AF386338

AN INTRODUCTION TO FOUR-DIMENSIONAL CHERN-SIMONS THEORY

AN INTRODUCTION TO FOUR-DIMENSIONAL CHERN-SIMONS THEORY

Meer Ashwinkumar

University of Bern, Switzerland

World Scientific

NEW JERSEY · LONDON · SINGAPORE · BEIJING · SHANGHAI · HONG KONG · TAIPEI · CHENNAI · TOKYO

Published by

World Scientific Publishing Co. Pte. Ltd.
5 Toh Tuck Link, Singapore 596224
USA office: 27 Warren Street, Suite 401-402, Hackensack, NJ 07601
UK office: 57 Shelton Street, Covent Garden, London WC2H 9HE

Library of Congress Control Number: 2024949782

British Library Cataloguing-in-Publication Data
A catalogue record for this book is available from the British Library.

AN INTRODUCTION TO FOUR-DIMENSIONAL CHERN-SIMONS THEORY

ISBN 978-981-12-9516-4 (hardcover)
ISBN 978-981-12-9517-1 (ebook for institutions)
ISBN 978-981-12-9518-8 (ebook for individuals)

For any available supplementary material, please visit
https://www.worldscientific.com/worldscibooks/10.1142/13900#t=suppl

Desk Editor: Muhammad Ihsan Putra

Typeset by Stallion Press
Email: enquiries@stallionpress.com

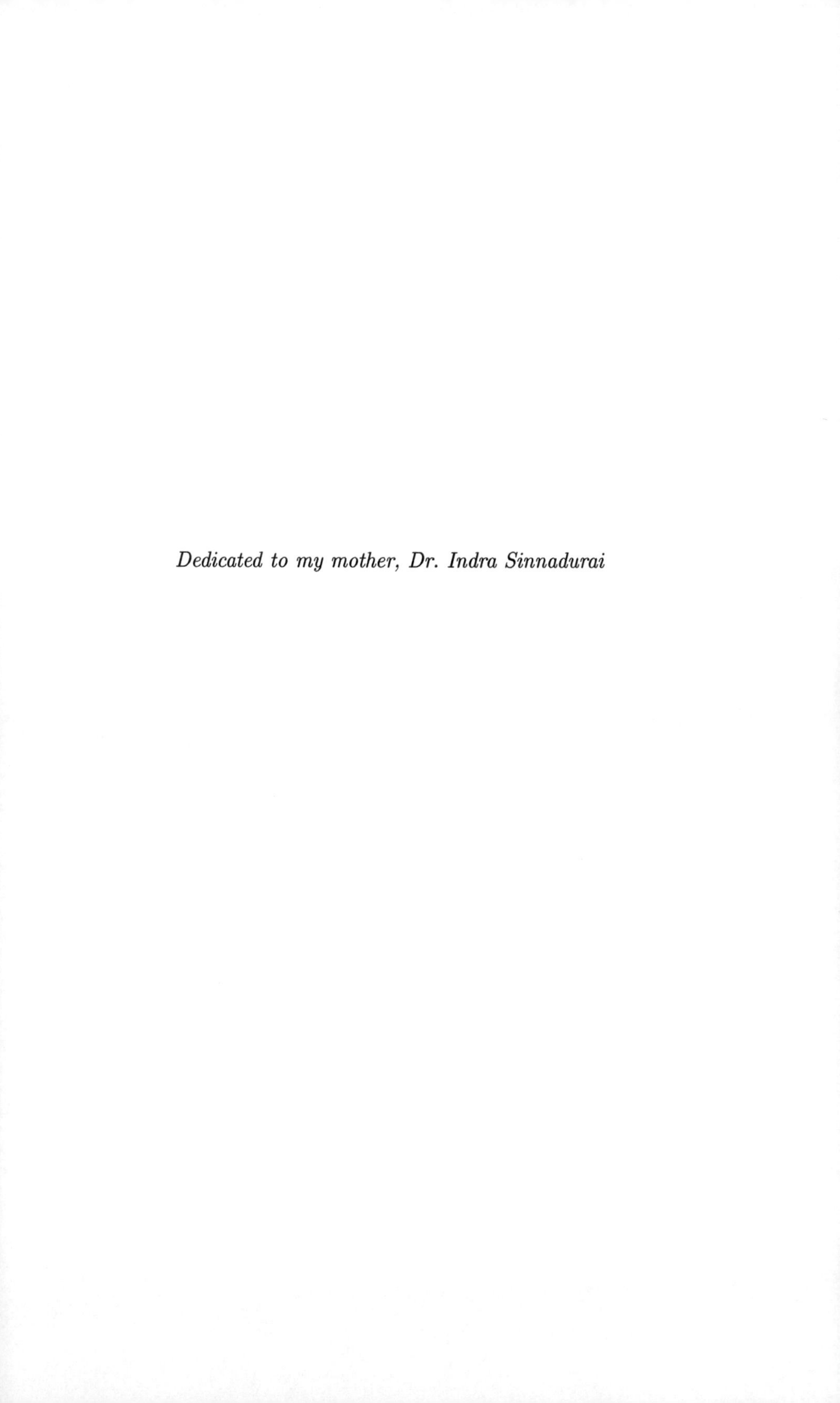

Dedicated to my mother, Dr. Indra Sinnadurai

Preface

Integrability is an ubiquitous concept in theoretical physics. Since one of the aims of physics is to understand nature in depth, systems that exhibit integrability are of particular interest, as their rich symmetry structures often allow for considerable analytical progress in exploring their dynamics. These symmetries, manifested in a large number of conserved quantities, are often related to interesting mathematical structures.

Although integrable systems are traditionally associated with classical dynamics and statistical mechanics, in recent years, an organizing principle for various integrable systems has emerged in the form of a quantum field theory, which is a four-dimensional variant of Chern–Simons gauge theory. It may be rather surprising that quantum field theory, the framework used to describe the fundamental particles and interactions of nature, would somehow know something about integrable systems.

To whet the reader's appetite in this regard, let us take a look at the diagram in Figure 1, that depicts an ice crystal. Here, one notes that each oxygen atom is bonded to four hydrogen atoms. In addition, each bond contains an atom of hydrogen, that has a preference of oxygen atom to which it has closer proximity.

The physics of an ice crystal of this sort can be captured via the language of statistical mechanics, in particular, the six-vertex model, which is a classical integrable lattice model. Drawing an arrow in the direction of preference of oxygen atom for each hydrogen atom in the diagram, the corresponding six-vertex model can be depicted as in Figure 2. The reason for the nomenclature of this model is the six possible configurations at each vertex, each associated with unique Boltzmann weights which contribute to the partition function of the system.

Now, let us pick orientations for the vertical and horizontal lines in this diagram. For the vertical lines, we shall pick an upward orientation,

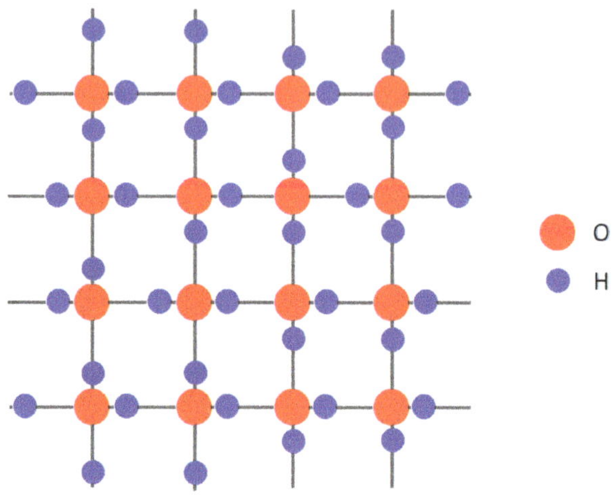

Figure 1.   An ice crystal, where each bond contains one hydrogen atom.

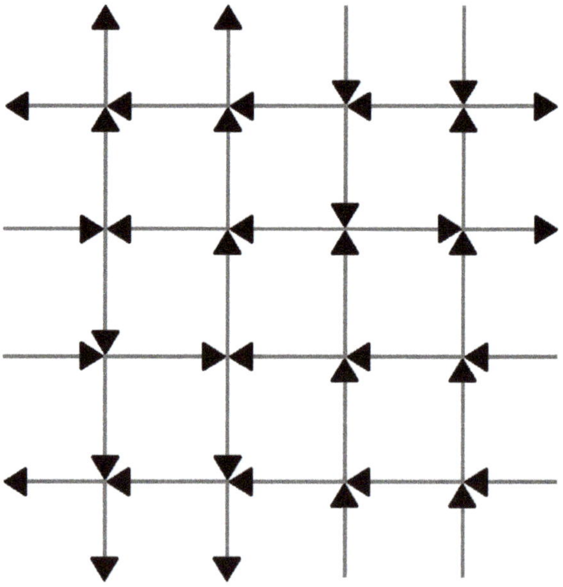

Figure 2.   The corresponding six-vertex model.

whereas for horizontal lines, we shall pick an orientation to the right. Then, for arrows that agree with the orientation, we shall replace it with a blue line, and otherwise we replace it with a grey line. We then obtain the diagram shown in Figure 3. Now, this diagram appears similar to a network of Wilson lines in a quantum gauge theory, more specifically, an $SU(2)$ gauge theory with Wilson lines in the fundamental representation, where each edge has been labeled by a basis vector of the representation. This is also the reason that squiggly red lines denoting gauge bosons mediating interactions between the Wilson lines have been included. The question one can ask at this point is as follows: Is there indeed a gauge theory that can capture the physics of the six-vertex model? The answer is **yes**, and that the gauge theory is four-dimensional Chern–Simons theory. In fact, as we shall see in the main text, a *correlation function* of Wilson lines in four-dimensional Chern–Simons theory corresponds to the partition function of an integrable lattice model, such as one that describes crystal ice.

In fact, similar Wilson line configurations lead to the realization of integrable *quantum* spin chains, which describe ferromagnets, via four-dimensional Chern–Simons theory. This is just the tip of the iceberg, and many more integrable systems can be shown to arise from this theory.

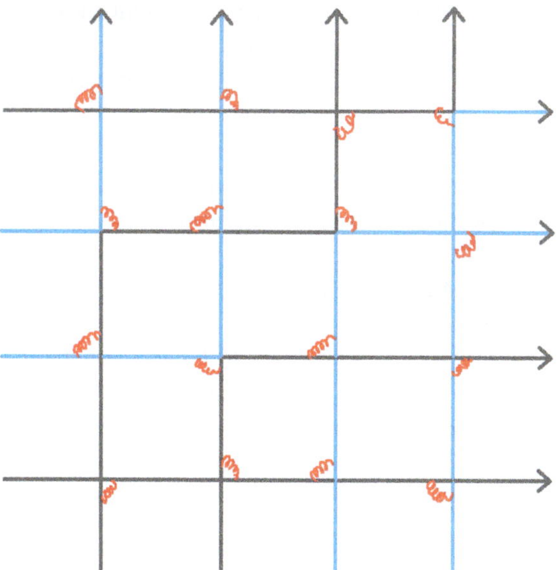

Figure 3. A network of Wilson lines in an $SU(2)$ gauge theory, each in the fundamental representation.

The following is a chapter-by-chapter overview of the content of this book. The first chapter deals with the relation of four-dimensional Chern–Simons theory to integrable lattice models, including how it realizes the Yang–Baxter equation with spectral parameter, solved by the R-matrix whose elements are the Boltzmann weights for statistical mechanical systems analogous to the aforementioned six-vertex model, as well as the mathematical structure that underlies such models, namely, the Yangian algebra. The focus will mostly be on rational solutions to the Yang–Baxter equation. I shall also review the framing anomaly of 4d Chern–Simons theory that restricts the manifold on which it can be defined. In addition, I review 't Hooft lines in 4d Chern–Simons theory and their relation to Baxter's TQ relations.

In the second chapter, I will review how 4d Chern–Simons theory gives rise to integrable field theories, realized by both order and disorder surface defects. In the latter case, I will present the unifying action that facilitates the derivation of integrable field theories. I will also briefly describe how the Poisson bracket of Lax connections can be obtained from Dirac brackets in 4d Chern–Simons theory. I then describe the relation of 4d Chern–Simons theory to 6d Chern–Simons theory on twistor space and 4d integrable field theories associated with the anti-self-dual Yang-Mills equation.

In the third chapter, I will describe the embedding of 4d Chern–Simons theory in topological string theory, and its relation to Chern–Simons theories in other dimensions. In addition, I will describe how 4d Chern–Simons theory can be derived from supersymmetric gauge theories that can be understood as worldvolume theories of D-branes in string theory. In particular, I review how 4d Chern–Simons theory can be derived from a stack of D5-branes subject to an Ω-deformation, or, alternatively, from a D4-NS5 system. Notably, such a derivation provides an integration cycle that ensures that the 4d Chern–Simons path integral is well defined beyond perturbation theory.

In the fourth chapter, I will discuss 4d Chern–Simons theory on manifolds with boundaries, and holographic duals of 4d Chern–Simons theory. I will discuss the dual of 4d Chern–Simons theory in the context of twisted holography, namely, 2d BF theory coupled to quantum mechanics, which realizes the Yangian algebra. In addition, I shall describe an alternative holographic dual, namely the 3d WZW model, which realizes rational R-matrices in a holographic manner. Finally, I describe a constrained version of the 3d WZW model that is holographically dual to 4d Chern–Simons theory, known as 3d Toda theory.

In the fifth chapter, I will discuss progress towards quantization and dualities of integrable field theories from the perspective of 4d Chern–Simons theory. I first review the derivation of the 1-loop RG flow of integrable field theories realized using disorder surface defects in 4d Chern–Simons theory. I then review the discretization of integrable field theories in terms of integrable lattice models (or equivalently, how the thermodynamic limit of integrable lattice models leads to integrable field theories). Moreover, I will explain how the thermodynamic limit of these lattice models can be interpreted in terms of the polarization of D-branes in string theory. I then describe various dualities of integrable field theories that can be derived from 4d Chern–Simons theory. In particular, I describe dualities that can be deduced from discretization of the integrable field theories, as well as generalizations of well-known bosonization dualities, that have been employed as a stepping stone towards quantization of said integrable field theories.

The aim of this book is to provide a pedagogical introduction to 4d Chern–Simons theory, intended as a reference both for postgraduate students as well as seasoned researchers interested in this nascent area of research. In this regard, I have strived to be as pedagogical as possible, while summarizing important results in 4d Chern–Simons theory to date.

I have benefited greatly from discussions with many mentors and colleagues who have worked on this subject. In particular, I would like to thank my Ph.D. supervisor, Meng-Chwan Tan, with whom I first collaborated on this topic. I am also very grateful to my postdoctoral mentor, Masahito Yamazaki, from whom I learned much about this subject. I am also grateful to my postdoctoral mentors, Matthias Blau and Susanne Reffert, for their support, and discussions on 4d Chern–Simons theory and related topics. I would also like to thank my collaborators and other researchers who work on 4d Chern–Simons theory and related topics, with whom I have had fruitful exchanges, including Roland Bittleston, Kevin Costello, Tudor Dimofte, Ben Hoare, Nafiz Ishtiaque, Sylvain Lacroix, Nat Levine, Marc Magro, Seyed Faroogh Moosavian, Domenico Orlando, Jitendra Pal, Kee-Seng Png, Surya Raghavendran, Jun-ichi Sakamoto, David M. Schmidtt, Daniel C. Thompson, Benoit Vicedo, Junya Yagi, Kentaroh Yoshida, Qin Zhao, and Yehao Zhou.

Meer Ashwinkumar
Advanced Postdoctoral Researcher
University of Bern
Bern, Switzerland
July, 2025

Contents

Chapter 1

Integrable Lattice Models from 4d Chern–Simons Theory

In this chapter, we shall describe the fundamentals of 4d Chern–Simons theory, with a focus on the realization of integrable lattice models in this theory. We shall elucidate its properties, and the realization of integrable lattice models using Wilson lines. We also review the derivation of the Yangian algebra from 4d Chern–Simons theory, as well as the framing anomalies of the theory. We also describe 't Hooft lines in 4d Chern–Simons theory, that are responsible for realizing Baxter's TQ relations.

1.1 Origins of 4d Chern–Simons Theory

The quantum field theory defined on three-manifolds known as Chern–Simons theory (which we shall refer to as 3d Chern–Simons theory in what follows) has a long history with numerous applications, including to condensed matter physics and mathematics. Of particular significance is its application to the mathematical theory of knots by Witten [1], where the Jones polynomial was interpreted as a correlation function of a Wilson loop in 3d Chern–Simons theory. In subsequent works [2, 3], Witten showed that this theory could describe certain integrable models of two-dimensional lattice statistical mechanics (namely critical IRF/vertex models), whose underlying algebraic structure is described by quantum groups.

The integrability of the models that arise from gauge theory in this way was understood at that time to be related to one of the Reidemeister moves of knot theory. These arise when knots are described in terms of projections to a plane, i.e., two such projections describe the same knot if they can be related by certain transformations known as Reidemeister moves. The most important of such moves for integrability depicted in Figure 1.1. This

1

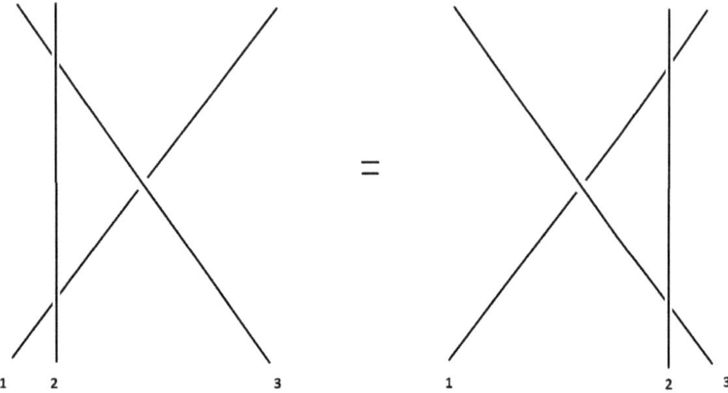

Figure 1.1.   A Reidemeister move in knot theory.

is in fact nothing but a diagrammatic representation of the Yang–Baxter equation

$$R_{12}R_{13}R_{23} = R_{23}R_{13}R_{12}, \qquad (1.1.1)$$

(obtained by assigning a matrix, known as an R-matrix, to each crossing).

However, Witten's work did not include a gauge-theoretic realization of an important generalization of the Yang-Baxter equation that involves a functional dependence of the R-matrices on additional parameters known as spectral parameters, i.e., the equation is generalized to

$$R_{12}(z_1, z_2)R_{13}(z_1, z_3)R_{23}(z_2, z_3) = R_{23}(z_2, z_3)R_{13}(z_1, z_3)R_{12}(z_1, z_2).$$
$$(1.1.2)$$

Such a Yang–Baxter equation is encountered in solving certain well-known classical integrable lattice models, such as the six-vertex and eight-vertex models, which in turn are related to the XXZ and XYZ integrable quantum spin chains describing ferromagnets.

A natural question is then whether the Yang–Baxter equation with spectral parameters can be realized via a Chern–Simons gauge theory. The answer to this question comes in the form of a four-dimensional variant of Chern–Simons theory. This theory first appeared in the PhD thesis of Nikita Nekrasov [4], and was rediscovered in 2013 by Kevin Costello [5, 6], who described its relation to integrable lattice models as well as the infinite-dimensional quantum group known as the Yangian in a mathematically rigorous manner. The theory was then elucidated and further developed by Costello, Witten and Yamazaki [7, 8] in a language more familiar to physicists, and as a result, integrable lattice models of rational, trigonometric and elliptic type (and their underlying mathematical structures) were shown to be realizable via standard QFT techniques in 4d Chern–Simons theory.

1.2 Properties of 4d Chern–Simons Theory

The starting point for the definition of four-dimensional Chern–Simons theory is the following classical action

$$S = \frac{1}{2\pi} \int_{\Sigma \times C} \mathrm{d}z \wedge \mathrm{CS}(A) \,, \tag{1.2.1}$$

where, for the time being, we shall set $\Sigma = \mathbb{R}^2$ and $C = \mathbb{C}$, and $\mathrm{CS}(A)$ is the Chern–Simons three-form, defined as

$$\mathrm{Tr}\left(A \wedge \mathrm{d}A + \frac{2}{3} A \wedge A \wedge A \right) = \varepsilon^{ijk} \mathrm{Tr}\left(A_i \partial_j A_k + \frac{2}{3} A_i A_j A_k \right) \mathrm{d}x \wedge \mathrm{d}y \wedge \mathrm{d}\bar{z} \,, \tag{1.2.2}$$

where the indices $i, j, \ldots$ run over x, y and $\bar{z}$ (the convention used here for the totally anti-symmetric tensor is $\varepsilon^{xy\bar{z}} = 1$), where x and y are Cartesian coordinates on $\Sigma = \mathbb{R}^2$, and z and $\bar{z}$ are complex coordinates on $C = \mathbb{C}$.

It is important to note that we are defining this Chern–Simons theory on a product 4-manifold, i.e., $\mathbb{R}^2 \times \mathbb{C}$. The complex structure on the second factor is crucial, as the fundamental field of the theory is in fact a gauge field without a z component, or, in mathematical parlance, a *partial* connection. That is, it takes the form

$$A = A_x \mathrm{d}x + A_y \mathrm{d}y + A_{\bar{z}} \mathrm{d}\bar{z}, \tag{1.2.3}$$

where the components depend non-trivially on all coordinates of the underlying 4-manifold.

No reality condition can be placed on such a connection, so the gauge group we are concerned with is a complex Lie group that we denote G, and each component of the gauge field is complex-valued. The notation Tr refers to an invariant and non-degenerate bilinear form on $\mathfrak{g}$, which can be identified with a trace in a suitable representation when $\mathfrak{g}$ is semisimple or reductive.

More generally, when $\mathfrak{g}$ is the direct sum of simple Lie algebras, Tr refers to the Killing form of $\mathfrak{g}$ which is invariant and non-degenerate. Picking an orthonormal basis t_a of $\mathfrak{g}$ with respect to the Killing form, the conventional normalization chosen is $\mathrm{Tr}(t_a t_b) = \delta_{ab}$.

Unlike three-dimensional Chern–Simons gauge theory which has complete diffeomorphism symmetry (since its action is defined without a metric), the classical action at hand is not invariant under four-dimensional diffeomorphisms. Nevertheless, two-dimensional orientation-preserving diffeomorphisms of $\mathbb{R}^2$ *do* leave the action invariant, and, as we shall see, this is a crucial feature of 4d Chern–Simons theory that enables it to describe integrable lattice models.

Since the action is holomorphic in the complex gauge field components, one immediately notices that the path integral is not a convergent one. Nevertheless, perturbation theory in the loop counting parameter $\hbar$ can be performed formally, that is, where one restricts $\hbar$ to be close to 0 (as we shall see in Chapter 4, realizing 4d Chern–Simons theory in string theory provides a suitable path integral analog of an integration contour from complex analysis that ensures convergence, and thereby a non-perturbative definition of the theory for any value of $\hbar$).

This leads us to yet another difference from 3d Chern–Simons theory, that is, the fact that $\hbar$ does not need to be quantized to ensure gauge invariance of the path integral. In fact, the action itself is gauge-invariant up to boundary terms (that can be removed via appropriate boundary conditions). To see this, we use integration by parts to cast the action in the form

$$S = -\frac{1}{2\pi} \int_{\mathbb{R}^2 \times \mathbb{C}} z \operatorname{Tr} F \wedge F, \tag{1.2.4}$$

where $F := \mathrm{d}A + A \wedge A$, which is invariant under the gauge transformation

$$A_i \mapsto -g A_i g^{-1} - \partial_i g g^{-1}, \tag{1.2.5}$$

where $i = x, y, \bar{z}$.

In a manner analogous to the case of 3d Chern–Simons theory, we can derive the classical equations of motion of 4d Chern–Simons theory by varying the classical action, whereby we find

$$\frac{1}{\pi} \int_{\Sigma \times C} \mathrm{d}z \wedge \operatorname{Tr}(\delta A \wedge F). \tag{1.2.6}$$

The resulting equations of motion are

$$F_{xy} = 0 \,, \quad F_{x\bar{z}} = 0 \,, \quad F_{y\bar{z}} = 0 \,. \tag{1.2.7}$$

The 4d Chern–Simons gauge field is thus a connection on a flat bundle on $\mathbb{R}^2$, which varies holomorphically as one moves along $\mathbb{C}$. These equations in fact imply that there are no local gauge-invariant observables in the theory.

One notes at this point that the loop-counting parameter $\hbar$ has dimensions of length, since the theory is invariant if we rescale z and $\hbar$ by a common factor. This in fact implies that the theory is unrenormalizable by power-counting, since naively one expects any perturbation series to diverge at high energies. However, since all possible gauge-invariant counterterms vanish on-shell via (1.2.7) the theory is in fact renormalizable and therefore quantizable in perturbation theory, as shown by Costello [5, 6].

The property of being unrenormalizable by power counting is actually related to a crucial property of the theory, that is, when constructed in perturbation theory, it is infrared-free. As we shall see, this enables us to analyze interactions between Wilson line operators in the theory locally, which will allow us to easily recover the Yang–Baxter equation with spectral parameters and its solutions from 4d Chern–Simons theory.

1.2.1 General choices of Σ and C

There are more general choices that one can make for the four-manifold on which 4d Chern–Simons is defined. In what follows, we shall focus on the case where the four-manifold is a product of two 2-manifolds, $\Sigma \times C$, where Σ is referred to as the "topological plane" and C is referred to as the "holomorphic plane".[1] The action in this case can be written as

$$S = \frac{1}{2\pi} \int_{\Sigma \times C} \omega \wedge \mathrm{CS}(A) \,, \tag{1.2.8}$$

where ω is, in general, a closed meromorphic one-form defined on C. With appropriate boundary conditions imposed, the equations of motion are now

$$\omega \wedge F = 0. \tag{1.2.9}$$

Moreover, as long as ω is closed, there is manifest gauge invariance up to total derivatives that are not important in perturbation theory.

In this chapter, we shall not consider choices of ω which contain zeroes (these, however, play an important role in the next chapter where classical integrable field theories are derived from 4d Chern–Simons theory). The reason for this is because the action involves the ratio $\omega/\hbar$, and therefore a zero of ω corresponds to the limit $\hbar \to \infty$. However, since we are restricted

---

[1] More general choices of four-manifolds are possible, which require the structure of a two-dimensional integrable foliation [7]

to perturbation theory and small values of $\hbar$, it is not obvious how the theory works near such zeroes. On the other hand, poles of ω are perfectly sensible within perturbation theory, as long as one imposes appropriate boundary conditions for the gauge fields and gauge transformation parameters at such poles. These are necessary since ω is generally not closed near such a pole.

Only simple and double poles are relevant, as one sees if one considers the possibilities for a complex Riemann surface C with a holomorphic one-form ω that is allowed to have poles but not zeroes.

Indeed, according to the Riemann-Roch theorem, for any meromorphic differential, the following equality is satisfied

$$\#\text{poles} - \#\text{zeroes} = 2 - 2g. \tag{1.2.10}$$

Since ω is assumed to not have any zeroes, we find that we are immediately restricted to $g = 0, 1$. For $g = 0$, there are two possibilities

- A single pole with multiplicity 2 corresponding to $C = \mathbb{C}$ with $\omega = dz$ (Picking coordinates $z' = 1/z$ close to ∞ reveals the double pole there).
- Two simple poles corresponding to $C = \mathbb{C}^\times = \mathbb{C}/\mathbb{Z}$ with $\omega = dz/z$ (which has simple poles at 0 and ∞).

The $g = 1$ case has no poles, and corresponds to $C = \mathbb{C}/(\mathbb{Z} + \tau\mathbb{Z})$, i.e., an elliptic curve with $\omega = dz$. Although the choice of ω is only unique up to an overall constant, this constant can be removed by rescaling $\hbar$.

The three possibilities for C correspond to three classes of possible solutions to the Yang–Baxter equation with spectral parameter, which are known as quasi-classical R-matrices, that take the form

$$R_\hbar(z) = I + \hbar r(z) + O(\hbar^2). \tag{1.2.11}$$

Here, $r(z)$ is known as a classical r-matrix, and satisfies the classical Yang–Baxter equation

$$\begin{aligned}
&[r_{12}(z_1 - z_2), r_{13}(z_1 - z_3)] + [r_{12}(z_1 - z_2), r_{23}(z_2 - z_3)] \\
&+ [r_{13}(z_1 - z_3), r_{23}(z_2 - z_3)] = 0 \,.
\end{aligned} \tag{1.2.12}$$

Solutions to this equation were classified by Belavin and Drinfeld [9], modulo trivial equivalences, under certain assumptions. It was found that these solutions were divided into three classes in which $r(z)$ takes the form of either a rational, trigonometric or elliptic function. These correspond to the aforementioned three possibilities for the holomorphic plane C:

$$
\begin{aligned}
\text{Rational} \quad &: \quad C = \mathbb{C}\,, & \omega &= \mathrm{d}z\,, \\[6pt]
\text{Trigonometric} &: \quad C = \mathbb{C}^\times\,, & \omega &= \frac{\mathrm{d}z}{z}\,, \\[6pt]
\text{Elliptic} \quad &: \quad C = \mathbb{C}/(\mathbb{Z} + \tau\mathbb{Z})\,, & \omega &= \mathrm{d}z\,.
\end{aligned}
\tag{1.2.13}
$$

Now, we shall turn to the possible choices of Σ. Classically, there is no restriction on Σ, but quantum mechanically, there is a framing anomaly analogous to that which occurs in 3d Chern–Simons theory [1]. We shall review a derivation of this anomaly in Section 1.6. This anomaly requires that Σ must be equipped with a framing, i.e., its tangent bundle must be trivialized.

A framing can be equivalently thought of as a pair of everywhere linearly independent vector fields v_1 and v_2 on the surface Σ. With the choice of a framing, one can pick a metric such that v_1 and v_2 are everywhere orthonormal, and then one can define angles between lines on Σ. This will be necessary for the analysis of another framing anomaly, that is, for Wilson lines, reviewed in Section 1.4. The condition that Σ is framed is quite restrictive; for example, if a framed 2-manifold is compact, it must have Euler characteristic equal to zero and thus must be of genus 1, i.e., a torus. We shall also consider the allowed choice of $\Sigma = \mathbb{R}^2$ in what follows.

1.2.2 *Wilson lines*

The simplest non-trival gauge-invariant operators in 4d Chern–Simons theory are Wilson line operators. Recall that no gauge-invariant local operators exist as they all vanish by the equations of motion.

We shall consider two classes of Wilson line operators. The operators in the first class take a form familiar from gauge theory:

$$
W_\rho(K) = P \exp\left(\int_K A_i^\rho(x^1, x^2, z, \bar{z}) \mathrm{d}x^i \right),
\tag{1.2.14}
$$

where we have used the notation $x^1 = x$ and $x^2 = y$. This is the path-ordered exponential of the integral of the connection along a curve K that lies entirely in Σ. Note that we do not take a trace, but instead implement gauge invariance by taking $\Sigma = \mathbb{R}^2$, and insisting that the gauge field vanishes at infinity along Σ. The reason we do this is that the theory is infrared-free, allowing us to consider the gauge field to vanish at infinity along $\Sigma = \mathbb{R}^2$. It is therefore the holonomy, rather than its trace, that is physically meaningful.

The gauge field here is associated with a particular representation, ρ, of the Lie algebra $\mathfrak{g}$. If one thinks of the Wilson operator as the worldline of a particle in two-dimensional spacetime, ρ plays the role of the vector space of internal states of the particle. In addition, the Wilson operator is labeled by a spectral parameter, which is the point on C at which it is located.

The Wilson line operators in the second class are not associated with representations of $\mathfrak{g}$, but instead finite-dimensional representations of an infinite-dimensional Lie algebra denoted $\mathfrak{g}[[z]]$, if we place such a Wilson line at $z = 0$. This is the Lie algebra $\prod_{n\geq0}(\mathfrak{g} \otimes z^n)$ of series in z whose coefficients are in $\mathfrak{g}$ (there is no dependence on the choice of C here, as the present considerations are local along C). If we instead place the Wilson line at $z = z_0$, it would be associated with a representation of $\mathfrak{g}[[z-z_0]]$. Given a finite-dimensional representation $\widehat{\rho}$ of $\mathfrak{g}[[z]]$, the corresponding Wilson line is defined as

$$W_{\widehat{\rho}}(K) = P\exp\left(\int_K A_i^{\widehat{\rho}}(x^1, x^2, z, 0)\mathrm{d}x^i\right), \qquad (1.2.15)$$

where $\bar{z}$ has formally been set to zero. The precise meaning of this is that one defines

$$A_i(x^1, x^2, z, 0) := \sum_{k\geq0} \frac{z^k}{k!} \frac{\partial^k}{\partial z^k} A_i(x^1, x^2, z, \bar{z})\bigg|_{z=\bar{z}=0}. \qquad (1.2.16)$$

To ensure convergence of the series, we restrict ourselves to representations of $\mathfrak{g}[[z]]$ wherein the series terminates after finitely many terms.

1.3   Computing an *R*-matrix from 4d Chern–Simons Theory

The first testament to the power of 4d Chern–Simons theory in describing integrable lattice models is its ability to provide us with solutions of the Yang–Baxter equation with spectral parameters, via the computation of correlation functions of crossed Wilson lines. This will turn out to be rather simple, in comparison to the computation of R-matrices in 3d Chern–Simons theory [10–16].

The fundamental reason for the simple description of R-matrices in 4d Chern–Simons theory is the infrared freedom of the theory, coupled with its diffeomorphism invariance. In a typical gauge theory, gluon exchange

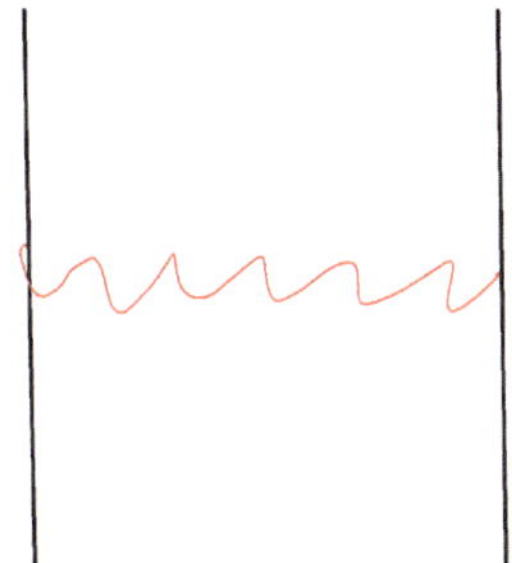

Figure 1.2.   Gluon exchange between non-intersecting Wilson lines.

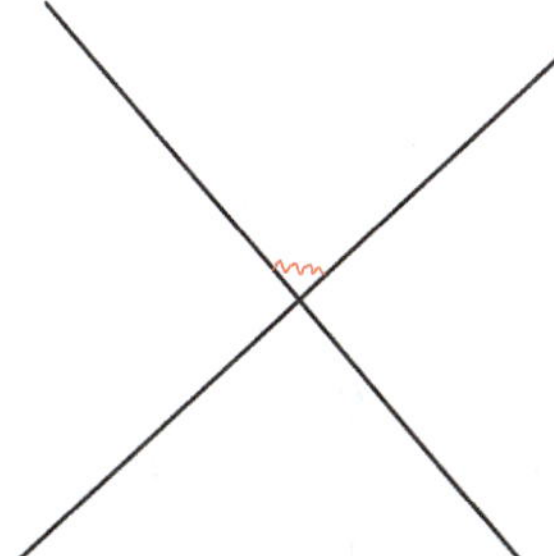

Figure 1.3.   Gluon exchange between crossed Wilson lines.

between Wilson lines as depicted in Figure 1.2 cannot be ignored. However, since 4d Chern–Simons theory is infrared-free, one can scale up the metric of Σ (that becomes relevant to the description of the theory after gauge fixing), and the gluon exchange becomes irrelevant.

The only non-trivial contributions arise from gluon exchange close to *crossings* of Wilson lines, as depicted in Figure 1.3. This is why 4d Chern–Simons theory is able to describe integrable lattice models: all interactions are confined to intersections of Wilson lines, and as we see, these interactions can be computed to be Boltzmann weights (R-matrix elements) of the lattice models. In fact, the topological invariance along Σ ensures that the two diagrams in Figure 1.4 are equivalent, which in turn implies the sought-after Yang–Baxter equation that ensures the integrability of these lattice models.

Note that one does not obtain more general versions of the Yang-Baxter equation, such as the dynamical Yang-Baxter equation, where the regions between the Wilson lines are also labeled by certain parameters. The reason

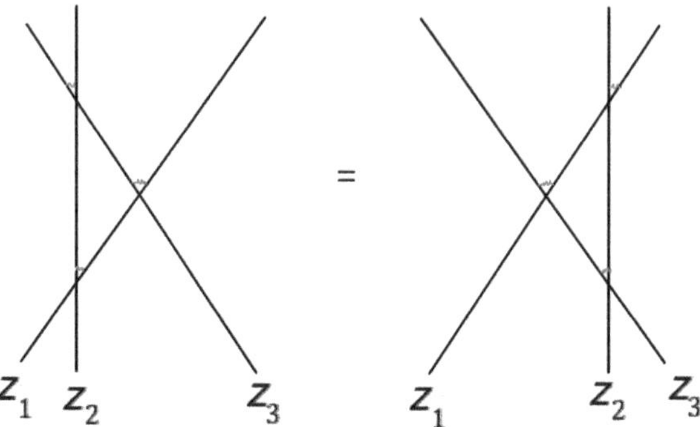

Figure 1.4.   The Yang–Baxter equation.

for this is that one chooses a boundary condition such that there is only one classical solution to expand around in perturbation theory, that does not have any continuous deformations. The regions between the Wilson lines would generally be labeled by a basis vector of the Hilbert space, $\mathcal{H}$, of quantum states of the theory. But since the space of classical solutions modulo gauge transformations is a point, $\mathcal{H}$ is one-dimensional, and there are no labels in between Wilson lines.

Nevertheless, due to the infrared-freedom of the theory, away from crossings, each Wilson line is labeled by a state (i.e., a basis vector) in its associated representation. This is what leads us to configurations in 4d Chern–Simons theory such as that depicted in Figure 3 of the Preface, where all quantum interactions are localized to crossings of Wilson lines, and build up Boltzmann weights (that depend on the adjacent basis vectors) such that the theory describes an integrable lattice model.

Let us now verify that the correlation function of crossed Wilson lines indeed has the expected behaviour at order $\hbar$. We shall evaluate the Feynman diagram for gluon exchange between the two crossed Wilson lines depicted in Figure 1.5, which are associated to representations ρ and ρ' of $\mathfrak{g}$ and supported at $z = z_1$ and $z = z_2$, respectively. Here, we have scaled up the metric on Σ by a large factor such that Σ becomes $\mathbb{R}^2$ near the crossing and the crossed Wilson lines become straight.

1.3.1   *4d Chern–Simons propagator*

We shall now compute the propagator of 4d Chern–Simons theory, focusing mainly on the example of $C = \mathbb{C}$, corresponding to the case of rational

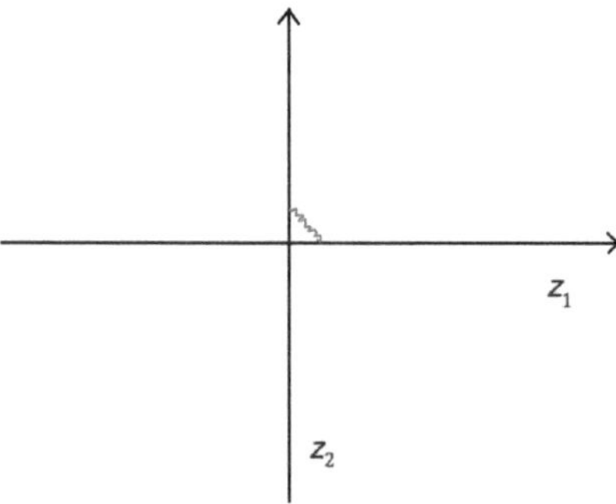

Figure 1.5. Feynman diagram for gluon exchange at order $O(\hbar)$ between crossed Wilson lines. The arrows indicate the orientations of the Wilson lines.

integrable models. At infinity on C, we impose the boundary condition that all fields go to zero, which ensures that we only have a trivial classical solution. We shall use the Cartesian coordinates x and y to parameterize $\Sigma = \mathbb{R}^2$. As in [7], we shall evaluate the propagator in the gauge

$$\tilde{\partial}_i A^i = 0, \tag{1.3.1}$$

where $\tilde{\partial}_i = (\partial_x, \partial_y, 4\partial_z)$. The reason the factor of four is included in this definition is that if a gauge field A satisfies (1.3.1) as well as the linearized equations of motion $\mathrm{d}z \wedge \mathrm{d}A = 0$, then each component of the gauge field is harmonic for the choice of metric

$$\mathrm{d}s^2 = \mathrm{d}x^2 + \mathrm{d}y^2 + \mathrm{d}z\mathrm{d}\bar{z} \tag{1.3.2}$$

that we pick on $\mathbb{R}^2 \times \mathbb{C}$, i.e., they each satisfy

$$g^{\mu\nu}\frac{\partial}{\partial x^\mu}\frac{\partial}{\partial x^\nu}A_i = \left(\frac{\partial^2}{\partial x^2} + \frac{\partial^2}{\partial y^2} + 4\frac{\partial}{\partial z}\frac{\partial}{\partial \bar{z}}\right)A_i = 0, \tag{1.3.3}$$

where $g^{\mu\nu}$ has components $g^{xx} = 1$, $g^{yy} = 1$ and $g^{z\bar{z}} = 2$.

We shall now compute the propagator of 4d Chern–Simons theory in the aforementioned gauge. It is useful to write the propagator as a two-form on two copies of $\mathbb{R}^2 \times \mathbb{C}$. We shall also employ a propagator two-form with adjoint indices removed, i.e.,

$$P^{ab}(x, y, z, \bar{z}) = \delta^{ab} P(x, y, z, \bar{z}). \tag{1.3.4}$$

This propagator two-form satisfies

$$\frac{i}{2\pi}\mathrm{d}z \wedge \mathrm{d}P(x, y, z, \bar{z}) = \delta_{x,y,z,\bar{z}=0} \tag{1.3.5}$$

and

$$\left(\partial_x \iota_{\partial_x} + \partial_y \iota_{\partial_y} + 4\partial_z \iota_{\bar z}\right) P(x,y,z,\bar z) = 0, \tag{1.3.6}$$

where $\delta_{x,y,z,\bar z=0}$ indicates a delta-function distribution 4-form.

Defining $x = x' - x''$, $y = y' - y''$, $z = z' - z''$, $\bar z = \bar z' - \bar z''$, the propagator two-form is

$$
\begin{aligned}
P^{ab}(x,y,z,\bar z) &:= \frac{1}{2}\langle A_i^a(x',y',z',\bar z')A_j^b(x'',y'',z'',\bar z'')\rangle \mathrm{d}x^i \wedge \mathrm{d}x^j \\[2mm]
&= -\frac{\delta^{ab}}{4\pi}\left(\mathrm{d}y \wedge \mathrm{d}\bar z \frac{\partial}{\partial x} + \mathrm{d}\bar z \wedge \mathrm{d}x \frac{\partial}{\partial y} + 4\mathrm{d}x \wedge \mathrm{d}y \frac{\partial}{\partial z}\right) \\[2mm]
&\quad \times \frac{1}{(x^2 + y^2 + z\bar z)} \\[2mm]
&= \frac{\delta^{ab}}{2\pi}\left(x\mathrm{d}y \wedge \mathrm{d}\bar z + y\mathrm{d}\bar z \wedge \mathrm{d}x + 2\bar z\mathrm{d}x \wedge \mathrm{d}y\right)\frac{1}{(x^2 + y^2 + z\bar z)^2}\,.
\end{aligned}
\tag{1.3.7}
$$

This can be verified as follows. Away from the origin, the RHS of (1.3.5) ought to be zero. We can check that $\mathrm{d}z \wedge \mathrm{d}P$ is indeed equal to

$$
\begin{aligned}
\mathrm{d}z \wedge &\left[\frac{1}{2\pi}\frac{4\mathrm{d}x \wedge \mathrm{d}y \wedge \mathrm{d}\bar z}{(x^2 + y^2 + |z|^2)^2} - \frac{1}{\pi}\frac{(2x\mathrm{d}x + 2y\mathrm{d}y + \bar z\mathrm{d}z + z\mathrm{d}\bar z)}{(x^2 + y^2 + |z|^2)^3}\right.\\[2mm]
&\left. \wedge (x\mathrm{d}y \wedge \mathrm{d}\bar z + y\mathrm{d}\bar z \wedge \mathrm{d}x + 2\bar z\mathrm{d}x \wedge \mathrm{d}y)\right]
\end{aligned}
\tag{1.3.8}
$$

$$= 0.$$

Moreover, we can verify (1.3.6), i.e., that $(\partial_x \iota_{\partial_x} + \partial_y \iota_{\partial_y} + 4\partial_z \iota_{\bar z})P(x,y,z,\bar z)$ is equal to

$$
\begin{aligned}
\partial_x &\left(\frac{-y\mathrm{d}\bar z + 2\bar z\mathrm{d}y}{(x^2 + y^2 + |z|^2)^2}\right) + \partial_y\left(\frac{x\mathrm{d}\bar z - 2\bar z\mathrm{d}x}{(x^2 + y^2 + |z|^2)^2}\right) + 4\partial_{\bar z}\left(\frac{-x\mathrm{d}y + y\mathrm{d}x}{(x^2 + y^2 + |z|^2)^2}\right)\\[2mm]
&= -2\frac{(-2xy\mathrm{d}\bar z + 4x\bar z\mathrm{d}y)}{(x^2 + y^2 + |z|^2)^3} - 2\frac{(2yx\mathrm{d}\bar z - 4y\bar z\mathrm{d}x)}{(x^2 + y^2 + |z|^2)^3} - 2(4)\frac{(-\bar z x\mathrm{d}y + \bar z y\mathrm{d}x)}{(x^2 + y^2 + |z|^2)^3}
\end{aligned}
$$

$$= 0. \tag{1.3.9}$$

The normalization of the propagator is fixed by checking (1.3.5) at the origin. The first step in this direction is the observation that the propagator two-form restricted to the unit three-sphere takes the form

$$P(x, y, z, \bar{z}) = \frac{1}{2\pi}(x \, dy \wedge d\bar{z} - y \, dx \wedge d\bar{z} + 2\bar{z} \, dx \wedge dy). \qquad (1.3.10)$$

If we were to integrate the propagator over a four-ball of radius 1 (which has volume $\pi^2/2$), then Stokes' theorem tells us that we can use (1.3.10) in the computation, i.e.,

$$\begin{aligned}
&\frac{i}{2\pi} \int_{x^2+y^2+z\bar{z}\leq 1} dz \wedge dP(x, y, z, \bar{z}) \\
&= \frac{i}{2\pi} \frac{1}{2\pi} \int_{x^2+y^2+z\bar{z}\leq 1} 4 \, dx \, dy \, dz \, d\bar{z} \\
&= \frac{i}{2\pi} \frac{1}{2\pi} \int_{x^2+y^2+z\bar{z}\leq 1} (-8i) dx \, dy \, du \, dv \\
&= \frac{i}{2\pi} \frac{1}{2\pi} (-8i) \frac{\pi^2}{2} = 1,
\end{aligned} \qquad (1.3.11)$$

where $z = u + iv$.

1.3.2 Rational R-matrix

Now, denote by $t_{a,\rho}$ the matrix whereby the Lie algebra element t_a acts in the representation ρ. The gauge field that appears in a Wilson line in such a representation can be written as $A = \sum_a A^a t_{a,\rho}$. The propagator between the two Wilson lines depicted in Figure 1.5 is then

$$I_1 = \hbar \left((t_{a,\rho} \otimes t_{b,\rho'})\right) \int dx dy' \, P^{ab}(x - x', y - y', z_1 - z_2, \bar{z}_1 - \bar{z}_2). \qquad (1.3.12)$$

Evaluating this integral gives us

$$\begin{aligned}
I_1 &= \hbar \, c_{\rho,\rho'} \frac{1}{2\pi} \int_{-\infty}^{\infty} dx \int_{-\infty}^{\infty} dy \frac{2(\bar{z}_1 - \bar{z}_2)}{(x^2 + y^2 + |z_1 - z_2|^2)^2} \\
&= \hbar \, c_{\rho,\rho'} \frac{1}{4} \int_{-\infty}^{\infty} dy \frac{2(\bar{z}_1 - \bar{z}_2)}{(y^2 + |z_1 - z_2|^2)^{\frac{3}{2}}} \\
&= \hbar \, c_{\rho,\rho'} \frac{(\bar{z}_1 - \bar{z}_2)}{|z_1 - z_2|^2} \\
&= \frac{\hbar \, c_{\rho,\rho'}}{z_1 - z_2},
\end{aligned} \qquad (1.3.13)$$

where

$$c_{\rho,\rho'} = \sum_a t_{a,\rho} \otimes t_{a,\rho'} \,. \tag{1.3.14}$$

is the tensor Casimir. This agrees with the known semi-classical expansion of the rational quasi-classical R-matrix

$$R = 1 + \frac{\hbar\, c_{\rho,\rho'}}{z_1 - z_2} + \mathcal{O}(\hbar^2) \,. \tag{1.3.15}$$

In fact, given the leading-order term, general theorems (see [17, p. 814] or [18, p. 418]) imply that the *full* rational R-matrix is determined by the general properties that it obeys, up to an overall prefactor.

Also note that the R-matrix is a function of the difference $z_1 - z_2$, and this is because the action that we started with has a symmetry under the translation $z \to z + a$. The above computation can also be generalized to case of non-perpendicular, straight Wilson lines, and one obtains the same result at order $O(\hbar)$. However, the result will not be the same at order $O(\hbar^2)$ and higher, due to the framing anomaly which we describe in the next section.

We would also like to fix the normalization factor for the R-matrix using 4d Chern–Simons theory. To this end, we note that the unitarity relation

$$R_{21}(z_2 - z_1)R_{12}(z_1 - z_2) = 1 \tag{1.3.16}$$

on the R-matrix can also be derived from the topological invariance along Σ, as depicted in Figure 1.6. This unitarity relation can be used to

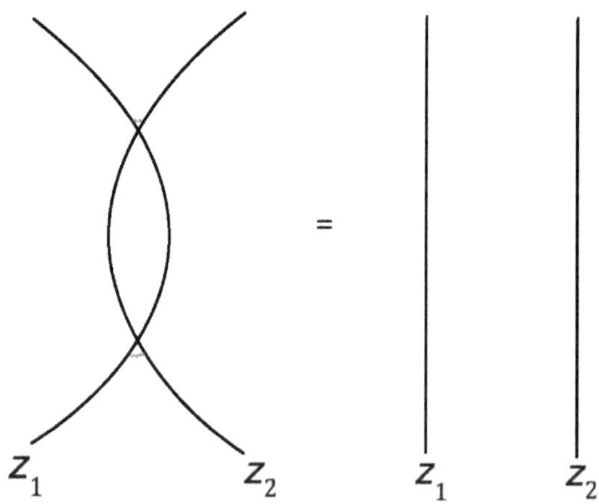

Figure 1.6. The unitarity relation.

constrain the normalization factor. Consider the example of the R-matrix for fundamental representations of $G = GL(N)$, which can be computed to be

$$R_{12}(z_1 - z_2) = c(z_1 - z_2)\left(\mathbb{1} + \frac{\hbar}{z_1 - z_2}P_{12}\right), \qquad (1.3.17)$$

where P_{12} is the permutation operator, and $c(z_1 - z_2)$ is the normalization factor to be fixed. Substituting (1.3.17) into the unitarity relation (1.3.16), we find

$$\begin{aligned}
c(z_2 - z_1)c(z_1 - z_2)&\left(\mathbb{1} + \frac{\hbar}{z_2 - z_1}P_{21}\right)\left(\mathbb{1} + \frac{\hbar}{z_1 - z_2}P_{12}\right) \\
&= c(z_2 - z_1)c(z_1 - z_2)\left(\mathbb{1} + \frac{\hbar^2}{(z_1 - z_2)^2}\mathbb{1}\right) = \mathbb{1},
\end{aligned} \qquad (1.3.18)$$

(where we used the identity $P_{21}P_{12} = \mathbb{1}$) from which we obtain two possible solutions for the normalization factor,

$$c(z_1 - z_2) = \frac{1}{1 \pm \frac{\hbar}{z_1 - z_2}}. \qquad (1.3.19)$$

One can likewise use the unitarity relation to constrain the normalization factor for R-matrices for other gauge groups; see, for example, the discussion in Appendix A of [7].

1.4   The Framing Anomaly for Wilson Lines

We have thus far considered straight Wilson lines in 4d Chern–Simons theory. An interesting quantum effect arises when one considers *curved* Wilson lines. This is known as the framing anomaly, and it says that what is constant along a Wilson line is not z but the sum of z and a $\hbar$-dependent expression that is proportional to the angle, φ, between the tangent vector to the Wilson line and a fixed direction in Σ.

The relevant Feynman diagram from which we can deduce the framing anomaly is depicted in Fig. 1.7. In particular, its evaluation involves an extra Feynman rule that is the bulk interaction vertex, which can be read off from the action, and takes the form

$$\frac{i}{2\pi}f^{abc}dz. \qquad (1.4.1)$$

It is then straightforward to deduce the color factor of the Feynman diagram in Figure 1.7, as follows. Expanding the gauge field in terms

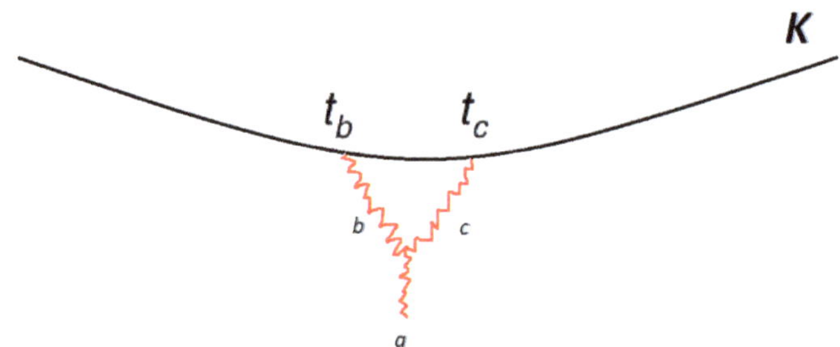

Figure 1.7.   The one-loop diagram that leads to a framing anomaly for Wilson operators.

of the generators of $\mathfrak{g}$ as $A = \sum_a A_a t_a$, and connecting the vertices via propagators, one finds the following coupling of an external gauge field A_a to the curved Wilson line operator

$$\sum_{a,b,c} A_a f_{abc} t_b t_c \,. \tag{1.4.2}$$

Then, employing the antisymmetry of f_{abc}, the product of matrices $t_b t_c$ can be replaced by $\frac{1}{2}[t_b, t_c] = \frac{1}{2} f_{bcd} t_d$. Via the identity $\sum_{b,c} f_{abc} f_{bcd} = 2 \mathsf{h}^\vee \delta_{ad}$, where $\mathsf{h}^\vee$ the dual Coxeter number of $\mathfrak{g}$, we then find that the external gauge field is coupled to to the Wilson line via a factor of

$$\mathsf{h}^\vee \sum_a A_a t_a. \tag{1.4.3}$$

In evaluating the diagram explicitly, one may use the diffeomorphism invariance of 4d Chern–Simons theory to scale Σ up such that the curve K has a very large radius of curvature. More specifically, one chooses coordinates such that K is close to the x-axis and is described by a curve $y = y(x)$, with y everywhere small. The curve K shall be located at $z = z_0$ (before quantum corrections are taken into account). It is then straightforward to compute the diagram explicitly (details of this can be found in [7]). The upshot is that the amplitude, I, one obtains is not a gauge-invariant quantity. Instead, performing a gauge transformation $A \to A + D\varepsilon$, the variation of I will be

$$\delta I = -\frac{\hbar\, \mathsf{h}^\vee}{2\pi} \int_K \mathrm{d}x \left(\frac{\mathrm{d}^2 y}{\mathrm{d}x^2} \partial_z \varepsilon(x, y(x), z_0) \right). \tag{1.4.4}$$

Having assumed that $y(x)$ is small in the region of interest of K that is close to the x-axis, one can replace $\varepsilon(x, y(x), z_0)$ by $\varepsilon(x, 0, z_0)$. This gauge anomaly can be cancelled in an interesting manner, that is, via a quantum correction to the classical form of the Wilson operator. This is achieved by replacing z_0 in the integral $\int_K dx A_x(x,0,z_0)$ that appears in the Wilson line by $z_0 - \hbar\, \mathsf{h}^\vee \frac{1}{2\pi} dy/dx$. Its gauge variation is then

$$-\frac{\hbar\, \mathsf{h}^\vee}{2\pi} \int_K dx \left(\frac{d^2 y}{dx^2} \partial_z \varepsilon(x,0,z_0) \right), \qquad (1.4.5)$$

which cancels the gauge anomaly (1.4.4).

Thus, for a curved Wilson line, canceling the gauge anomaly requires us to shift the position of the Wilson line in C by $-\frac{1}{2\pi}\hbar\, \mathsf{h}^\vee dy/dx$. Note that in deriving this formula, we have assumed that dy/dx is small. A more general formula can be obtained by re-expressing dy/dx in terms of the angle φ between the tangent vector to K and a fixed direction on $\mathbb{R}^2$, e.g., the x-axis.

Note that no higher-order corrections in $\hbar$ are necessary to cancel the anomaly. This is because requiring $z - \hbar\, \mathsf{h}^\vee \varphi/\pi$ to be constant along K is consistent with a symmetry of the action for $C = \mathbb{C}$, namely that it is invariant under a common rescaling of z and $\hbar$. For other choices of $\mathbb{C}$, we note that our analysis thus far has been local along C, so the analysis applies for these choices too as long as one picks local coordinates along C such that the action takes the same form as the $C = \mathbb{C}$ case. For the trigonometric case of $C = \mathbb{C}^\times$, this requires replacing z with $\log z$.

1.5 The Yangian Algebra

It is well known that the R-matrices that are rational solutions to the Yang-Baxter equation with spectral parameters are intimately related to the Yangian algebra. More precisely, such an R-matrix is an intertwiner for representations of the Yangian algebra. A natural question to ask is how one explicitly realizes the Yangian algebra in 4d Chern–Simons theory. Answering this question is the aim of this section.

Recall from Section 1.2 that there are two classes of Wilson lines relevant to 4d Chern–Simons theory. The second class is associated with representations of $\mathfrak{g}[[z]]$. More specifically, a Wilson line in this class is associated with a finite-dimensional vector space W, along with a sequence of operators

$$t_{a,n,W} : W \to W \qquad (1.5.1)$$

for $n \geq 0$ (we shall sometimes abbreviate these operators to $t_{a,n}$). These operators represent the action of $t_a z^n \in \mathfrak{g}[[z]]$. The gauge invariance of such a Wilson line relies on the commutation relation of $\mathfrak{g}[[z]]$, i.e.,

$$[t_{a,n,W}, t_{b,m,W}] = f_{ab}{}^c t_{c,n+m,W} \tag{1.5.2}$$

being satisfied. However, at the quantum level, gauge invariance requires a deformation of the algebra (1.5.2), or more precisely, its universal enveloping algebra, $U(\mathfrak{g}[[z]])$. There are three approaches to observe the Yangian structure, namely, computing operator product expansions (OPEs) of Wilson lines, realizing the "RTT" presentation of the Yangian, or explicitly computing the two-loop corrections to gauge invariance of Wilson lines [7], whereby one can deduce that the relevant deformation ought to correspond to the Yangian algebra. We shall first consider the first two of these approaches in what follows, and subsequently dedicate a separate subsection to the third approach, which is slightly more involved.

The OPE of Wilson lines along Σ can be evaluated by bringing two parallel Wilson lines along Σ close together. Naively, one might start with two ordinary Wilson lines in representations of $\mathfrak{g}$, and expect that bringing them close together will result in another Wilson line in a representation of $\mathfrak{g}$. It turns out that this is not the case in 4d Chern–Simons theory, and instead the OPE furnishes a Wilson line that is associated with a representation of a deformation of $U(\mathfrak{g}[[z]])$. To observe this, we need to evaluate the Feynman diagram for the interaction between two Wilson lines along the x direction, and separated by a distance ϵ along the y direction, depicted in Figure 1.8, which corresponds to the amplitude

$$\frac{i}{2\pi} \hbar \left(t^a \otimes t^b f_{abc} \right)$$

$$\times \int_{x_1,x_2} \int_{x,y,z,\bar{z}} P\left(x_1 - x, y, z\right) \wedge \mathrm{d}z \wedge A^c(x, y, z) \wedge P\left(x_2 - x, y - \epsilon, z\right).$$

$$\tag{1.5.3}$$

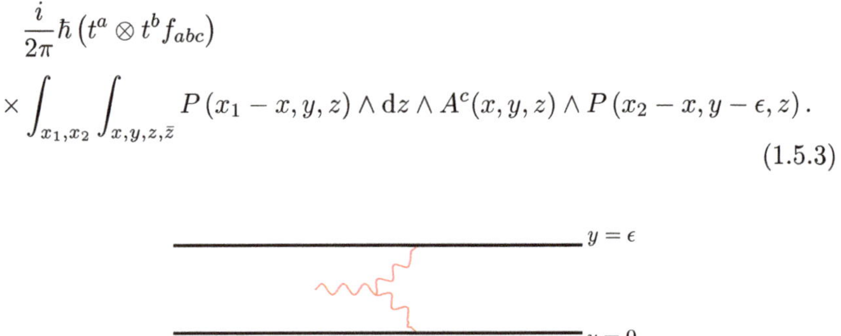

Figure 1.8. Feynman diagram corresponding to the order $\hbar$ contribution to the OPE of two Wilson lines. The coordinate along the Wilson line at $y = \epsilon$ is x_2, while the coordinate along the Wilson line at $y = 0$ is x_1.

The evaluation of this amplitude in the limit $\epsilon \to 0$ is elucidated clearly in [7], so we shall only state the result, which is

$$-\frac{\hbar}{2}\left(t^a \otimes t^b f_{abc}\right) \int \mathrm{d}x \partial_z A^c(x, y, z, \bar{z}). \tag{1.5.4}$$

The coupling to the holomorphic derivative of the gauge field indicates that we have obtained a Wilson line that is not in a representation of $\mathfrak{g}$. However, this Wilson line is not in a representation of $\mathfrak{g}[[z]]$ either, but rather a *deformation* of its universal enveloping algebra, $U(\mathfrak{g}[[z]])$. Indeed, the latter has a coproduct

$$\Delta : U(\mathfrak{g}[[z]]) \to U(\mathfrak{g}[[z]]) \otimes U(\mathfrak{g}[[z]]) \tag{1.5.5}$$

defined by

$$\Delta : t_{a,n} \mapsto t_{a,n} \otimes \mathbb{1} + \mathbb{1} \otimes t_{a,n}. \tag{1.5.6}$$

On the other hand there is a deformation of $U(\mathfrak{g}[[z]])$, namely the *Yangian*, which has a coproduct

$$\Delta_\hbar : Y(\mathfrak{g}) \to Y(\mathfrak{g}) \otimes Y(\mathfrak{g}), \tag{1.5.7}$$

defined by

$$\Delta_\hbar : t_{a,0} \mapsto t_{a,0} \otimes \mathbb{1} + \mathbb{1} \otimes t_{a,0} \tag{1.5.8}$$

$$\Delta_\hbar : t_{a,1} \mapsto t_{a,1} \otimes \mathbb{1} + \mathbb{1} \otimes t_{a,1} - \frac{\hbar}{2} f_a^{bc} t_{b,0} \otimes t_{c,0}. \tag{1.5.9}$$

We immediately recognize that the quantum correction to the OPE in (1.5.4) corresponds to the term that deforms the coproduct of the Yangian in (1.5.9). Costello used this result, together with the associativity and certain braiding properties of the OPE, to prove that in quantum 4d Chern–Simons theory, the line operators correspond to representations of the Yangian [5].

Now, we shall review another approach, from the work of Costello, Witten and Yamazaki [8], for uncovering the Yangian structure, that is able to provide the deformation of $\mathfrak{g}[[z]]$ to all orders in $\hbar$. We shall focus on the simplest case, that is the Yangian for the algebra $\mathfrak{g} = \mathfrak{gl}_N$.[2]

---

[2]The derivations for other examples of $\mathfrak{g}$ are slightly more involved as the corresponding Yangians include conditions on a "quantum determinant". This in turn requires the analysis of networks of Wilson lines [7], that lies outside of the scope of this book.

Firstly, take e^i_j to be the elementary $N \times N$ matrix, with the only non-zero entry being 1 in the (i, j) entry. This forms a basis of $\mathfrak{gl}_N$, whose generators we denote as t^i_j, The Lie algebra commutation relations are

$$[t^i_j, t^k_l] = \delta^k_j t^i_l - \delta^i_l t^k_j \,, \tag{1.5.10}$$

and the corresponding invariant bilinear form can be written as

$$(t^i_j, t^k_l) = \delta^i_l \delta^k_j \,. \tag{1.5.11}$$

We shall now consider a diagram analogous to that of Figure 1.5. However, we shall now replace the horizontal Wilson line by one that is in a representation $\mathfrak{gl}_N[[z]]$, such that $\frac{1}{k!} \partial^k_z A$ is coupled to an operator $t^i_{j,k,W}$. In addition, the vertical Wilson line will be placed at a point z in $\mathbb{C}$, and taken to be in the fundamental representation of $\mathfrak{gl}_N$, whose generators are e^i_j. Computing the order $\hbar$ contribution in an identical manner to that performed in Section 1.3, we find that it takes the form

$$\sum_{i,j} \sum_{k \geq 0} \frac{1}{k!} \partial^k_z \frac{1}{z} (t^i_{j,k} \otimes e^j_i)$$

$$= \sum_{i,j} \sum_{k \geq 0} (-1)^k \frac{1}{z^{k+1}} (t^i_{j,k} \otimes e^j_i) \,. \tag{1.5.12}$$

Now, let us place incoming and outgoing states on this vertical Wilson line denoted $\langle i|$ and $|j\rangle$, that is, above and below the horizontal Wilson line, and thereby extract the $(i, j)^{\text{th}}$ entry of this operator. This is permitted due to the infrared-freedom of the theory, due to which one can label each Wilson line by a state in its associated representation. In this way, one obtains the operator

$$T^i_j(z) : W \to W \tag{1.5.13}$$

known as a *transfer* matrix, as depicted in Figure 1.9. One can expand $T^i_j(z)$ in powers of z to obtain

$$T^i_j(z) = \delta^i_j \mathbb{1}_W + \hbar \sum_{n=0}^{\infty} (-1)^n \frac{t^i_{j,n}}{z^{n+1}}. \tag{1.5.14}$$

We can now retrieve the Yangian algebra, in a presentation known as the RTT relation, due to the topological invariance along Σ. This is done by considering a diagram analogous to that given in Figure 1.4, but where one of the diagonal Wilson lines is replaced by a Wilson line that is classically in a representation of $\mathfrak{gl}_N[[z]]$. This is depicted in Figure 1.10.

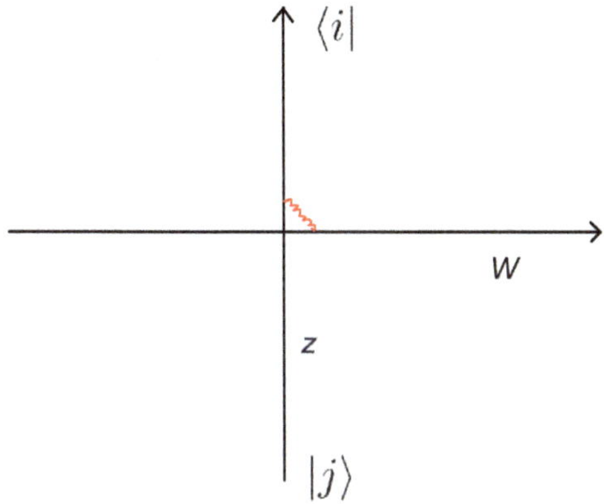

Figure 1.9.   Realization of a transfer matrix in 4d Chern-Simons theory.

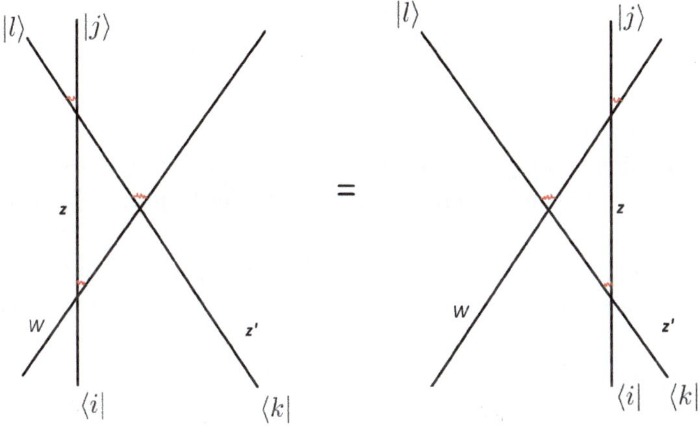

Figure 1.10.   Realization of the RTT relation in 4d Chern–Simons theory.

One immediately deduces from the diagram the following relation

$$\sum_{r,s} R^{ik}_{rs}(z - z')T^r_j(z')T^s_l(z) = \sum_{r,s} T^i_r(z)T^k_s(z')R^{rs}_{jl}(z - z') \,. \qquad (1.5.15)$$

Here, one must take note of the distinction between the entries $R^{ij}_{rs}(z-z')$ of the R-matrix which are scalar functions, and the entries T^r_j of the transfer matrix which are operators acting on the representation W. This is the sought-after RTT relation, which can be succinctly written as

$$R(z - z')T(z')T(z) = T(z)T(z')R(z - z') \,. \qquad (1.5.16)$$

This relation is known to give a presentation of the Yangian algebra for $\mathfrak{gl}_N$, with no further conditions necessary [17, 18].

Although we shall not provide details here, much of the analysis in this and previous subsections generalizes to the other two possibilities of meromorphic one-form described in Section 1.2.1, which realize trigonometric and elliptic lattice models, respectively. One can choose boundary conditions known as Manin triple boundary conditions in the trigonometric case, or restrict the gauge group in the elliptic case, such that there exists a unique classical solution [7]. This ensures that we get the simple Yang-Baxter equation, as opposed to its dynamical counterpart, and furthermore, ensures that we can realize lattice models such as the six-vertex and eight-vertex model. The underlying quantum groups can be shown, respectively, to be the quantum loop group and elliptic quantum group [8].

1.5.1 *The Yangian algebra from two-loop correction to gauge invariance of Wilson lines*

An alternate approach to deriving the Yangian algebra from 4d Chern–Simons theory is by studying quantum corrections to the gauge invariance of a Wilson line in a representation of $\mathfrak{g}[[z]]$. We shall only present the main aspects of this derivation here, leaving the reader to pursue further details in [7].

To motivate this approach, we recall that for a generic quantum field theory with gauge symmety, at tree level, there are three Feynman diagrams, depicted in Figure 1.11, that one must consider when investigating whether the matrices t_a, whereby gauge fields A^a couple to Wilson lines, satisfy the

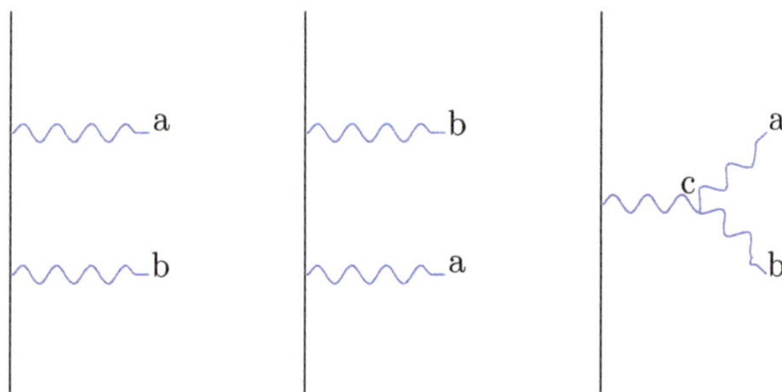

Figure 1.11.   Tree-level Feynman diagrams that are relevant to the coupling of two gauge bosons to a Wilson line.

usual Lie algebra relations. These diagrams are proportional to $t_a t_b, t_b t_a$, and $f^c_{ab} t_c$, where f_{abc} are the structure constants that appear in the cubic interaction vertex. It can be shown that the sum of the three diagrams is gauge-invariant and BRST-invariant if and only if the Lie algebra relation

$$[t_a, t_b] = f^c_{ab} t_c \tag{1.5.17}$$

is satisfied.

Four-dimensional quantum field theories often exhibit a problem where one-loop 3-point triangle diagrams introduce an anomaly that violates nonabelian gauge symmetry irreversibly, and analogous anomalies occur in other even dimensions as well. However, 4d Chern–Simons theory does not encounter this issue at order $\hbar$. Instead, we discover an anomaly at order $\hbar^2$. Nevertheless, this anomaly does not signify a total breakdown of gauge invariance, nor does it indicate that the theory is not well-defined. Instead, it signifies a *deformation* of the gauge symmetry algebra that takes the form of a quantum correction to the commutation relations of $\mathfrak{g}[[z]]$.

This interpretation is permissible in 4d Chern–Simons theory, but not in conventional gauge theories because the classification of semi-simple groups is discrete. However, the infinite-dimensional gauge algebra $\mathfrak{g}[[z]]$ is amenable to continuous deformations. Specifically, due to the anomalous diagrams involving multiple gauge bosons attached to the Wilson line (and therefore involving products of elements of $\mathfrak{g}[[z]]$), what we observe is a deformation of the universal enveloping algebra of $\mathfrak{g}[[z]]$, $U(\mathfrak{g}[[z]])$, rather than a deformation of $\mathfrak{g}[[z]]$ as a Lie algebra.

1.5.2 *The anomaly as a deformation of $U(\mathfrak{g}[[z]])$*

In [7], the possible Feynman diagrams that could contribute to the deformation of $\mathfrak{g}[[z]]$ were classified. It was shown that up permutations of vertices along the Wilson lines, there is only one diagram that is nontrivial, that is depicted in Figure 1.12.

As shown in [7], each of the Feynman diagrams depicted have the same color factor, up to contributions that can be canceled by counterterms. The color factor has the following nontrivial part :

$$f^{acf} f^{fgd} f^{geb} \{\rho(t^c), \rho(t^d), \rho(t^e)\} = ([[t^a, t^c], t^d], [t^b, t^e]) \{\rho(t^c), \rho(t^d), \rho(t^e)\} , \tag{1.5.18}$$

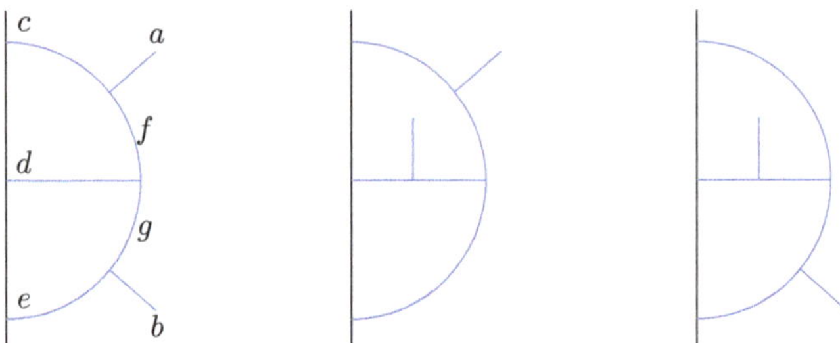

Figure 1.12.   The non-trivial Feynman diagrams contributing to the deformation of $\mathfrak{g}[[z]]$.

where a, b are indices for the external lines and

$$\{\rho(t^{a_1}), \rho(t^{a_2}), \rho(t^{a_3})\} = \tfrac{1}{3!} \sum_{\sigma \in S_3} \rho(t^{a_{\sigma(1)}})\rho(t^{a_{\sigma(2)}})\rho(t^{a_{\sigma(3)}}) \,, \qquad (1.5.19)$$

where the sum is over the permutations of the indices $1, 2, 3$.

It then remains to evaluate the numerical factor associated with the nontrivial Feynman diagrams. This is a rather involved computation, but given its importance, we have included the details of the computation in Appendix A. The result is that the numerical factor multiplied by the color factor of the anomaly is

$$\frac{\hbar^2}{12} f^{afc} f^{fgd} f^{gbe} \{\rho(t^c), \rho(t^d), \rho(t^e)\} \,. \qquad (1.5.20)$$

We can now demonstrate that in order for a Wilson line to be well defined at the quantum level, there must be quantum corrections to the Lie algebra satisfied by the operators coupled to the gauge field. As we shall see, the quantum-corrected relation determines that the Wilson line should be constructed from a representation of the Yangian, rather than the Lie algebra $\mathfrak{g}[[z]]$. The Yangian itself is a deformation of the universal enveloping algebra of $\mathfrak{g}[[z]]$, denoted $U(\mathfrak{g}[[z]])$.

Let us consider a Wilson line in a representation V of our gauge group. Assuming that the level one generators $t_{a,1}$ act through operators denoted as $\rho(t_{a,1})$, we now focus on analyzing the four two-loop diagrams depicted in Figure 1.13.

Here, a vertex labeled with the number 1 or 2 on the Wilson line indicates a coupling of the first or second z-derivative of the gauge field. It is important to note that we implicitly consider variations of these diagrams where the ordering of the vertices on the Wilson lines has been permuted.

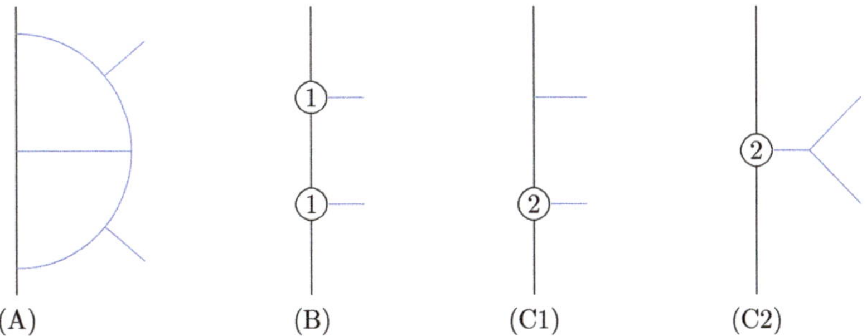

Figure 1.13. Four Feynman diagrams contributing to the anomaly.

It is observed that the amplitude of each of these diagrams may lack gauge invariance. Specifically, for the first diagram (A), this non-invariance arises from the two-loop computation we have been discussing. On the other hand, for the remaining diagrams, the failure to be gauge invariant is much more straightforward, as we will now demonstrate.

Suppose that at the quantum level, the coupling of the first and second derivatives of the gauge field to the Wilson line is governed by operators $\rho_{a,1} : V \to V$ and $\frac{1}{2}\rho_{a,2} : V \to V$, respectively, where a is an adjoint index.

For diagram (B), the amplitude can be expressed as follows:

$$\int_{p_1 < p_2 \in \mathbb{R}} \partial_z A^a(p_1) \partial_z A^b(p_2) \rho_{a,1} \circ \rho_{b,1}. \tag{1.5.21}$$

If we perform a linearized BRST transformation $A^a \mapsto A^a + \mathrm{d}c^a$, and subsequently perform integration by parts and apply Stokes' theorem, we obtain:

$$\int_{p \in \mathbb{R}} \partial_z c^a(p) \partial_z A^b(p) [\rho_{a,1}, \rho_{b,1}] . \tag{1.5.22}$$

Likewise, for diagram (C1), the amplitude is

$$\frac{1}{2} \int_{p_1 < p_2 \in \mathbb{R}} (\partial_z^2 A^a(p_1)) A^b(p_2) \rho_{a,2} \circ \rho_{b,0} + \frac{1}{2} \int_{p_1 < p_2 \in \mathbb{R}} (A^a(p_1)) \partial_z^2 A^b(p_2) \rho_{a,0} \circ \rho_{b,2} . \tag{1.5.23}$$

The obstruction to its gauge invariance is given by

$$\frac{1}{2} \int_{p \in \mathbb{R}} (\partial_z^2 c^a(p)) A^b(p) [\rho_{a,2}, \rho_{b,0}] + \frac{1}{2} \int_{p \in \mathbb{R}} c^a(p) \partial_z^2 A^b(p) [\rho_{a,0}, \rho_{b,2}] . \tag{1.5.24}$$

Next, diagram (C2) can be evaluated to give

$$\tfrac{1}{4}\int_{p\in\mathbb{R},\,v=(x,y,z)} \mathrm{d}z\,\partial^2_{z_p} P(p,v)A^a(v)A^b(v)f_{ab}{}^c\rho_{c,2}\,. \tag{1.5.25}$$

In this context, v denotes the position of the internal vertex, and ∂_{z_p} indicates the application of a z-derivative to the p-coordinate of the propagator. The failure of this to be gauge-invariant is

$$-\tfrac{1}{2}\int_{p\in\mathbb{R}} \partial^2_z\left(c^a(p)A^b(b)\right)f_{ab}{}^c\rho_{c,2}\,, \tag{1.5.26}$$

where we have used the fact that $\mathrm{d}z_v \wedge \mathrm{d}P(p,v) = \delta_{p=v}$ and imposed the equations of motion for A.

Finally, recall that the anomaly for diagram (A) is

$$\tfrac{\hbar^2}{12}\int_{p\in\mathbb{R}} \partial_z c^a(p)\partial_z A^b(p)f^{afc}f^{fgd}f^{gbe}\{\rho(t^c),\rho(t^d),\rho(t^e)\}\,. \tag{1.5.27}$$

The sum of the anomalies for the diagrams (A), (B) and (C) must be zero. This implies the equations

$$[\rho(t_{a,2}),\rho(t_{b,0})] = f_{ab}{}^c\rho(t_{c,2})\,, \tag{1.5.28}$$

$$[\rho(t_{a,1}),\rho(t_{b,1})] = f_{ab}{}^c\rho(t_{c,2}) - \tfrac{\hbar^2}{12}Q_{ab}\,, \tag{1.5.29}$$

where the first equation arises from the cancellation of the coefficients of $(\partial^2_z c^a(p))A^b(p)$ and $c^a(p)\partial^2_z A^b(p)$, and the second from those of $(\partial_z c^a(p))\partial_z A^b(p)$. In the equations above, we have written $\rho_{a,n}$ explicitly as $\rho(t_{a,n})$ and we have defined

$$Q_{ab} = f^{afc}f^{fgd}f^{gbe}\{\rho(t^c),\rho(t^d),\rho(t^e)\}\,. \tag{1.5.30}$$

The first equation is the classical commutation relation. The second equation implies that the operators $\rho(t_{a,1})$ do not commute to give $\rho(t_{c,2})$, as one might find classically, but rather a linear combination of $\rho(t_{c,2})$ and a particular cubic polynomial in the level 0 generators $\rho(t_{a,0})$.

By decomposing the exterior square $\wedge^2\mathfrak{g}$ into $\mathfrak{g} \oplus \wedge^2_0\mathfrak{g}$, where $\wedge^2_0\mathfrak{g}$ represents the kernel of the Lie bracket map from $\wedge^2\mathfrak{g}$ to $\mathfrak{g}$, we can express the second relation as a sum of two separate and independent relations.

$$f^{ab}{}_c[\rho(t_{a,1}),\rho(t_{b,1})] = \mathsf{h}^\vee\rho(t_{c,2}) - \tfrac{\hbar^2}{12}f^{ab}{}_c Q_{ab}\,, \tag{1.5.31}$$

$$\Lambda^{ab}[\rho(t_{a,1}),\rho(t_{b,1})] = -\tfrac{\hbar^2}{12}\Lambda^{ab}Q_{ab}\,, \tag{1.5.32}$$

where $\Lambda^{ab} \in \wedge^2 \mathfrak{g}$. The first relation can be satisfied via a redefinition of the operator $\rho(t_{c,2})$ to

$$\rho'(t_{c,2}) = \rho(t_{c,2}) - \frac{\hbar^2}{12h^\vee} f^{ab}{}_c Q_{ab} \,. \tag{1.5.33}$$

The second relation, however, cannot be satisfied in such a simple way.

This analysis reveals that to quantize a classical Wilson line modulo $\hbar^3$, it is necessary to define the operators $\rho(t_{a,1})$ in a manner that satisfies the relation (1.5.32). This relation belongs to the set of relations in the Yangian algebra, which demonstrates its nature as a non-trivial deformation of the universal enveloping algebra of $\mathfrak{g}[[z]]$. With some work, it can be shown that the relation obtained is equivalent to the known relation for the Yangian, given, e.g., in eqn. (4) in Theorem 12.1.1 of [18].

1.6 Framing Anomaly on Σ

We have previously discussed the framing anomaly of Wilson lines in 4d Chern–Simons theory. We shall now review another framing anomaly in 4d CS, namely the framing anomaly of the topological surface Σ. As we shall see, to avoid the anomaly, we shall require that Σ is a framed surface. Although this framing anomaly was discussed in the seminal works on 4d CS [5–7], we shall mostly follow the derivation of Khan [19].

The derivation of this anomaly involves the computation of the one-loop effective action of 4d CS as a functional of a background field that solves the 4d CS equations of motion, and performing a gauge variation of this effective action with respect to the aforementioned background field.

Let us decompose the 4d CS gauge field as

$$A + B \tag{1.6.1}$$

where A is a background field that solves the equations of motion, and B is a fluctuation field. The 4d CS action can then be written as

$$S[A+B] = S[A] + \frac{1}{2\pi} \int \omega \wedge \mathrm{Tr}\left(B \wedge \mathrm{d}_A B + \frac{1}{3} B \wedge [B, B] \right) \tag{1.6.2}$$

where $\mathrm{d}_A B = \mathrm{d}B + [A, B]$ with $[X, Y] := X \wedge Y - (-1)^{|X||Y|} Y \wedge X$. Given that this action is invariant under the gauge transformation

$$B \to B + \mathrm{d}_A \epsilon + [B, \epsilon], \tag{1.6.3}$$

we require a choice of gauge fixing to eliminate redundant degrees of freedom in the path integral. This shall be achieved using the Batalin-Vilkovisky (BV) formalism.

The BV formalism goes beyond the usual BRST formalism, as it requires not just the introduction of a ghost field c, but also an anti-field $B^\vee$ for the field B, and an anti-field $c^\vee$ of the ghost field c.

As we shall review below, the BV field space realizes a differential graded Lie algebra with an odd symplectic pairing. The BV field space is given by

$$\Omega^*_{\mathrm{CS}_4}(\Sigma \times C; \mathfrak{g}) = \bigoplus_{m+n=*} \Omega^m_{\mathrm{dR}}(\Sigma; \mathfrak{g}) \otimes \Omega^{(0,n)}_{\bar\partial}(C; \mathfrak{g}), \qquad (1.6.4)$$

where $\Omega^m_{\mathrm{dR}}(\Sigma; \mathfrak{g})$ denotes the space of $\mathfrak{g}$-valued m-forms on Σ, and $\Omega^{(0,n)}_{\bar\partial}(C; \mathfrak{g})$ denotes the space of $\mathfrak{g}$-valued $(0,n)$ forms on C. In addition, there is a gradation by form degree F, which is related to the ghost number/homological grading by

$$\mathrm{gh} = 1 - F. \qquad (1.6.5)$$

In particular, the ghost field

$$c \in \Omega^0_{\mathrm{CS}_4}(\Sigma \times C; \mathfrak{g}), \qquad (1.6.6)$$

has ghost number 1, while the fluctuation field

$$B_i \mathrm{d}x^i + B_{\bar z} \mathrm{d}\bar z \in \Omega^1_{\mathrm{CS}_4}(\Sigma \times C; \mathfrak{g}), \qquad (1.6.7)$$

has ghost number 0. Moreover, the two form anti-field of the fluctuation field,

$$B^\vee_{ij} \mathrm{d}x^i \wedge \mathrm{d}x^j + B^\vee_{i\bar z} \mathrm{d}x^i \wedge \mathrm{d}\bar z \in \Omega^2_{\mathrm{CS}_4}(\Sigma \times C; \mathfrak{g}), \qquad (1.6.8)$$

has ghost number -1, while the three-form field antifield of the ghost field

$$c^\vee_{ij\bar z} \mathrm{d}x^i \wedge \mathrm{d}x^j \wedge \mathrm{d}\bar z \in \Omega^3_{\mathrm{CS}_4}(\Sigma \times C; \mathfrak{g}). \qquad (1.6.9)$$

has ghost number -2.

By combining the wedge product of forms with the Lie bracket on $\mathfrak{g}$, we arrive at a Lie algebra structure on the BV field space, with the Lie bracket having ghost number -1. The differential on the BV field space is given by

$$\mathrm{d}_{\mathrm{CS}_4} = \mathrm{d}_A(\Sigma) + \bar\partial_A(C), \qquad (1.6.10)$$

which is the sum of the twisted de Rham exterior derivative d_A along Σ, and the twisted Dolbeault exterior derivative $\bar\partial_A$ along C.

The aforementioned odd symplectic pairing on the BV field space is given by

$$\langle \alpha, \beta \rangle := \int_{\Sigma \times C} \omega \wedge \mathrm{Tr}(\alpha \wedge \beta). \qquad (1.6.11)$$

Then, for a field $\mathcal{B} \in \Omega^*_{\mathrm{CS}_4}(\Sigma \times C; \mathfrak{g})$ in the BV field space, which we can formally write as

$$\mathcal{B} = c + B + B^{\vee} + c^{\vee}, \tag{1.6.12}$$

the BV action is the generalized Chern–Simons action defined using (1.6.11):

$$S_{\mathrm{BV}}[\mathcal{B}] = \frac{1}{2\pi} \langle \mathcal{B}, \mathrm{d}_{\mathrm{CS}_4}\mathcal{B} + \frac{1}{3}[\mathcal{B}, \mathcal{B}] \rangle, \tag{1.6.13}$$

(where $[\alpha, \beta] := \alpha \wedge \beta + \beta \wedge \alpha$ on BV field space) which is equivalent to

$$S_{\mathrm{BV}} = \frac{1}{2\pi} \int_{\Sigma \times C} \omega \wedge \mathrm{Tr}(B \wedge \mathrm{d}_A B + \frac{1}{3} B \wedge [B, B])$$
$$+ \frac{1}{\pi} \int_{\Sigma \times C} \omega \wedge \mathrm{Tr}(B^{\vee} \wedge (\mathrm{d}_A c + [B, c])) + \frac{1}{2\pi} \int_{\Sigma \times C} \omega \wedge \mathrm{Tr}(c^{\vee}[c, c]). \tag{1.6.14}$$

Gauge-fixing in the language of the BV formalism is performed by picking a Lagrangian subspace L of the BV field space, in a way such that the quadratic part of the action becomes non-degenerate along this subspace. To define the gauge-fixing condition in the present context, one first picks a Riemannian metric along Σ, with components denoted g_{ij}, and a Kähler metric along C, with components denoted $g_{z\bar{z}}$.

We then define the operator

$$\delta_A = \left(\mathrm{d}_A(\Sigma)\right)^{\dagger} + 2\left(\overline{\partial}_A(C)\right)^{\dagger}, \tag{1.6.15}$$

where we have used the metric to define adjoints of the twisted de Rham and Dolbeault operators on Σ and C respectively, given by

$$\mathrm{d}_A(\Sigma)^{\dagger} = g^{ij} \iota_{\frac{\partial}{\partial x^i}} D_j, \tag{1.6.16}$$

$$\overline{\partial}_A(C)^{\dagger} = g^{z\bar{z}} \iota_{\frac{\partial}{\partial \bar{z}}} D_z = g^{z\bar{z}} \iota_{\frac{\partial}{\partial \bar{z}}} \partial_z, \tag{1.6.17}$$

with D_i defined with respect to both the background gauge field A and the Christoffel connection on Σ, while D_z is just the ordinary derivative ∂_z since we are working with partial connections and a Kähler metric on C.

The crucial property of the operator (1.6.15) is that its action on one-forms

$$\delta_A : \Omega^1_{\mathrm{CS}_4} \rightarrow \Omega^0_{\mathrm{CS}_4} \tag{1.6.18}$$

is

$$\delta_A(B_i\mathrm{d}x^i + B_{\bar{z}}\mathrm{d}\bar{z}) = g^{ij}D_iB_j + 2g^{z\bar{z}}\partial_z B_{\bar{z}}. \tag{1.6.19}$$

The Lagrangian subspace L can then be defined as the kernel of δ_A,

$$L = \{X \in \Omega^*_{\mathrm{CS}_4} | \delta_A X = 0\}, \tag{1.6.20}$$

which is equivalently the analog of the Lorentz gauge for B, i.e.,

$$g^{ij}D_iB_j + 2g^{z\bar{z}}\partial_z B_{\bar{z}} = 0, \tag{1.6.21}$$

where, as before, D_i is the covariant derivative with respect to both the gauge field components along Σ, and the Christoffel connection on Σ.

The path integral with integration restricted to L

$$Z[A] = \int_{X \in L} DX\, e^{-S[X]} \tag{1.6.22}$$

is (formally) non-degenerate as a result, and we are then in a position to compute the quantum effective action

$$\Gamma[A] = -\log Z[A]. \tag{1.6.23}$$

To compute the one-loop effective action, we only need to compute the (super)-determinant of the operator d_A. This can be shown to be equivalent to the square-root of a "holomorphic-topological" torsion, defined analogously to the standard topological torsion encountered in 3d Chern–Simons theory. As we shall see, the upshot here is that this torsion, computed via one-loop diagrams shown in Figure 1.14, is not an invariant of the gauge equivalence class of the background field A, as long as Σ has non-trivial curvature. Instead, the amplitudes associated with these diagrams, and thus the one-loop effective action, vary nontrivially under gauge transformations.

To simplify analysis, in what follows, we shall restrict ourselves to $C = \mathbb{C}$, with the metric $g_{z\bar{z}} = \frac{1}{2}$ and $\omega = \mathrm{d}z$. We recall an additional crucial property of the operator δ_A now, namely, that it can be used to define a generalized Hodge Laplacian. To be precise, the operator

$$\Delta(A, g) := \delta_A\mathrm{d}_A + \mathrm{d}_A\delta_A, \tag{1.6.24}$$

becomes the standard Hodge Laplacian with respect to the metric on $\Sigma \times C$, when $A = 0$ and $\mathfrak{g} = \mathbb{C}$.

To compute the one-loop diagrams of Figure 1.14, we can utilize the relevant propagators, as done in earlier sections. It shall be convenient,

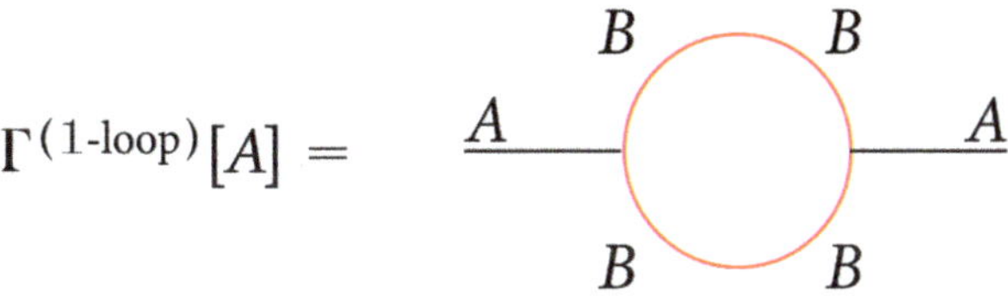

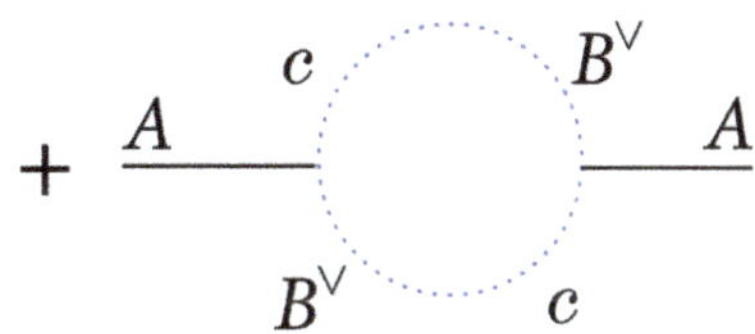

Figure 1.14. Feynman diagrams that contribute to the one-loop effective action as a functional of the classical solution A.

however, to express each propagator, P, in terms of a Green's function of the Hodge Laplacian acting on $L^\perp = \mathrm{Ker}\, \mathrm{d}_A \subset \Omega^*_{\mathrm{CS}_4}$, i.e.,

$$P(x,y) = \pi\delta_{A,x}G(x,y). \tag{1.6.25}$$

Explicitly, the two-point function of the fluctuation field B is

$$P_{BB}(x,y) = \pi\delta_{A,x}G(x,y)|_{\Omega^1_{\mathrm{CS}_4}}, \tag{1.6.26}$$

while the propagator involving the fields c and $B^\vee$ is given by

$$P_{cB^\vee}(x,y) = \pi\delta_{A,x}G(x,y)|_{\Omega^0_{\mathrm{CS}_4}}. \tag{1.6.27}$$

Let us first consider the amplitude whose internal fields only involve the fluctuation field B. This is given by

$$\frac{1}{8\pi^2}\int_{(\Sigma\times C)^2} f_{abc}f_{def}\,\mathrm{d}z_1\,\mathrm{d}z_2\wedge A^a(x_1)\wedge P^{bc}_{BB}(x_1,x_2)\wedge P^{de}_{BB}(x_2,x_1)\wedge A^f(x_2).$$

In order to regularize the UV divergent amplitude, we express the Green's function in terms of the heat kernel of the Hodge Laplacian $\Delta(A,g)$ acting on $\Omega^*_{\mathrm{CS}_4}(\Sigma\times\mathbb{C};\mathfrak{g})$, i.e.,

$$G(x,y) = \int_0^\infty \mathrm{d}\tau\, K_{\Omega^*_{\mathrm{CS}_4}}(x,y;\tau), \tag{1.6.28}$$

where the heat kernel is defined as

$$K_{\Omega^*_{\mathrm{CS}_4}}(x,y;\tau) = \sum_n e^{\lambda_n\tau}\big(*\,\omega_n(x)\big)\wedge\big(\omega_n(y)\big) \tag{1.6.29}$$

where $\{\omega_n\}$ denotes a basis of eigenforms of $\Delta(A, g)$ with respective eigenvalues $\{\lambda_n\}$, and where the Hodge star is defined with respect to $\epsilon^{xy\bar{z}} = 1$.

It is sufficient to consider the amplitude to quadratic order in the background field A. Given that there are already two external A-fields, one can employ the heat kernel for $\Delta(A, g)$ where $A = 0$. This is given explicitly by

$$K^{ab}_{\Omega^1_{\mathrm{CS}_4}(\Sigma \times C; \mathfrak{g})}(x, y; \tau)\big|_{A=0} = \delta^{ab} K_{\Omega^1_{\mathrm{CS}_4}(\Sigma \times C)}(x, y; \tau), \qquad (1.6.30)$$

the second factor being the usual heat kernel for the Hodge Laplacian acting on one-forms. The amplitude thus takes the form

$$\frac{1}{4} \mathrm{h}^\vee \delta_{ab} \int \mathrm{d}\tau_1 \mathrm{d}\tau_2 \mathrm{d}z_1 \mathrm{d}z_2 \, A^a(x_1) \wedge \delta_{x_1} K_{\Omega^1_{\mathrm{CS}_4}(\Sigma \times C)}(x_1, x_2, \tau_1)$$
$$\wedge \delta_{x_2} K_{\Omega^1_{\mathrm{CS}_4}(\Sigma \times C)}(x_2, x_1, \tau_2) \wedge A^b(x_2). \qquad (1.6.31)$$

For the sake of simplifying the computation, we shall further specify the background field. We shall specify it to be a solution of the equations of motion of the form

$$A = A_{\bar{z}}(z, \bar{z}) \mathrm{d}\bar{z}, \qquad (1.6.32)$$

which solves the equations of motion since it is constant along Σ. Note that this solution is more general than the trivial solution $(A = 0)$ used in the computation of rational R-matrices.

We now use the factorization property of the heat kernel

$$K_{\Omega^l(M_1 \times M_2)}\big((x_1, x_2), (y_1, y_2); \tau\big)$$
$$= \sum_{m+n=l} K_{\Omega^m(M_1)}(x_1, y_1; \tau) \wedge K_{\Omega^n(M_2)}(x_2, y_2; \tau), \qquad (1.6.33)$$

whereby (1.6.31) can be written as

$$\frac{\mathrm{h}^\vee}{4} \delta_{ab} \int \mathrm{d}\tau_1 \mathrm{d}\tau_2 \mathrm{d}z_1 \mathrm{d}z_2 \, A^a(x_1)$$

$$\wedge \delta_{x_1} \bigg(K_{\Omega^1(\Sigma)}\big((x_1, y_1), (x_2, y_2), \tau_1\big) K_{\Omega^0(\mathbb{C})}(z_1, z_2, \tau_1)$$

$$+ K_{\Omega^0(\Sigma)}\big((x_1, y_1), (x_2, y_2), \tau_1\big) K_{\Omega^1(\mathbb{C})}(z_1, z_2, \tau_1) \bigg) \qquad (1.6.34)$$

$$\wedge \delta_{x_2} \bigg(K_{\Omega^1(\Sigma)}\big((x_2, y_2), (x_1, y_1), \tau_2\big) K_{\Omega^0(\mathbb{C})}(z_2, z_1, \tau_2)$$

$$+ K_{\Omega^0(\Sigma)}\big((x_2, y_2), (x_1, y_1), \tau_2\big) K_{\Omega^1(\mathbb{C})}(z_2, z_1, \tau_2) \bigg) \wedge A^b(x_2),$$

where x and y denote coordinates on Σ. Now, note that when δ_{x_1} and δ_{x_2} act on heat kernels associated with Σ in (1.6.34), the resulting expressions always contain a factor of either $d\bar{z}_1$ or $d\bar{z}_2$. This follows since, from standard properties of heat kernels, we know that $K_{\Omega^0(\mathbb{C})}(z_1, z_2, \tau_1)$ and $K_{\Omega^1(\mathbb{C})}(z_2, z_1, \tau_2)$ are proportional to $d\bar{z}_1$, while $K_{\Omega^0(\mathbb{C})}(z_2, z_1, \tau_2)$ and $K_{\Omega^1(\mathbb{C})}(z_1, z_2, \tau_1)$ are proportional to $d\bar{z}_2$. The wedge product of each of these expressions with the background gauge fields is thus equal to zero.

Next, consider the first term of the factorized heat kernel in each parenthesis in (1.6.34). The action of δ_{x_1} and δ_{x_2} on $K_{\Omega^0(\mathbb{C})}(z_1, z_2, \tau_1)$ and $K_{\Omega^0(\mathbb{C})}(z_2, z_1, \tau_2)$ results in 0-forms.

Moreover, $K_{\Omega^1(\Sigma)}((x_1, y_1), (x_2, y_2), \tau_1)$ and $K_{\Omega^1(\Sigma)}((x_2, y_2), (x_1, y_1), \tau_2)$ are each $(1,1)$ forms on $\Sigma \times \Sigma$. This implies that the first term in each parenthesis conspires to produce a top form on $\Sigma \times \Sigma$.

On the other hand, δ_{x_1} and δ_{x_2} annihilate $K_{\Omega^1(\mathbb{C})}(z_1, z_2, \tau_1)$ and $K_{\Omega^1(\mathbb{C})}(z_1, z_2, \tau_2)$, respectively. Hence, we find that (1.6.34) is

$$\frac{\mathsf{h}^\vee}{4}\delta_{ab} \int d\tau_1 d\tau_2 dz_1 dz_2 \, A^a(x_1)$$
$$\wedge K_{\Omega^1(\Sigma)}((x_1, y_1), (x_2, y_2), \tau_1)\delta_{x_1} K_{\Omega^0(\mathbb{C})}(z_1, z_2, \tau_1) \tag{1.6.35}$$
$$\wedge K_{\Omega^1(\Sigma)}((x_2, y_2), (x_1, y_1), \tau_2)\delta_{x_2} K_{\Omega^0(\mathbb{C})}(z_2, z_1, \tau_2) \wedge A^b(x_2).$$

Next, we use the composition property of the heat kernel

$$\int_M K_{\Omega^*(M)}(x, y, \tau) \wedge K_{\Omega^*(M)}(y, z, \tau') = K_{\Omega^*(M)}(x, z, \tau + \tau'), \tag{1.6.36}$$

whereby (1.6.35) becomes

$$\frac{\mathsf{h}^\vee}{4}\delta_{ab} \int d\tau_1 d\tau_2 dz_1 dz_2 \, A^a(x_1) \wedge K_{\Omega^1(\Sigma)}((x_1, y_1), (x_1, y_1), \tau_1 + \tau_2)$$
$$\times \delta_{x_1} K_{\Omega^0(\mathbb{C})}(z_1, z_2, \tau_1)\delta_{x_2} K_{\Omega^0(\mathbb{C})}(z_2, z_1, \tau_2) \wedge A^b(x_2), \tag{1.6.37}$$

Using (1.6.19) and $g_{z\bar{z}} = \frac{1}{2}$, we find

$$\delta_{x_1} K_{\Omega^0(\mathbb{C})}(z_1, z_2, \tau_1)\delta_{x_2} K_{\Omega^0(\mathbb{C})}(z_2, z_1, \tau_2)$$
$$= \partial_{z_1} K_{\mathbb{C}}(z_1, z_2, \tau_1)\partial_{z_2} K_{\mathbb{C}}(z_2, z_1, \tau_2) \tag{1.6.38}$$

where

$$K_{\mathbb{C}}(z_1, z_2; \tau) = \frac{1}{4\pi\tau}e^{-\frac{|z_1 - z_2|^2}{4\tau}}. \tag{1.6.39}$$

We also make use of the short-time asymptotics [20] of the one-form heat kernel on the surface Σ

$$K_{\Omega^1(\Sigma)}(x, x; \tau) = \frac{dV_\Sigma}{4\pi\tau} - \frac{R_\Sigma dV_\Sigma}{12\pi} + O(\tau), \tag{1.6.40}$$

where R_Σ denotes the Ricci scalar of Σ and dV_Σ denotes the volume form on Σ. This identity leads us to the following form for the amplitude:

$$- 4h^\vee \frac{\delta_{ab}}{12\pi} \int d\tau_1 d\tau_2 d^2 z_1 d^2 z_2 \left(\frac{dV_\Sigma}{4\pi(\tau_1 + \tau_2)} - \frac{R_\Sigma dV_\Sigma}{12\pi} + O(\tau_1 + \tau_2) \right)$$

$$\times A_{\bar{z}}^a(z_1, \bar{z}_1) A_{\bar{z}}^b(z_2, \bar{z}_2) \partial_{z_1} K_\mathbb{C}(z_1, z_2, \tau_1) \partial_{z_2} K_\mathbb{C}(z_2, z_1, \tau_2). \qquad (1.6.41)$$

We shall focus on the term that depends on R_Σ in what follows. As we shall see, the other terms will cancel with their respective contributions from the ghost-antifield loop.

We can now use (1.6.39) and perform the (τ_1, τ_2) integrals using $\int_0^\infty d\tau \frac{1}{\tau^2} e^{-\frac{a}{\tau}} = \frac{1}{a}$ to give

$$4h^\vee \frac{\delta_{ab}}{12\pi} \left(\frac{1}{4\pi} \right)^2 \int dV_\Sigma\, R_\Sigma(x) d^2 z_1 d^2 z_2 A_{\bar{z}}^a(z_1, \bar{z}_1) A_{\bar{z}}^b(z_2, \bar{z}_2) \frac{1}{(z_1 - z_2)^2}.$$
$$(1.6.42)$$

We then perform a gauge transformation $A_{\bar{z}} \to D_{\bar{z}}\varepsilon$ and keep only the linear term in A since the other term that is generated may be canceled via addition of a counterterm.

Then, upon integration by parts, the standard complex analysis lemma

$$\partial_{\bar{z}} \frac{1}{z^2} = -2\pi \partial_z \delta^{(2)}(z, \bar{z}) \qquad (1.6.43)$$

implies that the contribution to anomaly from the BB loop is

$$\mathcal{A}_{BB} = \frac{h^\vee}{12\pi^2} \int_{\Sigma \times \mathbb{C}} dV_\Sigma\, d^2 z\, R_\Sigma \mathrm{Tr}(\varepsilon\, \partial_z A_{\bar{z}}). \qquad (1.6.44)$$

The evaluation of the Feynman diagram where the ghost field and gauge anti-field propagate in the loop is analogous to the computation above, the main difference being the use of the short-time asymptotics of the diagonal heat kernel of the Hodge Laplacian acting on *zero-forms* on Σ:

$$K_{\Omega^0(\Sigma)}(x, x; \tau) = \frac{dV_\Sigma}{4\pi\tau} + \frac{R_\Sigma dV_\Sigma}{12\pi} + O(\tau). \qquad (1.6.45)$$

The contribution to the anomaly from the $cB^\vee$ loop is $\frac{1}{2}$ times the BB-anomaly

$$\mathcal{A}_{cB^\vee} = \frac{\mathsf{h}^\vee}{24\pi^2} \int_{\Sigma \times \mathbb{C}} dV_\Sigma \, d^2z \, R_\Sigma \mathrm{Tr}(\varepsilon \, \partial_z A_{\bar{z}}), \qquad (1.6.46)$$

Notably, the singular terms in $\tau_1 + \tau_2$ cancel when the BB and $cB^\vee$ contributions to the anomaly are summed. The final expression for the total anomaly

$$\mathcal{A} = \mathcal{A}_{BB} + \mathcal{A}_{cB^\vee}$$

is therefore given by

$$\mathcal{A} = \frac{\mathsf{h}^\vee}{8\pi^2} \int_{\Sigma \times \mathbb{C}} dV_\Sigma \, d^2z \, R_\Sigma \mathrm{Tr}(\varepsilon \, \partial_z A_{\bar{z}}). \qquad (1.6.47)$$

This indicates the presence of a gauge anomaly due to the curvature of Σ in 4d Chern–Simons theory.

One obvious way to cure this anomaly is to insist that the curvature of Σ vanishes, i.e., by requiring the tangent bundle of Σ to be trivialized, meaning that we require Σ to be a *framed* surface. This is why the gauge anomaly derived above is usually referred to as the framing anomaly, although as we shall see shortly, there are other means to cancel this anomaly.

1.6.1 *Anomaly cancelation using holomorphic surface operators*

Another class of operators that 4d CS admits is surface defects wrapped along the complex surface C. These surface defects are usually defined as two-dimensional quantum field theories that depend on the complex structure of C. Such surface defects in 4d CS have been studied by Khan in [19], and the reader is encouraged to refer to this work for more details on them.

Our main aim in this subsection is to illustrate how such surface defects can be used as an alternate approach to cancel the gauge anomaly derived above. To this end, let us consider a particular surface defect wrapped along C, namely, a gauged $\beta - \gamma$ system, with the action

$$\frac{1}{2\pi} \int_C d^2z (\beta_i \partial_{\bar{z}} \gamma^i + A_{\bar{z}}^a J_a) \qquad (1.6.48)$$

where

$$J_a = \beta_i (T_a)^i{}_j \gamma^j \qquad (1.6.49)$$

is a current that generates an affine Kac–Moody algebra, with T_a denoting a particular representation of the Lie algebra of G under which γ^j transforms. This algebra can be derived using the standard $\beta - \gamma$ operator product expansion

$$\beta_i(z)\gamma^j(z') \sim -\frac{\delta_i{}^j}{z - z'} + \text{regular terms.} \tag{1.6.50}$$

Such a surface operator suffers from a gauge anomaly, and this holds in general for a surface operator supporting an affine Kac–Moody algebra. It is known that if one performs a gauge transformation of the effective action of such a surface operator, one obtains

$$\frac{k}{2\pi} \int_C \mathrm{d}^2 z \,\mathrm{Tr}(\varepsilon \partial_z A_{\bar z}), \tag{1.6.51}$$

where k is the level of the algebra.[3] A derivation of this result is given in Appendix B.

Now, consider such a surface operator in 4d CS located at a point $w = w_0$ on Σ, where w is shorthand for the local coordinates on Σ. Let us further specify the curvature of Σ to be localized at w_0

$$\sqrt{g_\Sigma} R_\Sigma = 4\pi \delta^{(2)}(w - w_0). \tag{1.6.52}$$

This corresponds to Σ having the topology of a cigar, as depicted in Figure 1.15. Now, we see that if the level of the affine Kac–Moody algebra

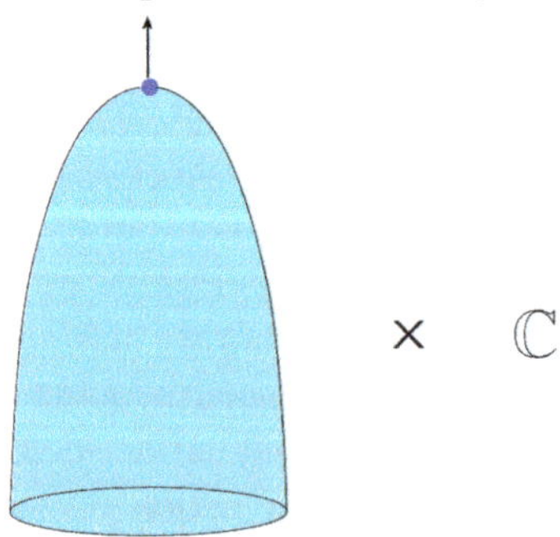

Figure 1.15.   Anomaly cancelation between gauge anomalies of 4d Chern–Simons theory and holomorphic surface defect on $C = \mathbb{C}$.

---

[3]This anomaly can be shown to satisfy the Wess-Zumino consistency condition.

on the surface defect is critical

$$k + \mathsf{h}^\vee = 0, \tag{1.6.53}$$

then the four-dimensional framing anomaly and the two-dimensional gauge anomaly of the surface operator cancel. As a result, we find that the 4d-2d coupled system is gauge anomaly-free and well defined, at least to one loop. Gauge anomaly cancellation is expected to hold to all loops, since 4d Chern–Simons theory and the theory on the holomorphic surface operator both admit a one-loop exact quantization [5] (see also the work of Gwilliam and Williams [141]).

1.7   Baxter's TQ Relations and 't Hooft Lines

Although we have focused on integrable lattice models thus far, it is straightforward to realize integrable spin chains using Wilson lines in 4d Chern–Simons theory. Given Cartesian coordinates, x and y, on $\Sigma = \mathbb{R}^2$, we can place a finite number, say K, of Wilson lines along the y-direction, separated along the x-direction, and at $z = 0$. If we then place a Wilson line along the x direction, at some point $z = \tilde{z}$, we can then realize the transfer matrix associated with a spin chain with K sites.

A seminal development in the study of integrable systems, in particular, integrable spin chains, is the Bethe ansatz, which is essential in the derivation of the spectrum of a spin chain. It has been known for some time that the Bethe ansatz equations can be restated as an equation involving the monodromy operator T, as well as a new operator, denoted Q, known as Baxter's TQ relation. A discussion of the relation between these equations can be found, for example, in [21].

It is thus a crucial endeavor to interpret this Q-operator in 4d Chern–Simons theory. Seminal work in this direction was carried out by Costello, Gaiotto and Yagi [22], where the Q-operator was interpreted in terms of 't Hooft lines. In the following, we shall review some basic aspects of the derivation of the TQ-relations in 4d CS, and leave the reader to pursue further details in the original reference [22].

1.7.1   *'t Hooft lines in 4d Chern–Simons theory*

't Hooft lines are widely studied in the context of four-dimensional gauge theory. For example in 4d $U(1)$ Maxwell gauge theory, a line operator with no electric charge but with magnetic charge k is defined to be a 't Hooft line. Here, the magnetic charge is defined by evaluating the Chern class of the gauge field on the two-sphere that surrounds the line, where it is the connection of a non-trivial $U(1)$ principal bundle.

To satisfy the equations of motion, the field strength generated by the 't Hooft line is that of a Dirac monopole singularity,

$$F \sim \frac{\epsilon_{ijk} x^i \, \mathrm{d}x^j \wedge \, \mathrm{d}x^k}{\|x\|^3} + \text{regular terms}, \tag{1.7.1}$$

with x^i parameterizing the directions normal to the 't Hooft line.

Since we are interested in a non-abelian 4d gauge theory, we should try to understand how an operator of this type can be constructed in 4d Yang–Mills theory with gauge group G. To this end, we need to embed the $U(1)$ Dirac monopole singularity into G, which amounts to a choice of cocharacter $\rho : U(1) \to G$. The field strength in this case has the form

$$F \sim \frac{\rho \, \epsilon_{ijk} x^i \, \mathrm{d}x^j \wedge \, \mathrm{d}x^k}{\|x\|^3} + \text{regular terms}, \tag{1.7.2}$$

where ρ is regarded as an element of the Lie algebra, $\mathfrak{g}$. The magnetic charge of this 't Hooft line can be defined by associating a $U(1)$ bundle to the G-bundle defined on the two-sphere surrounding the 't Hooft line via ρ, and evaluating the Chern class of the $U(1)$ bundle.

When defining 't Hooft lines in the setting of the topological-holomorphic theory that is 4d Chern–Simons theory, it is more convenient to work with cylinders instead of two-spheres. Let us set the topological surface Σ to be $\mathbb{R}^2$ and parameterize it by coordinates x and y, and set C to be $\mathbb{C}$, parameterized, as usual, by z and $\bar{z}$. We are now interested in a 't Hooft line along the x direction, and will describe it in terms of a holomorphic bundle on the boundary of the solid cylinder where $|z| \le \epsilon$ and $|y| \le \epsilon$, for infinitesimal ϵ, as depicted in Figure 1.16. The boundary is divided into the following regions :

$$y = -\epsilon, \ |z| \le \epsilon, \tag{1.7.3}$$

$$-\epsilon \le y \le \epsilon, \ |z| = \epsilon, \tag{1.7.4}$$

$$y = \epsilon, \ |z| \le \epsilon. \tag{1.7.5}$$

Since the equations of motion do not include a flatness condition along C, up to gauge transformations, a solution to these equations can be described by a holomorphic bundle on the discs $|z| \le \epsilon$ located at the points $y = -\epsilon$ and $y = \epsilon$, with an isomorphim between the two bundles at the boundary $|z| = \epsilon$.

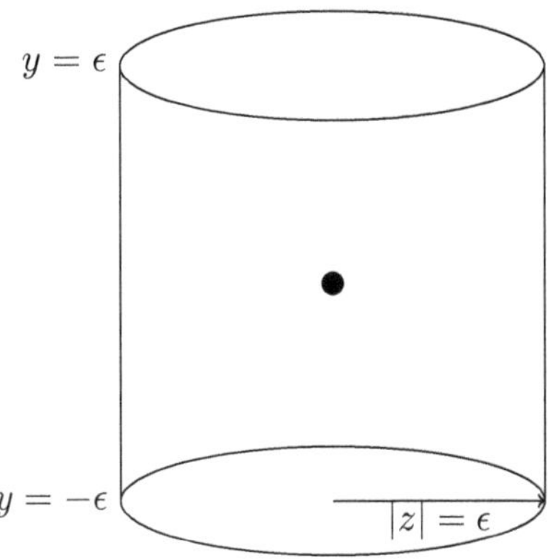

Figure 1.16. The 't Hooft singularity of interest, surrounded by a cylindrical surface. The vertical direction is parameterized by the y coordinate. The horizontal plane is identified with $C = \mathbb{C}$, with the center of the cylinder identified with the origin of $\mathbb{C}$.

The aforementioned isomorphism can be determined via parallel transport in the y-direction from $y = -\epsilon$ to $y = \epsilon$. To this end , we trivialize the holomorphic bundles at these two points. From (1.7.4), we determine that parallel transport between the two points would then be an element of the loop group LG. The on-shell definition of the 't Hooft line therefore amounts to declaring that the solutions to the equations of motion are locally equivalent to the Dirac singularity determined by the coweight, ρ, corresponding to the point

$$z^\rho \in LG \tag{1.7.6}$$

in the loop group. For example, given a $G = SL(2, \mathbb{C})$ coweight

$$\rho = \begin{pmatrix} \frac{1}{2} & 0 \\ 0 & -\frac{1}{2} \end{pmatrix} \tag{1.7.7}$$

we have

$$z^\rho = \begin{pmatrix} z^{\frac{1}{2}} & 0 \\ 0 & z^{-\frac{1}{2}} \end{pmatrix}. \tag{1.7.8}$$

So far, we have picked a choice of trivialization at the points ϵ and $-\epsilon$ on the cylinder, and we still have freedom to change these trivializations, with these changes at $-\epsilon$ and ϵ corresponding respectively to left and right multiplication by elements of the group of loops which extend to holomorphic maps from a disc to G, denoted L_+G. This implies that the moduli space of solutions to the 4d CS equations of motion on the boundary of the solid cylinder is

$$L_+G\backslash LG/L_+G. \tag{1.7.9}$$

The space $L_+G\backslash LG/L_+G$ is closely related to an object known in the mathematical literature as the affine Grassmannian, and both spaces agree if we consider a very small cylinder around the 't Hooft line, whereby one can replace LG by the group $G((z))$ of Laurent series valued in G.

One can also interpret the 't Hooft line as arising from a singular gauge transformation, with the singularity located at $z = 0$ and $y = 0$. In 4d CS, all classical solutions are locally trivial, so we can engineer the 't Hooft line's classical configuration by gluing two trivial gauge fields by the gauge transformation z^μ, for some coweight μ. The parallel transport of the gauge field from $y < 0$ to $y > 0$ therefore takes the form

$$g_1(z)z^\mu g_2(z) \tag{1.7.10}$$

with $g_1(z)$ and $g_2(z)$ denoting arbitrary G-valued holomorphic functions of z that are regular at the point 0. Notably, two 't Hooft lines associated to coweights related by a Weyl action are equivalent, since the corresponding singularities are related by gauge transformations.

For our purposes, we ought to also place 't Hooft lines at infinity on C. This amounts to a modification of the boundary conditions at this point. The upshot here is that due to the Dirichlet boundary conditions at $z = \infty$ away from the 't Hooft line breaking the gauge symmetry there, two 't Hooft lines at $z = \infty$ associated to coweights in the same Weyl orbit are not equivalent.

This is crucial for the goal of identifying Q-operators with 't Hooft lines, as the precise identification is that between a Q-operator with an 't Hooft line of charge μ in the bulk together with an 't Hooft line of charge $-\mu$ at infinity. Any such configuration is labeled by a coweight as opposed to a Weyl orbit of a coweight. Q-operators are likewise labeled by coweights and not Weyl orbits of coweights.

1.7.2   *TQ relations from the Witten effect*

We shall focus on deriving the TQ relations using 't Hooft and Wilson lines in 4d Chern–Simons theory with gauge group $SL(2,\mathbb{C})$.

The integrable system of interest is a spin chain with N sites and periodic boundary conditions, which can be modified by a twist. This involves the application of an element of the Cartan subgroup H, denoted $e^{i\phi}$ for $\phi \in \mathfrak{h}$, of the gauge group when the endpoints of the spin chain are identified.

The TQ relation for this system is well known to be[4]

$$\tilde{\mathbf{T}}(z,\phi)\mathbf{Q}_{\pm}(z,\phi) = \left(z - \frac{1}{2}\hbar\right)^{N} \mathbf{Q}_{\pm}(z+\hbar,\phi) + \left(z + \frac{1}{2}\hbar\right)^{N} \mathbf{Q}_{\pm}(z-\hbar,\phi).$$

$$(1.7.13)$$

It can be shown using suitable normalizations that this equation can be brought to the form

$$\mathbf{T}(z,\phi)\mathbf{H}_{\frac{1}{2}}(z,\phi) = \mathbf{H}_{\frac{1}{2}}(z+\hbar,\phi) + \mathbf{H}_{\frac{1}{2}}(z-\hbar,\phi), \qquad (1.7.14)$$

where $\tilde{\mathbf{T}}$ and $\mathbf{Q}_{\pm}$ are related to $\mathbf{T}$ and $\mathbf{H}_{\pm}$ respectively. It turns out that these are precisely the normalizations that arise in 4d Chern–Simons theory; details on this can be found in [22].

We shall now go on to review the derivation of (1.7.14) from the Witten effect in 4d Chern–Simons theory. Recall that the usual θ-angle in a four-dimensional gauge theory is defined so that the term

$$-\frac{\theta}{8\pi^2} \int \mathrm{Tr}\,(F \wedge F), \qquad (1.7.15)$$

is included in the action. The Witten effect [23] can be simply stated as the fact that if the magnetic charge of a line operator is non–zero while θ is also non–zero, then the electric charge of this line operator must be shifted by an amount proportional to the product of the magnetic charge and the θ angle. In particular, it means that purely magnetic line operators become dyonic in the presence of a non–zero θ angle. Moreover, if one varies θ to θ' continuously, one finds that the electric charge shifts by an amount proportional to the magnetic charge multiplied by $(\theta' - \theta)$.

Let us state the Witten effect more precisely. If T is the maximal torus of the gauge group G, then we denote the coweight lattice where magnetic charges live as $\Gamma = \mathrm{Hom}(U(1), T)$ and the weight lattice where electric

---

[4]There are additional relations of relevance here. If $\mathbf{T}_j^+(z,\phi)$ is the transfer matrix for a Verma module of highest weight j, one has the QQ relation

$$2i\sin(\phi/2)\tilde{\mathbf{T}}_{j-\frac{1}{2}}^+(z) = \mathbf{Q}_+(z+\hbar j)\mathbf{Q}_-(z-\hbar j). \qquad (1.7.11)$$

In addition, Q_+ and Q_- are related by the Weyl group of $SL(2,\mathbb{C})$:

$$w\mathbf{Q}_+(z,\phi)w = \mathbf{Q}_-(z,-\phi). \qquad (1.7.12)$$

The QQ relation has a derivation from 4d Chern–Simons theory as well [22].

charges live as $\Gamma^\vee = \mathrm{Hom}(T, U(1))$. The Killing form, κ, furnishes a map $\kappa : \Gamma \to \Gamma^\vee$ between these lattices.

In the case of $SL(2, \mathbb{C})$ that we are interested in, both these lattices are $\mathbb{Z}$ but the map from magnetic to electric charge lattices involves multiplication by 2.

Given a four-dimensional gauge theory with gauge group G and a line operator with magnetic charge $m \in \Gamma$, the Witten effect implies that the electric charge of the line operator is no longer in $\Gamma^\vee$, but rather in

$$\Gamma^\vee + \frac{\kappa(m)\theta}{2\pi}. \tag{1.7.16}$$

Varying θ to θ' shifts the electric charge of a line defect with magnetic charge m by

$$\frac{\kappa(m)(\theta' - \theta)}{2\pi}. \tag{1.7.17}$$

Now, we recall that the 4d Chern–Simons action can be rewritten (with the coupling constant factor of $1/\hbar$ included explicitly) as

$$S = -\frac{1}{2\pi\hbar} \int_{\mathbb{R}^2 \times \mathbb{C}} z \, \mathrm{Tr}\, F \wedge F, \tag{1.7.18}$$

whereby it takes the form of (1.7.15), but with a position–dependent θ-angle. Hence, we conclude that in 4d Chern–Simons theory, if the position of a line defect with magnetic charge m shifts from z to $z + \hbar$, its electric charge shifts from e to

$$e + \kappa(m)\frac{((z + \hbar) - z)}{\hbar} = e + \kappa(m). \tag{1.7.19}$$

Let us consider a fractional 't Hooft line at z with magnetic charge $\frac{1}{2}$, denoted $\mathbf{H}_{\frac{1}{2}}(z)$. The Witten effect described above implies that if we shift z to $z + \hbar$, the resulting line operator is equivalent to a dyonic line operator at z with magnetic charge $\frac{1}{2}$ and electric charge $\kappa(\frac{1}{2}) = 1$.

This is essentially how we arrive at the TQ relation, i.e., by fusing a Wilson and 't Hooft line with appropriate respective electric and magnetic charges, such that the fused line operator is another 't Hooft line at a shifted point. We shall consider Wilson lines in the fundamental representation of the gauge group in what follows. However, there are further subtleties to take into account before we can obtain the TQ relation explicitly.

On the 't Hooft line, the gauge symmetry is broken to a Borel subgroup, B of the gauge group. This implies that fusing the Wilson and 't Hooft lines couples the fundamental representation of the gauge group to the Borel gauge symmetry. However, as a representation of B, this fundamental representation is an extension of the charge 1 and -1 representations of the Cartan subgroup. Thus, the Witten effect implies that

$$\mathbf{T}(z)\mathbf{H}_{\frac{1}{2}}(z,\phi) = \mathbf{H}_{\frac{1}{2}}(z + \hbar) + \mathbf{H}_{\frac{1}{2}}(z - \hbar), \tag{1.7.20}$$

which is precisely the TQ relation we are after.

One can also verify that the Witten effect leads to the TQ-relation via a Feynman diagram analysis. Denote a Wilson-'t Hooft line of electric charge e and magnetic charge $1/2$ by $\mathbf{H}_{(e,\frac{1}{2})}(z)$. The Witten effect we have studied so far implies that

$$\mathbf{H}_{(e,\frac{1}{2})}(z) = \mathbf{H}_{\frac{1}{2}}(z + e\hbar), \tag{1.7.21}$$

and we would like to derive this to leading order in $\hbar$.

Place a 't Hooft line at $y = 0, z = 0$. As we have seen before, the 't Hooft line field configuration can be understood as a singular gauge transformation of the form $\mathrm{Diag}\left(z^{\frac{1}{2}}, z^{-\frac{1}{2}}\right)$ at $y = 0$, gluing trivial field configurations at $y \leq 0$ and $y \geq 0$.

The effect of this 't Hooft line on a Wilson line along $x = 0$ and z is thus the matrix

$$M(z) = \begin{pmatrix} z^{1/2} & 0 \\ 0 & z^{-1/2} \end{pmatrix} (1 + O(\hbar/z)). \tag{1.7.22}$$

If we were to now add electric charge e to the 't Hooft line, we would equivalently be including a term

$$e \int_{y=0,z=0} A_h \tag{1.7.23}$$

in the Lagrangian of the theory.

We now have a crossing of a dyonic line and a Wilson line. The leading order in $\hbar$ contribution to the interaction between these lines can then be evaluated in a similar manner to the R-matrix computation in Section 1.3. The main difference is that only the Cartan component of the gauge field from the Wilson-'t Hooft line enters the interaction, due to the presence of the 't Hooft line breaking the gauge symmetry.

Since the component of the quadratic Casimir involving $h \in \mathfrak{sl}_2(\mathbb{C})$ is $\frac{1}{2}h \otimes h$, where $h = \mathrm{Diag}(1, -1))$, the introduction of the electric charge e to the 't Hooft line modifies the matrix M to

$$M(z) \mapsto \left(1 + \frac{e\hbar}{2z}\begin{pmatrix} 1 & 0 \\ 0 & -1 \end{pmatrix}\right) M(z) + O\left(\hbar^2\right). \tag{1.7.24}$$

It then follows that

$$e\hbar\partial_z M(z) = \frac{e\hbar}{2z}hM(z) + O\left(\hbar^2\right).$$

which confirms the Witten effect and TQ relation to leading order in $\hbar$:

$$\mathbf{H}_{\left(1,\frac{1}{2}\right)}(z) = \mathbf{H}_{\frac{1}{2}}(z + e\hbar) + O\left(\hbar^2\right). \tag{1.7.25}$$

To show that this result holds to all orders in $\hbar$, one needs to use the fact that the 't Hooft line has only one parameter, which corresponds to its position in the z-plane (see Section 13 of [22] for details). This means that

$$\mathbf{H}_{\left(e,\frac{1}{2}\right)}(z) = \mathbf{H}_{\frac{1}{2}}(z + g(\hbar, e)) \tag{1.7.26}$$

for an undetermined function $g(\hbar, e)$. We recall that there is a symmetry of the 4d CS action under a common rescaling of $\hbar$ and z, which means that we must have $g(\hbar, e) = e\hbar$.

Chapter 2

Integrable Field Theories from 4d Chern–Simons Theory

As first shown by Costello and Yamazaki [24], the second class of integrable systems that can be realized using 4d Chern–Simons theory consists of two-dimensional integrable field theories (IFTs), which can be interpreted as 4d Chern–Simons coupled to surface defects along Σ. These systems can be further divided into two subclasses, depending on the type of surface defects inserted:

- **Order defects**, where new degrees of freedom are introduced on the defect, that are moreover coupled to the bulk gauge theory in a gauge invariant manner, and
- **Disorder defects**, where the four-dimensional Chern–Simons gauge field is required to have some kind of singularity.

In this chapter, we shall consider both cases in turn, focusing on classical aspects of these theories. We shall also review the relation of 4d Chern–Simons theory with disorder defects to 6d holomorphic Chern–Simons theory on twistor space, that was first discovered by Bittleston and Skinner [37], and Penna [142].

2.1 Integrable Field Theories from Order Surface Defects

Let us investigate the coupling of order surface defects to 4d Chern–Simons theory. The action for the coupled $4\mathrm{d} - 2\mathrm{d}$ system takes the form

$$S_{4\mathrm{d}-2\mathrm{d}} = \frac{1}{2\pi\hbar} \int_{\mathbb{R}^2 \times C} \mathrm{d}z \wedge \mathrm{CS}(A) + \sum_{\alpha=1}^{n} \frac{1}{\hbar} \int_{\mathbb{R}^2 \times z_\alpha} \mathscr{L}_\alpha \left(\phi_\alpha; A_w|_{z_\alpha}, A_{\bar{w}}|_{z_\alpha} \right).$$

$$(2.1.1)$$

45

Note that the components of the gauge field are defined with respect to the complex coordinates $w = x + iy$ and $\bar{w} = x - iy$ are employed. In the study of integrable field theories using 4d Chern–Simons theory, it is usually the case that topological invariance along Σ is explicitly broken, and the complex coordinates w and $\bar{w}$ are employed.

In the context of integrable lattice models studied in the previous chapter, the analogue of the Lorentz gauge

$$\partial_{\bar{w}} A_w + \partial_w A_{\bar{w}} + g^{z\bar{z}} \partial_z A_{\bar{z}} = 0 \tag{2.1.2}$$

was used. We shall be concerned with the limit where the volume of C becomes small, corresponding to the limit $g^{z\bar{z}} \to \infty$, whereby the gauge-fixing condition simplifies to

$$g^{z\bar{z}} \partial_z A_{\bar{z}} = 0. \tag{2.1.3}$$

Moreover, with appropriate boundary conditions imposed on C, this gauge can simplify further to the holomorphic gauge

$$A_{\bar{z}} = 0. \tag{2.1.4}$$

For example, in the rational case, $C = \mathbb{CP}^1$, taking into account the Dirichlet boundary condition, $A = 0$, on the gauge fields at infinity, the gauge fixing condition is further reduced to (2.1.4) since global anti-holomorphic functions on a connected, compact Riemann surface are constants.

Note that using the gauge (2.1.4), the propagator only involves the components A_w and $A_{\bar{w}}$, and does not involve $A_{\bar{z}}$. For example, in the rational case, the only nontrivial propagator is of the form

$$\langle A_w^a(w, z) A_{\bar{w}}^b(w', z') \rangle \propto \frac{\delta^{ab}}{z - z'}. \tag{2.1.5}$$

Here, we have made use of an orthonormal basis, $\{T_a\}$, of the Lie algebra $\mathfrak{g}$ with respect to the Killing form. Since the cubic interaction of the gauge theory incorporates all three components $A_w, A_{\bar{w}}, A_{\bar{z}}$ of the gauge field, we observe that the bulk vertex associated with $A_{\bar{z}}$ does not contribute to the Feynman diagram analysis.

For general choices of C, the propagator takes the form

$$\int_{\Sigma} d^2w \, \langle A_{a\bar{w}}(z) A_{bw}(z') \rangle = r_{ab}(z, z'). \tag{2.1.6}$$

The expression on the right-hand side is nothing but the classical r-matrix that solves the classical Yang-Baxter equation

$$[r_{12}(z_{12}), r_{13}(z_{13})] + [r_{12}(z_{12}), r_{23}(z_{23})] + [r_{13}(z_{13}), r_{23}(z_{23})] = 0, \tag{2.1.7}$$

where $z_{ij} := z_i - z_j$.

We can obtain an effective two-dimensional theory on Σ either by integrating out the gauge field along the complex manifold C in the path integral, or by solving the classical equations of motion along C in 4d CS theory. We shall elucidate the first of these techniques below. Before doing so, let us discuss the form of the order surface defects that we shall utilize.

Given that we work in a gauge where $A_{\bar{z}}$ vanishes, it is natural to consider the coupling between chiral and anti-chiral defects which only depend on A_w and $A_{\bar{w}}$, respectively. We shall be interested in surface defects with global G-symmetry (where G is the gauge group of 4d CS). A large class of integrable field theories (IFTs) of interest arise from a combination of such chiral and antichiral defects, which support currents associated with the G symmetry, denoted as $J_w^\alpha = J_w^{\alpha a} T_a$ and $\overline{J}_{\bar{w}}^\beta = \overline{J}_{\bar{w}}^{\beta b} T_b$, respectively. The Lagrangians of these defects have the form

$$\begin{aligned}
\mathscr{L}_\alpha(\phi_\alpha; A_{\bar{w}}) &= \mathscr{L}_\alpha(\phi_\alpha) + J_w^\alpha A_{\bar{w}} \quad (\alpha = 1, \dots n_+) \\
\overline{\mathscr{L}}_\beta(\phi_\beta; A_w) &= \overline{\mathscr{L}}_\beta(\phi_\beta) + \bar{J}_{\bar{w}}^\beta A_w \quad (\beta = 1, \dots n_-),
\end{aligned} \tag{2.1.8}$$

for some fields, denoted ϕ, supported on the defects.

As we shall review below, one can obtain a 2d field theory whose Lagrangian has an interaction term proportional to

$$\sum_{\alpha=1}^{n_+} \sum_{\bar{\alpha}=1}^{n_-} \hbar r_{ab}(z_\alpha - z_{\bar{\alpha}}') J_a^\alpha \overline{J}_b^{\bar{\alpha}}. \tag{2.1.9}$$

Here, the classical r-matrix can be either rational, trigonometric or elliptic. In the rational case, for instance, the classical r-matrix can be expressed as:

$$r_{ab}(z, z') = \frac{\delta_{ab}}{z - z'}. \tag{2.1.10}$$

Let us derive the integrable field theory at the path integral level explicitly in the rational case, for chiral and anti-chiral defects located at z_+ and z_-, respectively, as depicted in Figure 2.1. Modulo the kinetic terms of the defects, the 4d-2d coupled system in the holomorphic gauge

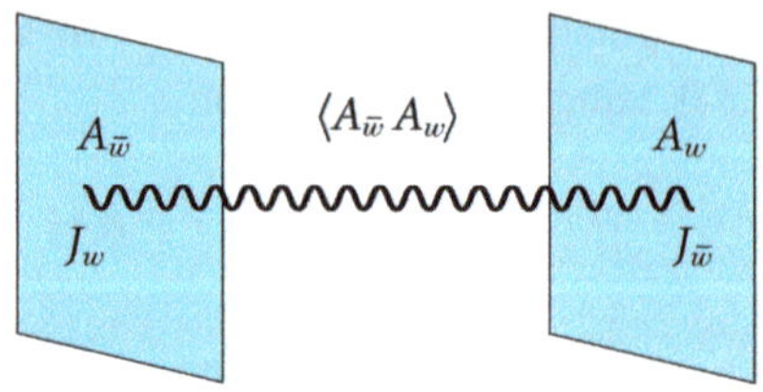

Figure 2.1. Coupling between chiral and anti-chiral defects.

(2.1.4) is

$$\int DA e^{\frac{i}{2\pi\hbar}\int_{\Sigma\times C} dz d\bar{z} dw d\bar{w} 2 A^a_{\bar{w}}\partial_{\bar{z}} A_{aw} + i\int_{\Sigma\times z_+} dw d\bar{w} A^a_{\bar{w}} J_{wa} + i\int_{\Sigma\times z_-} dw d\bar{w} A^a_w J_{\bar{w}a}}.$$

(2.1.11)

We now shift the fields A_- and A_+ to $A_- + A_{-0}$ and $A_+ + A_{+0}$ respectively. Writing the 4d measure as d^4x and the 2d measure along Σ as d^2w, this gives

$$\int DA e^{\left(\frac{i}{2\pi\hbar}\int_{\Sigma\times C} d^4x 2(A^a_{\bar{w}} + A^a_{\bar{w}0})\partial_{\bar{z}}(A_{aw} + A_{aw0}) + i\int_{\Sigma\times z_+} d^2w(A^a_{\bar{w}} + A^a_{\bar{w}0}) J_{wa} + i\int_{\Sigma\times z_-} d^2w(A^a_w + A^a_{w0}) J_{\bar{w}a}\right)}.$$

(2.1.12)

Let

$$\partial_{\bar{z}} A^a_{w0} = -\delta^{(2)}(z - z_+) J^a_w$$
$$\partial_{\bar{z}} A^a_{\bar{w}0} = \delta^{(2)}(z - z_-) J^a_{\bar{w}}.$$

(2.1.13)

Then, the path integral reduces to

$$\int DA e^{\frac{i}{\pi\hbar}\int_{\Sigma\times C} d^4x (A^a_{\bar{w}}\partial_{\bar{z}} A_{wa} + A^a_{\bar{w}0}\partial_{\bar{z}} A_{w0a})} e^{i\int_{\Sigma\times z_+} d^2w(A_{\bar{w}0a} J^a_w)} e^{i\int_{\Sigma\times z_-} d^2w(A_{w0a} J^a_{\bar{w}})}$$

$$= \int DA e^{\frac{i}{\pi\hbar}\int_{\Sigma\times C} d^4x (A^a_-\partial_{\bar{z}} A_{wa})} e^{i\int_{\Sigma\times z_-} d^2w(A_{w0a}(z_-) J^a_{\bar{w}}(z_-))}.$$

(2.1.14)

Using the delta function identity

$$\frac{1}{2\pi i}\partial_{\bar{z}}\frac{1}{z} = -\delta^{(2)}(z, \bar{z})$$

(2.1.15)

we have

$$A^a_{w0} = \frac{\hbar}{2i}\frac{J^a_w(z_+)}{z - z_+}.$$

(2.1.16)

Employing this expression, the path integral becomes

$$\int DA\, e^{\frac{i}{\pi\hbar}\int_{\Sigma\times C} d^4x(A^a_{\bar{w}}\partial_{\bar{z}}A_{wa})}\, e^{i\int_{\Sigma} d^2w\left(\frac{\hbar}{2i}\frac{J^a_w J_{a\bar{w}}}{z_- - z_+}\right)}$$

$$= \frac{1}{\mathcal{N}} e^{i\int_{\Sigma} d^2w\left(\frac{i\hbar}{2}\frac{J^a_w J_{a\bar{w}}}{z_+ - z_-}\right)}, \tag{2.1.17}$$

where $\mathcal{N}$ is a constant. Incorporating the kinetic terms from the order surface operators, we arrive at an effective two-dimensional integrable field theory. For an arbitrary number of chiral and anti-chiral order surface defects, the effective two-dimensional integrable field theory has the form,

$$S^{\text{eff}}_{2\,\text{d}} = \frac{1}{\hbar}\int_{\mathbb{R}^2}\left[\sum_{\alpha=1}^{n_+}\mathcal{L}^c_{\alpha}(\phi_{\alpha}) + \sum_{\beta=1}^{n_-}\mathcal{L}^a_{\beta}(\phi_{\beta}) + \sum_{\alpha=1}^{n_+}\sum_{\beta=1}^{n_-} r_{ab}\left(z_{\alpha} - z'_{\beta}\right) J^{\alpha}_a \bar{J}^{\beta}_b\right], \tag{2.1.18}$$

where we have performed a convenient rescaling of r_{ab}.

Let us now understand why such theories are integrable. The integrability of two-dimensional IFTs follows from the existence of quantity known as a Lax operator or Lax connection, which satisfies a condition known as the zero-curvature equation. Let us observe how this quantity arises naturally in the 4d CS setup. Pick a point $z \in C$ away from the surface defects. We then define a $\mathfrak{g}$-valued one-form $\mathcal{L}(z)$ on the plane $\mathbb{R}^2 \times \{z\}$ by the formula

$$\mathcal{L}(z) = A_w(z)\mathrm{d}w + A_{\bar{w}}(z)\mathrm{d}\bar{w}.$$

Here, z is treated as a parameter for the one-form on $\mathbb{R}^2 \times \{z\}$, considering it as one of the coordinates along the curve C. Recall that the equation of motion obtained by integrating out the four-dimensional gauge field $A_{\bar{z}}$ is

$$\partial_{\bar{w}}A_w - \partial_w A_{\bar{w}} + [A_{\bar{w}}, A_w] = 0. \tag{2.1.19}$$

This immediately implies the zero-curvature equation for the one-form $\mathcal{L}(z)$ on the plane $\mathbb{R}^2 \times \{z\}$ given by:

$$\mathrm{d}\mathcal{L}(z) + \mathcal{L}(z) \wedge \mathcal{L}(z) = 0. \tag{2.1.20}$$

Hence, we can identify $\mathscr{L}(z)$ as the two-dimensional Lax operator. The match to the conventional definition where it is holomorphic follows from the remaining equations of motion since we are in the gauge $A_{\bar{z}} = 0$.

The advantage of the 4d CS formulation of integrable field theories is that the Lax operator can be *derived*, whereas it had to be guessed in previous studies of IFTs. The Lax operator can be computed explicitly via a straightforward Feynman diagram computation, depicted in Figure 2.2.

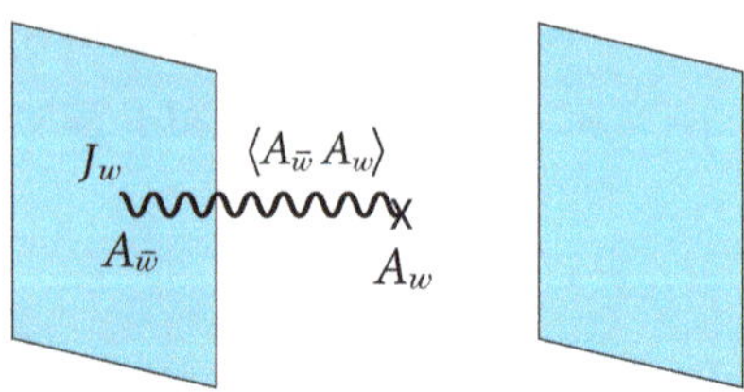

Figure 2.2. Lax operator from 4d Chern–Simons propagator.

For example, let us consider the case with n_+ chiral defects and n_- anti-chiral defects. In this case, the one-form Lax operator has the form

$$\mathcal{L}(z) = \sum_{\alpha=1}^{n_+} J^\alpha\left(z_\alpha\right) r\left(z_\alpha, z\right) \mathrm{d}w + \sum_{\beta=1}^{n_-} \bar{J}^\beta\left(z_\beta\right) r\left(z, z_\beta\right) \mathrm{d}\bar{w}. \tag{2.1.21}$$

For simplicity, let us focus on the case $n_+ = n_- = 1$, with chiral (anti-chiral) defect located at $z = 1 (z = -1)$, and restrict ourselves to the rational case $(C = \mathbb{C})$. We then find

$$\mathcal{L}(z) = \frac{J(1)\mathrm{d}w}{1 - z} + \frac{\bar{J}(-1)\mathrm{d}\bar{w}}{z + 1} = \frac{j + z \star j}{1 - z^2}, \tag{2.1.22}$$

where we defined the $\mathfrak{g}$-valued one-form j (current) by

$$j := J(1)\mathrm{d}w + \bar{J}(-1)\mathrm{d}\bar{w}$$

and $\star$ is the Hodge-star operator in the two-dimensions such that $\star j = J(1)\mathrm{d}w - \bar{J}(-1)\mathrm{d}\bar{w}$.

Via the relation (2.1.22), the zero-curvature equation for the Lax operator corresponds to two equations for the current one-form, namely, the current conservation and flatness equations for the current:

$$\mathrm{d} \star j = 0$$

$$\mathrm{d}j + j \wedge j = 0.$$

Indeed, in the literature on integrable field theories, one often constructs the Lax operator using a conserved and flat current present in such a theory. In the present context of 4d CS, however, currents do not always play a primary role.

2.1.1 *Line defects and the lax operator*

The existence of a Lax operator that satisfies the zero-curvature (flatness) condition eventually leads to the integrability of the associated integrable field theory. This follows since it allows us to define infinitely many non-local conserved charges for the classical field theory at hand, which would imply integrability as long as they are Poisson commuting. Let us review why the flatness condition implies the existence of an infinite number of conserved charges, before interpreting this in 4d Chern–Simons theory.

Given a two-dimensional IFT with a Lax operator for the Lie algebra $\mathfrak{g}$ satisfying the zero-curvature equation, we can define it on a cylinder $\mathbb{R} \times S^1$ with coordinates $t, \theta \in [0, 2\pi]$. In addition, we pick a representation, V, of $\mathfrak{g}$. Now, using the Lax operator, we construct the non-local operator

$$W(t_0, z, V) = \mathrm{Tr}_V \, P \exp \int_{\theta=0}^{2\pi} \mathcal{L}(z) \tag{2.1.23}$$

as the trace over V of the path-ordered exponential of the Lax operator along the circle $t = t_0$ in the cylinder. This is known as the monodromy operator.

The monodromy operator $W(t_0, V)$ is in fact independent of the value of t_0:

$$\partial_{t_0} W(t_0, z, V) = 0 \,, \tag{2.1.24}$$

and this follows from the zero-curvature equation.[1] Put differently, $W(t_0, z, V)$ is a conserved quantity, as depicted in Figure 2.3. Then, by expanding in z around any point on C, we discover that the Lax operator furnishes an infinite number of conserved, non-local charges acting on the

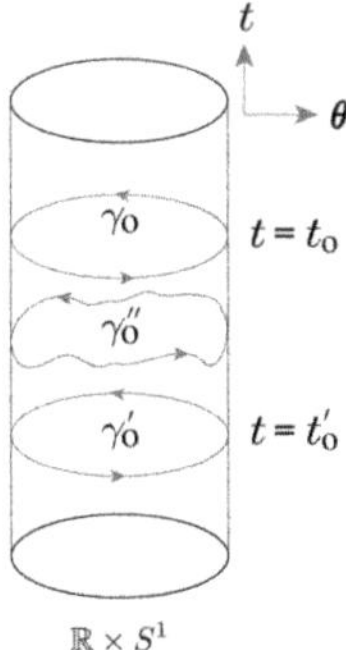

Figure 2.3. The IFT is studied on a cylinder $\mathbb{R}_t \times S_\theta^1$, and the monodromy operator is considered around a contour γ_0 at a fixed radial coordinate $t = t_0$ as in eqn. (2.1.23). The trace of the monodromy does not change under continuous deformations of the contour.

---

[1] For details, see, for example, page 20 of [25].

Hilbert space of the theory on a circle. For example, expanding around $z = \infty$ results in

$$W(t_0, z, V) = \exp\left(\sum_{n=0}^{\infty} \frac{Q_n}{z^n}\right). \tag{2.1.25}$$

We would now like to understand the physical origin of these conserved non-local charges in 4d Chern–Simons theory.

As one might expect, the path-ordered exponential of the four-dimensional Lax operator corresponds to the Wilson line of the four-dimensional gauge theory, at some point z in the holomorphic plane and on some path in the topological plane:

$$W(z, V) = \mathrm{Tr}_V\, P \exp \int_{\gamma \times \{z\}} A \,, \tag{2.1.26}$$

where γ is a winding cycle around the cylinder.

Non-local conserved charges are thus obtained by inserting a Wilson line into 4d Chern–Simons theory, at some value of z distinct from the location of the surface defects. The fact that 4d Chern–Simons theory is topological in the $w, \overline{w}$ plane away from the locations of the surface defects means that moving the Wilson line in this plane has no effect. This therefore implies that the non-local conserved charges obtained from it, by expanding in powers of z, are conserved quantities.

2.1.2　*Examples*

We shall now discuss some examples of integrable field theories that can be realized using order surface operators in 4d Chern–Simons theory.

2.1.2.1　*The Faddeev–Reshetikhin model*

The first example that we shall consider is the Faddeev–Reshetikhin model. This model arose in the quantum inverse scattering method approach to quantizing integrable field theories, as first utilized by Faddeev and Reshetikhin [26] to quantize the $SU(2)$ principal chiral model (PCM), with the action

$$S_{\mathrm{PCM}} = \int_{\Sigma} \mathrm{d}^2 x\, \mathrm{Tr}(g^{-1}\partial_\mu g g^{-1}\partial^\mu g)\,, \tag{2.1.27}$$

where Σ is 2d Minkowski spacetime and g is a $SU(2)$-valued field (precisely this model will later be realized using disorder defects in 4d CS).

Faddeev and Reshetikhin's approach to quantization required modifying the Lagrangian without changing the form of the equations of motion of the theory.[2] The equation of motion arising from (2.1.27) gives a conservation equation for the current $j_\mu = g^{-1}\partial_\mu g$, i.e.,

$$\partial_\mu j^\mu = 0\,, \tag{2.1.28}$$

where j_μ also satisfies the flatness condition $\partial_\mu j_\nu - \partial_\nu j_\mu + [j_\mu, j_\nu] = 0$. Let us make use of lightcone coordinates $\sigma^\pm = \tau \pm \sigma$ (where τ is a time coordinate and σ is a space coordinate), on Σ, which are obtained from $w, \overline{w}$ via analytic continuation. The equation of motion combined with the flatness condition then gives

$$\partial_\mp j_\pm = \pm \frac{1}{2}[j_+, j_-]\,. \tag{2.1.29}$$

and as we shall see, one can define a different theory whose equations of motion can be put in this form [28].

$$\mathcal{L}_\pm(z) = \frac{j_\pm}{1 \mp z}, \tag{2.1.30}$$

where we have made use of lightcone coordinates $\sigma^\pm = \tau \pm \sigma$ on Σ, which are obtained from $w, \overline{w}$ via analytic continuation.

The action for this theory is

$$S_{\mathrm{FR}}\left[g_{(\pm)}\right]$$

$$= 2 \int_\Sigma \mathrm{Tr}\left(\Lambda g_{(+)}^{-1}\partial_- g_{(+)} + \Lambda g_{(-)}^{-1}\partial_+ g_{(-)} - \frac{1}{2\nu}g_{(+)}\Lambda g_{(+)}^{-1}g_{(-)}\Lambda g_{(-)}^{-1}\right)\mathrm{d}^2\sigma\,, \tag{2.1.31}$$

where $\int_\Sigma \mathrm{d}^2\sigma = \int_\Sigma \mathrm{d}\tau \wedge \mathrm{d}\sigma = -\frac{1}{2}\int_\Sigma \mathrm{d}\sigma^+ \wedge \mathrm{d}\sigma^-$, and is referred to as the Faddeev–Reshetikhin model or linear chiral model. Here, both $g_{(+)}$ and $g_{(-)}$

---

[2]The reason for doing so is that the Poisson bracket of the Lax connection of this model suffers from "non-ultralocality", which is the presence of a derivative of a Dirac delta function. This is an obstacle to the lattice discretization of this model. In fact, for general non-ultralocal theories, it is not known how to define a lattice Lax matrix that gives a well-defined Poisson bracket without ambiguity (for more details on this issue, see, for example, Ref. [27]). Faddeev and Reshetikhin's approach to avoiding this pathology was to explicitly drop the term in the Poisson brackets involving the derivative of the delta function. The motivation for doing this was the idea that there exists a 'false vacuum' in the quantum theory, and the semiclassical description of the theory around such a vacuum does not involve the non-ultralocal term.

From the perspective of 4d CS, we shall discuss non-ultralocal Poisson brackets in Section 2.3, and lattice discretization and quantization of integrable field theories in Section 5.2.

are $SU(2)$-valued fields on Σ, while Λ is the Cartan generator of $\mathfrak{su}(2)$. The Lax connection for this model is

$$\mathcal{L}_\pm(z) = \frac{\mathcal{J}_\pm}{\nu \mp z}, \quad \mathcal{J}_{(\pm)} := g_{(\pm)}\Lambda g_{(\pm)}^{-1}, \tag{2.1.32}$$

where the equations of motion can be stated as

$$\partial_\mp \mathcal{J}_\pm = \mp \frac{1}{2\nu}[\mathcal{J}_+, \mathcal{J}_-], \tag{2.1.33}$$

which has the form of (2.1.29) for $\nu = -1$.

The derivation of the Faddeev–Reshetikhin model from 4d CS theory was first performed by Fukushima *et al.* [30]. This derivation involves coupling 4d CS theory to two order surface operators, each supporting fields valued in the complexified Lie group $G^{\mathbb{C}}$, denoted $\mathcal{G}_{(+)}$ and $\mathcal{G}_{(-)}$. The 4d-2d action is given by

$$S[A, \mathcal{G}_{(\pm)}] = \frac{1}{2\pi\hbar}\int_{\Sigma \times C} dz \wedge \mathrm{CS}(A) + \int_{\Sigma \times \{z_+\}} \mathrm{Tr}\left(\Lambda \mathcal{G}_{(+)}^{-1} D_- \mathcal{G}_{(+)}\right) d\sigma^+ \wedge d\sigma^-$$
$$+ \int_{\Sigma \times \{z_-\}} \mathrm{Tr}\left(\Lambda \cdot \mathcal{G}_{(-)}^{-1} D_+ \mathcal{G}_{(-)}\right) d\sigma^+ \wedge d\sigma^- , \tag{2.1.34}$$

where Λ is an element of the real Lie subalgebra of the complex Lie algebra $\mathfrak{g}^{\mathbb{C}}$, and z_+ and z_- are points on $C = \mathbb{CP}^1$. In the case where $G^{\mathbb{C}} = SL(2, \mathbb{C})$, we take $\Lambda = -\frac{i}{2}\sigma^3$, where σ^3 is the Pauli matrix,

$$\sigma^3 = \begin{pmatrix} 1 & 0 \\ 0 & -1 \end{pmatrix}. \tag{2.1.35}$$

Integrating out the 4d CS gauge fields in the path integral for the action (2.1.34) allows us to retrieve the Faddeev–Reshetikhin model in the form given in (2.1.31). Suitable reality conditions can be imposed on the fields of the 4d-2d system we started with if one wants to obtain the Faddeev–Reshetikhin model for $G = SU(2)$ [30].

2.1.2.2　*Massless Thirring and Gross–Neveu models*

Another set of important models arise from the consideration of chiral and anti-chiral defects that support fermionic degrees of freedom. These give rise to either the massless Thirring model or chiral Gross–Neveu model. If $\mathfrak{g} = \mathfrak{so}_n$ and the fermions are in its vector representation, we obtain the Gross–Neveu model, while if $\mathfrak{g} = \mathfrak{sl}_n$ and the fermions are in the sum of the fundamental anti-fundamental representation, we find the (massless) Thirring model.

To define the surface operators, we first introduce some notation for gamma matrices. The anti-commutation relation of gamma matrices, γ^μ ($\mu = 0, 1$), is normalized as

$$\{\gamma^\mu, \gamma^\nu\} = 2\eta^{\mu\nu}, \tag{2.1.36}$$

where $\eta^{\mu\nu} = \mathrm{diag}(-1, 1)$. The relation (2.1.36) can be satisfied by picking the gamma matrices to be

$$\gamma^0 = i\sigma^1, \qquad \gamma^1 = -\sigma^2, \tag{2.1.37}$$

where σ^i ($i = 1, 2, 3$) are Pauli matrices. We can also introduce

$$\gamma^5 = \sigma^3, \tag{2.1.38}$$

and then a Dirac fermion Ψ can be decomposed into eigenstates Ψ_L, Ψ_R of γ^5 with eigenvalues ± 1,

$$\Psi_L = \frac{1 + \gamma^5}{2}\Psi = \frac{1}{\sqrt{2}}\begin{pmatrix} \psi_L \\ 0 \end{pmatrix}, \qquad \Psi_R = \frac{1 - \gamma^5}{2}\Psi = \frac{1}{\sqrt{2}}\begin{pmatrix} 0 \\ \psi_R \end{pmatrix}. \tag{2.1.39}$$

The light-cone coordinates $\sigma^\pm$ on Σ are defined as before, and the associated metric $\eta_{\mu\nu}$ has the off-diagonal components $\eta_{+-} = \eta_{-+} = -1/2$.

Now we consider the 4d CS action coupled with the 2d fermion systems

$$S_f[A, \Psi] = S_{\mathrm{CS}}[A] + S_L[A_-, \Psi_L] + S_R[A_+, \Psi_R], \tag{2.1.40}$$

where $A = A^a T_a \in \mathfrak{g}$ and T_a are generators of $\mathfrak{g}$. The second and third terms are gauge-invariant 2d surface operator actions constructed from the free massless Weyl fermions coupled to the gauge field

$$S_+^f[A_-, \psi_L] = \frac{1}{\hbar_{2d}} \int_{\Sigma \times \{z_+\}} \mathrm{d}^2\sigma \, \psi_L^* i \left(\partial_- + A_-\right) \psi_L,$$

$$\tag{2.1.41}$$

$$S_-^f[A_+, \psi_R] = \frac{1}{\hbar_{2d}} \int_{\Sigma \times \{z_-\}} \mathrm{d}^2\sigma \, \psi_R^* i \left(\partial_+ + A_+\right) \psi_R,$$

with the fermionic fields ψ_L and ψ_R transforming in the fundamental representation, ρ.

We may integrate out the gauge fields to obtain the 2d action of the associated 2d integrable field theory:

$$S_{\text{Th}}[\psi] = \tag{2.1.42}$$

$$\frac{1}{\hbar_{2d}} \int_\Sigma \mathrm{d}^2\sigma \left(\psi_L^* i\partial_- \psi_L + \psi_R^* i\partial_+ \psi_R \right.$$

$$\left. + \frac{1}{4\hbar_{2d}} \frac{i\hbar}{z_+ - z_-} \left(\psi_L^* \rho\left(T_a\right) \psi_L \right) \left(\psi_R^* \rho\left(T^a\right) \psi_R \right) \right). \tag{2.1.43}$$

Note that the coupling constant of the massless Thirring/Gross–Neveu models can be identified with the inverse of the distance between two order defects, and the Lax operator is

$$\mathcal{L} = \frac{\hbar}{4\hbar_{2d}} \frac{\psi_L^* \rho\left(T^a\right) \psi_L \rho\left(T_a\right)}{z - z_+} \mathrm{d}\sigma^+ - \frac{\hbar}{4\hbar_{2d}} \frac{\psi_R^* \rho\left(T^a\right) \psi_R \rho\left(T_a\right)}{z - z_-} \mathrm{d}\sigma^-. \tag{2.1.44}$$

2.1.2.3   *Sigma models on odd-dimensional spheres*

As shown in Ref. [24], the coupling of chiral and anti-chiral free $\beta\gamma$ defects to 4d CS theory with gauge group $G = GL(N, \mathbb{C})$ can be studied, with the total action given by

$$S_{\text{4d-2d}} = \frac{1}{2\pi\hbar} \int_{\mathbb{R}^2 \times \mathbb{C}} \mathrm{d}z \wedge \mathrm{CS}(A) + \frac{1}{\hbar_{2d}} \int_{\mathbb{R}^2 \times \{z_+\}} \mathrm{d}w \mathrm{d}\bar{w} \, \beta^i \bar{\partial}_A \gamma_i$$

$$+ \frac{1}{\hbar_{2d}} \int_{\mathbb{R}^2 \times \{z_-\}} \mathrm{d}w \mathrm{d}\bar{w} \, \bar{\beta}_i \partial_A \bar{\gamma}^i. \tag{2.1.45}$$

We show that this gives rise to a sigma model on S^{2N-1}, coupled to a free boson. The chiral-free $\beta\gamma$ system has the explicit form

$$\int_{\mathbb{R}^2 \times \{z_+\}} \mathrm{d}w \mathrm{d}\bar{w} \, \beta^i \bar{\partial}_A \gamma_i = \int_{\mathbb{R}^2 \times \{z_+\}} \mathrm{d}w \mathrm{d}\bar{w} \left(\beta^i \partial_{\bar{w}} \gamma_i + A_{\bar{w}}^a \beta^i \rho(T_a)_i{}^j \gamma_j \right),$$

$$\tag{2.1.46}$$

while the anti-chiral free $\beta\gamma$ system has the explicit form

$$\int_{\mathbb{R}^2 \times \{z_-\}} \mathrm{d}w \mathrm{d}\bar{w} \, \bar{\beta}^i \partial_A \bar{\gamma}_i = \int_{\mathbb{R}^2 \times \{z_-\}} \mathrm{d}w \mathrm{d}\bar{w} \left(\bar{\beta}^i \partial_w \bar{\gamma}_i + A_w^a \bar{\beta}^i \rho(T_a)_i{}^j \bar{\gamma}_j \right).$$

$$\tag{2.1.47}$$

Here $\rho(T_a)$ denote the fundamental representation of the $\mathfrak{gl}(N, \mathbb{C})$ generators T_a. The associated 2d sigma model is derived by integrating out the gauge field A, which leads to the classical action

$$S_{2d}^{\text{eff}} = \frac{1}{\hbar_{2d}} \int_{\mathbb{R}^2} \left(\beta^i \partial_{\bar{w}} \gamma_i + \bar{\beta}_i \partial_w \bar{\gamma}^i + \frac{i}{2} \frac{\hbar}{z_+ - z_-} \beta^i \gamma_j \bar{\beta}_i \bar{\gamma}^j \right) \mathrm{d}w \mathrm{d}\bar{w}, \quad (2.1.48)$$

where the identity $(\rho(T^a) \otimes \rho(T_a))_{ij,kl} = \delta_{ik}\delta_{jl}$ is used.

Integrating out the auxiliary fields β^i or $\bar{\beta}_i$ results in a sigma model with target space $\mathbb{C}^N\backslash\{0\} \cong S^{2N-1} \times \mathbb{R}$. To be precise, if we define coordinates on $\mathbb{C}^n$ to be $u_i, \bar{u}_i$, and work on the open subset where $u_i \neq 0$, then this action is that of the σ-model on $\mathbb{C}^n\backslash 0$ with the metric

$$\frac{z_+ - z_-}{\|u\|^2} \sum \mathrm{d}u_i \, \mathrm{d}\bar{u}_i. \quad (2.1.49)$$

Employing this metric, there is an isometry

$$\mathbb{C}^n\backslash\{0\} \cong S^{2n-1} \times \mathbb{R}. \quad (2.1.50)$$

We observe that this construction has engineered the σ-model on S^{2n-1}, together with a single free boson.

2.1.2.4 *Sigma models on Kähler target spaces*

We now consider chiral curved $\beta\gamma$ defects, which are associated with a Kähler manifold, X, that enjoys a holomorphic G-action. The defect is defined in terms of fields $\gamma : \mathbb{C} \to X$ and $\beta \in \Omega^{1,0}(\mathbb{C}, \gamma^*T^*X)$ (here, T^*X is the holomorphic cotangent bundle of X). To define the coupling to 4d CS gauge fields, we employ the map $\rho : \mathfrak{g} \to \text{Vect}(X)$ which is the Lie algebra homomorphism from $\mathfrak{g}$ to the Lie algebra of holomorphic vector fields on X. These holomorphic vector fields generate the infinitesimal G-action on X. Picking local holomorphic coordinates $u_1, \ldots, u_n$ on X, and a basis T_a of $\mathfrak{g}$, the holomorphic vector fields can be expressed as

$$\rho(T_a) = \sum \rho_{a,i}(u)\partial_{u_i}, \quad (2.1.51)$$

where $\rho_{a,i}(u)$ are holomorphic functions of u_j. The explicit, local form of the surface defect action is

$$\int_{\Sigma \times \{z_+\}} \beta^i \bar{D} \, \gamma_i \, \mathrm{d}w \mathrm{d}\bar{w} \equiv \int_{\Sigma \times \{z_+\}} \left(\beta^i \partial_{\bar{w}} \gamma_i + A_{a,\bar{w}} \beta^i \rho_{a,i}(\gamma) \right) \mathrm{d}w \mathrm{d}\bar{w}. \quad (2.1.52)$$

To realize integrable field theories, [24] also introduced a complex-conjugate $\bar{\beta}\bar{\gamma}$ system located at another point on C, where $\bar{\gamma}$ is a map to $\bar{X}$ (defined to be X with the opposite complex structure) and $\bar{\beta} \in \Omega^{0,1}\left(\mathbb{C}, \bar{\gamma}^* T^* \bar{X}\right)$. The explicit surface defect action is

$$\int_{\Sigma \times \{z_-\}} \bar{\beta}^i D\, \bar{\gamma}_i \, \mathrm{d}w \mathrm{d}\bar{w} \equiv \int_{\Sigma \times \{z_-\}} \left(\bar{\beta}^i \partial_w \bar{\gamma}_i + A_{a,w} \bar{\beta}^i \bar{\rho}_{a,i}(\bar{\gamma})\right) \mathrm{d}w \mathrm{d}\bar{w}.$$

$$(2.1.53)$$

The full 4d-2d coupled system, with holomorphic and anti-holomorphic surface defects at points z_0 and z_1 on C, respectively, is

$$S_{\text{4d-2d}} = \frac{1}{2\pi\hbar} \int_{\mathbb{R}^2 \times \mathbb{C}} \mathrm{d}z \wedge \mathrm{CS}(A) + \int_{\mathbb{R}^2 \times \{z_+\}} \beta^i \bar{D} \gamma_i \, \mathrm{d}w \mathrm{d}\bar{w}$$

$$+ \int_{\mathbb{R}^2 \times \{z_-\}} \bar{\beta}_i D \bar{\gamma}^i \, \mathrm{d}w \mathrm{d}\bar{w}. \qquad (2.1.54)$$

Integrating out the gauge field gives us

$$\int_{\Sigma} \left(\beta^i \partial_{\bar{w}} \gamma_i + \bar{\beta}^i \partial_w \bar{\gamma}_i - \frac{i}{4} \frac{\hbar}{z_+ - z_-} \sum \beta^i \bar{\beta}^j \rho_{a,i}(\gamma) \bar{\rho}_{a,j}(\bar{\gamma})\right) \mathrm{d}w \mathrm{d}\bar{w},$$

$$(2.1.55)$$

and further integrating out β and $\bar{\beta}$ gives a sigma model with Kähler target space, X, so long as

$$g^{i\bar{j}}(u, \bar{u}) = \sum_a \rho^{a,i}(u) \bar{\rho}^{a,\bar{j}}(\bar{u})$$

is invertible. The metric tensor $g_{i\bar{j}}$ that enters the sigma model action is the inverse of this tensor $g^{i\bar{j}}$.

2.1.2.5   *Trigonometric and elliptic generalizations*

By generalizing the previous derivations, we can derive trigonometric and elliptic analogues of the aforementioned integrable field theories. For example, by taking $C = \mathbb{C}^\times$, the trigonometric analogue of the massless Thirring/Gross–Neveu model derived in Ref. [24] is

$$S_{2\,\text{d}}^{\text{eff}} = \frac{1}{\hbar} \Bigg[2 \int \psi_e \bar{\partial} \psi_f + 2 \int \psi_h \bar{\partial} \psi_h + 2 \int \bar{\psi}_e \partial \bar{\psi}_f + 2 \int \bar{\psi}_h \partial \bar{\psi}_h$$

$$+ \frac{4}{1 - z_1/z_2} \int \psi_f \psi_h \bar{\psi}_f \bar{\psi}_h + \frac{4}{1 - z_2/z_1} \int \psi_e \psi_h \bar{\psi}_e \bar{\psi}_h$$

$$+ \frac{z_2 + z_1}{z_2 - z_1} \int \psi_e \psi_f \bar{\psi}_e \bar{\psi}_f \Bigg], \qquad (2.1.56)$$

where the Lagrangian has been presented in the basis of fermions given by the e, f, h basis of $\mathfrak{sl}_2$, i.e., the fermionic field is expanded as

$\psi = \psi_e e + \psi_f f + \psi_h h$, with the invariant bilinear form normalized as $\langle e, f \rangle = 1$ and $\langle h, h \rangle = 2$. We can also derive a trigonometric generalization of the Faddeev–Reshetikhin model

$$
- \int_{\mathcal{M}} \mathrm{Tr}\left(\Lambda g_{(+)}^{-1} \partial_- g_{(+)} + \Lambda g_{(-)}^{-1} \partial_+ g_{(-)} + \frac{4}{1 - z_+/z_-} \mathcal{J}_+^e \mathcal{J}_-^f \right.
$$
$$
\left. - \frac{4}{1 - z_-/z_+} \mathcal{J}_+^f \mathcal{J}_-^e + \frac{z_- + z_+}{z_- - z_+} \mathcal{J}_+^h \mathcal{J}_-^h \right) d\sigma^+ \wedge d\sigma^- . \tag{2.1.57}
$$

Here we have, for example, $\mathcal{J}_+^e$ denoting the e-component of $\mathcal{J}_+ = g_{(+)} \Lambda g_{(+)}^{-1}$.

Likewise, by picking C to be an elliptic curve E, the elliptic analogue of the massless Thirring model

$$
S_{2\mathrm{d}}^{\mathrm{eff}} = \frac{1}{\hbar} \left[-\frac{1}{2} \sum_{i=1}^{3} \int \psi_i \bar{\partial} \psi_i - \frac{1}{2} \sum_{i=1}^{3} \int \bar{\psi}_i \partial \bar{\psi}_i \right.
$$
$$
\left. + \frac{1}{4} \sum_{i,j,k,l,m=1}^{3} w_i (z_0 - z_1)\, \epsilon_{ijk} \epsilon_{ilm} \int \psi_j \psi_k \bar{\psi}_l \bar{\psi}_m \right] \tag{2.1.58}
$$

was derived in Ref. [24], where we used $\psi = \sum_{i=1}^{3} \psi_i t_i$, where t_i form the Pauli matrix basis of $\mathfrak{sl}_2$, i.e., $[t_1, t_2] = t_3$. We can also derive an elliptic generalization of the Faddeev–Reshetikhin model

$$
- \int_{\mathcal{M}} \mathrm{Tr}\left(\Lambda g_{(+)}^{-1} \partial_- g_{(+)} + \Lambda g_{(-)}^{-1} \partial_+ g_{(-)} + \sum_{i=1}^{3} w_i(z_+ - z_-) \mathcal{J}_+^i \mathcal{J}_-^i \right) d\sigma^+ \wedge d\sigma^-
$$
$$
\tag{2.1.59}
$$

where we have, for example, $\mathcal{J}_+^1$ denoting the 1-component of $\mathcal{J}_+ = g_{(+)} \Lambda g_{(+)}^{-1}$.

More generally, one can consider integrable field theories arising from multiple defects at various points on C. Hence, we find that 4d Chern–Simons theory allows us to build a broad framework for integrable field theories that can be realized using order surface operators.

2.2 Integrable Field Theories from Disorder Surface Defects

We have so far considered integrable field theories where the meromorphic one-form, ω, is restricted to one of three choices, corresponding to rational, trigonometric or elliptic integrable field theories. In this section, we consider more general choices of meromorphic one-form on $C = \mathbb{CP}^1$, which include the possibility of higher-order poles, as well as zeroes. As we see, the latter can be identified with the location of disorder defects in 4d Chern–Simons theory, characterized by singular behavior of the gauge field.

Many of the models we can derive using disorder defects can be interpreted as sigma models with target space a Lie group, or some coset thereof. The simplest such model is the conformal WZW model, which corresponds to the case with no disorder defect insertions, but with boundary conditions that explicitly break the topological symmetry along Σ. We first study the derivation of this model from 4d Chern–Simons theory, before moving to more elaborate examples.

2.2.1   *Examples*

2.2.1.1   *Conformal WZW model*

The conformal WZW model arises by considering the meromorphic one-form

$$\omega = \frac{\mathrm{d}z}{z}. \tag{2.2.1}$$

The boundary conditions[3] picked on $\mathbb{CP}^1$ are depicted in Figure 2.4, where the relation to the usual derivation of the WZW models from 3d Chern–Simons theory is also depicted (as we elucidate in the following chapter, both Chern–Simons theories are related via dimensional reduction and a form of T-duality). Explicitly, these boundary conditions are

$$\begin{aligned}
A_{\bar{w}}\big|_{z=0} &= 0, \\
A_w\big|_{z=\infty} &= 0, \\
A_{\bar{z}}\big|_{z=0} &= 0, \\
A_{\bar{z}}\big|_{z=\infty} &= 0.
\end{aligned} \tag{2.2.2}$$

Now, in defining the 4d Chern–Simons theory Lagrangian, we consider a topologically-trivial G-bundle. A topologically-trivial complex bundle on $\mathbb{CP}^1$ is generically holomorphically trivial as well, allowing us to express

---

[3]The terminology "boundary condition" in this chapter refers to the conditions on the gauge fields at the poles of the one-form ω.

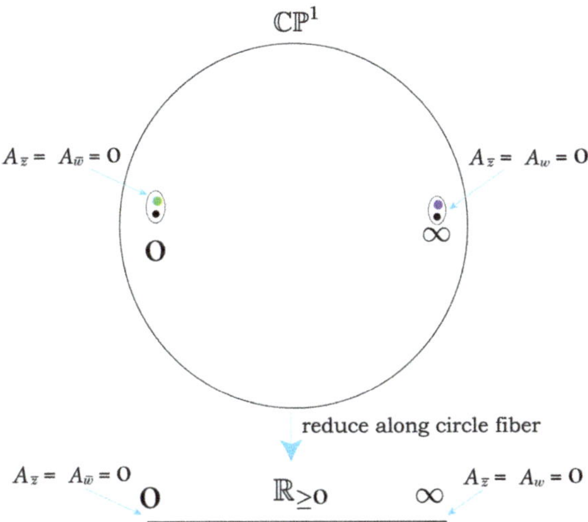

Figure 2.4. Boundary conditions realizing 2d conformal WZW model. Also depicted is the relation the 3d Chern–Simons theory realization of the WZW model, via shrinking of a circle fiber.

$A_{\bar{z}}$ as

$$A_{\bar{z}} = \widehat{g}^{-1}\partial_{\bar{z}}\widehat{g}, \tag{2.2.3}$$

where $\widehat{g} = g(w,\bar{w})$ at $z = \infty$ and $\widehat{g}$ is the identity, denoted 1, at $z = 0$, ensuring that the boundary conditions are obeyed. Without loss of generality, we choose $\widehat{g}$ to be invariant under the $U(1)$ action that rotates the z coordinate. The equations of motion $F_{w\bar{z}} = 0$ and $F_{\bar{w}\bar{z}} = 0$ can then be solved by setting

$$\begin{aligned}
A_{\bar{w}} &= \widehat{g}^{-1}\partial_{\bar{w}}\widehat{g}, \\
A_w &= \widehat{g}^{-1}\partial_w\widehat{g} - \widehat{g}^{-1}\partial_w g g^{-1}\widehat{g},
\end{aligned} \tag{2.2.4}$$

which satisfy the boundary conditions at $z = 0$ and $z = \infty$. Moreover these are the unique solutions to the equations of motion given a choice of $A_{\bar{z}}$, since if there was another solution to, e.g., $A_{\bar{w}}$, denoted $A'_{\bar{w}}$, the difference would satisfy

$$\partial_{\bar{z}}\left(A_{\bar{w}} - A'_{\bar{w}}\right) + [A_{\bar{z}}(\widehat{g}), A_{\bar{w}} - A'_{\bar{w}}] = 0, \tag{2.2.5}$$

i.e., it would be a holomorphic section of the adjoint bundle associated with the holomorphic bundle on $\mathbb{CP}^1$ which vanishes at one of the boundaries. Such sections do not exist.

In summary, setting $\widehat{A}_i = \widehat{g}^{-1}\partial_i\widehat{g}$, for $i = w, \bar{w}, \bar{z}$, we have

$$A = \widehat{A} + \widetilde{A}, \quad \widetilde{A} = -\widehat{g}^{-1}\partial_w gg^{-1}\widehat{g}\,\mathrm{d}w. \tag{2.2.6}$$

With the solutions for $A_{\bar{z}}$, $A_{\bar{w}}$ and A_w in hand, we can now insert them into the action. Now, the Chern–Simons three-form for the sum of two connections $\widehat{A}$ and $\widetilde{A}$ can be shown to be

$$\mathrm{CS}\left(\widehat{A} + \widetilde{A}\right) = \mathrm{CS}(\widehat{A}) + 2\,\mathrm{Tr}\left(F(\widehat{A})\widetilde{A}\right) - \mathrm{d}\,\mathrm{Tr}\left(\widehat{A}\widetilde{A}\right)$$
$$+ 2\,\mathrm{Tr}\left(\widehat{A}\widetilde{A}\widetilde{A}\right) + \mathrm{CS}\left(\widetilde{A}\right). \tag{2.2.7}$$

Firstly, note that $F(\widehat{A}) = 0$, and furthermore we have $\mathrm{CS}\left(\widetilde{A}\right) = \mathrm{Tr}\left(\widehat{A}\widetilde{A}\widetilde{A}\right) = 0$ since $\widetilde{A}$ only consists of a w component, which implies

$$\mathrm{CS}\left(A\right) = \mathrm{CS}\left(\widehat{A} + \widetilde{A}\right)$$
$$= \mathrm{CS}(\widehat{A}) - \mathrm{d}\,\mathrm{Tr}\left(\widehat{A}\widetilde{A}\right) \tag{2.2.8}$$
$$= \mathrm{CS}(\widehat{A}) + \mathrm{d}\,\mathrm{Tr}\left(\widehat{A}\widehat{g}^{-1}\partial_w gg^{-1}\widehat{g}\,\mathrm{d}w\right).$$

The 4d Chern–Simons action then requires us to integrate the wedge product of this expression with the one-form $\omega = \mathrm{d}z/z$ over $\mathbb{R}^2 \times \mathbb{CP}^1$. The integrand in the first term is the pure gauge Chern–Simons functional :

$$\mathrm{CS}(\widehat{A}) = -\frac{1}{3}\,\mathrm{Tr}\left(\widehat{g}^{-1}\,\mathrm{d}\widehat{g}\right)^3. \tag{2.2.9}$$

Because $\widehat{g}$ is picked to be invariant under the $U(1)$ rotation of circle fiber of $C = \mathbb{CP}^1$, the integral over the angular coordinate parametrizing the circle fiber results in a factor of $2\pi i$ (note that in polar coordinates where $z = r^{i\theta}$, $\frac{dz}{z} = \frac{dr}{r} + id\theta$):

$$\int_{\mathbb{R}^2 \times \mathbb{CP}^1} \mathrm{CS}(\widehat{A}) = -\frac{2\pi i}{3} \int_{\mathbb{R}^2 \times \mathbb{R}_{\geq 0}} \mathrm{Tr}\left(\widehat{g}^{-1}\mathrm{d}\widehat{g}\right)^3. \tag{2.2.10}$$

The remaining term in (2.2.8) is a total derivative in $\bar{z}$. We shall now employ the complex analysis lemma

$$\partial_{\bar{z}}\left(\frac{\mathrm{d}z}{z}\right) = 2\pi i\left(\delta_{z=0} - \delta_{z=\infty}\right), \tag{2.2.11}$$

which follows from the Cauchy-Pompeiu integral formula. Now, $\widehat{g} = 1$ near $z = 0$, and therefore $\widehat{A} = 0$ near $z = 0$. In addition, $\widehat{g} = g$ near $z = \infty$. This leads us to the action

$$\int_{\mathbb{R}^2 \times \mathbb{CP}^1} \frac{\mathrm{d}z}{z} \wedge \mathrm{d}\,\mathrm{Tr}\left(\widehat{A}\widehat{g}^{-1}\partial_w g g^{-1}\widehat{g}\mathrm{d}w\right)$$

$$= 2\pi i \int_{\mathbb{R}^2} \mathrm{Tr}\left(\widehat{A}\widehat{g}^{-1}\partial_w g g^{-1}\widehat{g}\mathrm{d}w\right)\Big|_{z=\infty} \tag{2.2.12}$$

$$= 2\pi i \int_{\mathbb{R}^2} \mathrm{Tr}\left(g^{-1}\partial_w g g^{-1}\partial_{\bar{w}}g\right).$$

As a result, we find the conformal WZW model

$$S_{\mathrm{WZW}} = \frac{k}{4\pi}\int_{\mathbb{R}^2}\mathrm{Tr}\left(g^{-1}\partial_w g g^{-1}\partial_{\bar{w}}g\right) - \frac{k}{12\pi}\int_{\mathbb{R}^2 \times \mathbb{R}_{\geq 0}}\mathrm{Tr}\left(\widehat{g}^{-1}\,\mathrm{d}\widehat{g}\right)^3$$

$$= \frac{k}{8\pi}\int_{\mathbb{R}^2}\mathrm{Tr}(j \wedge \star j) - \frac{k}{12\pi}\int_{\mathbb{R}^2 \times \mathbb{R}_{\geq 0}}\mathrm{Tr}(\widehat{j} \wedge \widehat{j} \wedge \widehat{j}),$$

where we have defined $j = g^{-1}\,\mathrm{d}g$ and $\widehat{j} = \widehat{g}^{-1}\,\mathrm{d}\widehat{g}$, and $k/(4\pi) = i/\hbar$.

As in the case with order surface defects, the Lax operator is the expectation value of the gauge field, evaluated at some point on $\mathbb{CP}^1$. To derive the Lax operator, we ought to apply an inverse gauge transformation by $\widehat{g}$, so that the gauge field becomes $A_w = 0$ and $A_{\bar{w}} = g^{-1}\partial_{\bar{w}}g = J_{\bar{w}}$. These values of the gauge field can be identified as the Lax operator, i.e.,

$$\mathcal{L}_{\bar{w}}(z) = J_{\bar{w}},$$
$$\mathcal{L}_w(z) = 0. \tag{2.2.13}$$

2.2.1.2 *Principal chiral model*

The conformal WZW model is a special case of the 2d principal chiral model (PCM) with Wess–Zumino term. The calculation just presented can in fact be generalized to derive the 2d PCM with Wess–Zumino term, using 4d Chern–Simons theory with the meromorphic one-form

$$\omega = \frac{(z - z_0)(z - z_1)}{z^2}\mathrm{d}z, \tag{2.2.14}$$

with boundary conditions specified in Figure 2.5, as well as the following singular behavior :

$$A_w \sim \frac{1}{z - z_0}$$

$$A_{\bar{w}} \sim \frac{1}{z - z_1}. \tag{2.2.15}$$

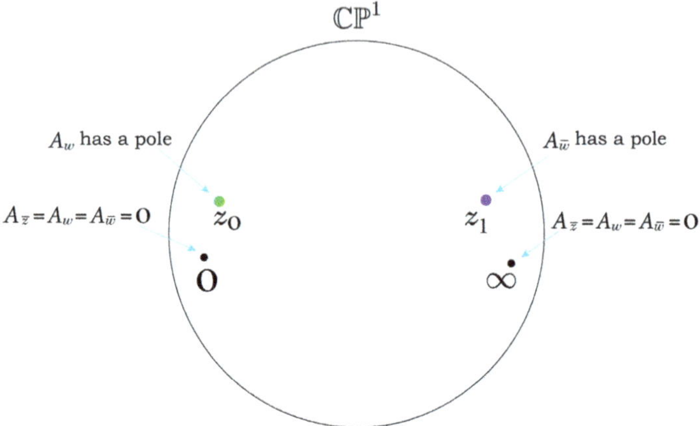

Figure 2.5. Boundary conditions realizing 2d principal chiral model with Wess–Zumino term.

These are examples of *disorder* surface defects in 4d Chern–Simons theory. As in the derivation of the conformal WZW model, we start by setting[4]

$$A_{\bar{z}} = \widehat{g}^{-1}\partial_{\bar{z}}\widehat{g}, \tag{2.2.16}$$

where, as before, $\widehat{g} = g(w, \overline{w})$ at $z = \infty$ and $\widehat{g} = 1$ at $z = 0$. Solving for the 4d Chern–Simons equations of motion gives

$$A_{\bar{w}} = \widehat{g}^{-1}\partial_{\bar{w}}\widehat{g} - \frac{z}{z - z_1}\widehat{g}^{-1}\partial_{\bar{w}}gg^{-1}\widehat{g}$$

$$A_w = \widehat{g}^{-1}\partial_w\widehat{g} - \frac{z}{z - z_0}\widehat{g}^{-1}\partial_w gg^{-1}\widehat{g}. \tag{2.2.17}$$

Substituting these solutions into the 4d Chern–Simons action and following steps analogous to the derivation of the conformal WZW model leads to the 2d action

$$S_{\text{PCM+WZ}} = \frac{k(z_1 - z_0)}{8\pi}\int_{\mathbb{R}^2}\text{Tr}(j \wedge \star j) - \frac{k(z_1 + z_0)}{12\pi}\int_{\mathbb{R}^2 \times \mathbb{R}_{\geq 0}}\text{Tr}(\widehat{j} \wedge \widehat{j} \wedge \widehat{j}), \tag{2.2.18}$$

where we have defined $j = g^{-1}\mathrm{d}g$ and $\widehat{j} = \widehat{g}^{-1}\,\mathrm{d}\widehat{g}$ and the level $k = 4\pi i/\hbar$ as in the example of the conformal WZW model. The conformal WZW model is obtained from this action by taking the limit $z_0 \to 0$ or $z_1 \to \infty$.

---

[4]This shall typically be the starting point for derivations of integrable sigma models from 4d Chern–Simons theory. The question of when this assumption holds for general meromorphic 1-forms ω is discussed in Remark 5.1 of [31].

2.2.1.3 *Coupled integrable field theories*

A natural generalization of the case of the 2d principal chiral model is to consider a meromorphic 1-form on $\mathbb{CP}^1$ with n second-order poles and $2n-2$ first-order zeroes, of the form

$$\omega = \frac{\prod_{i=1}^{n-1}\left(z - q_i^w\right)\prod_{j=1}^{n-1}\left(z - q_j^{\bar{w}}\right)}{\prod_{k=1}^{n}\left(z - p_k\right)^2}\,\mathrm{d}z. \tag{2.2.19}$$

The boundary conditions are that the gauge field is set to zero at all the second-order poles, p_i. Furthermore, at the zeroes q_i^w, we allow A_w to have a pole, and at $q_i^{\bar{w}}$, we allow $A_{\bar{w}}$ to have a pole, as shown in Figure 2.6. The field $A_{\bar{z}}$ defines a map $g : \mathbb{R}^2 \to G^n/G$ (up to gauge equivalence), where n copies of G correspond to the trivializations at p_i, and the right diagonal action is quotiented out. The constant gauge transformation is used to set the trivialization at p_n to correspond to the identity in G, such that $g = (g_1,\ldots,g_{n-1})$, where $g_i : \mathbb{R}^2 \to G$. We can then express $A_{\bar{z}}$ as

$$A_{\bar{z}} = \widehat{g}^{-1}\partial_{\bar{z}}\widehat{g}, \tag{2.2.20}$$

where $\widehat{g} = g_i$ near p_i and $\widehat{g}$ is the identity near p_n.

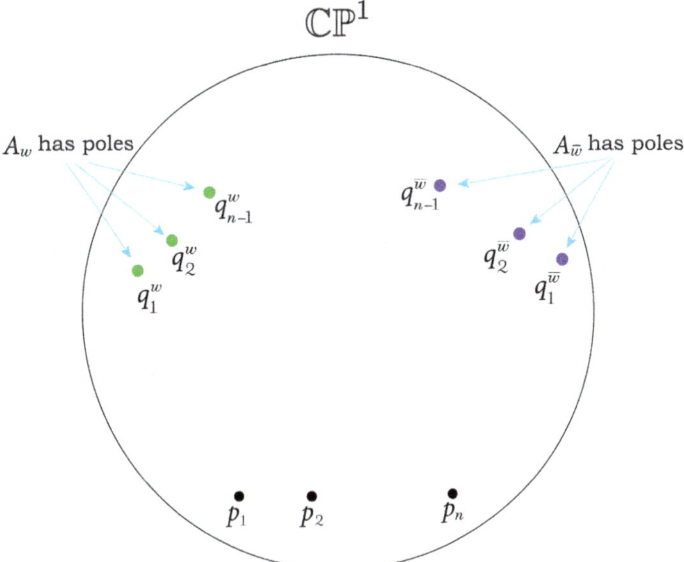

Figure 2.6. Boundary conditions realizing coupled integrable field theories.

We can now solve the equations of motion with

$$A_w = \widehat{A}_w + \widehat{g}^{-1} F(z) \widehat{g},$$
$$A_{\bar{w}} = \widehat{A}_{\bar{w}} + \widehat{g}^{-1} G(z) \widehat{g}, \tag{2.2.21}$$

where

$$F(z) = -\frac{(z - p_1) \ldots (z - p_n)}{(z - q_1^w) \ldots (z - q_{n-1}^w)}$$
$$\times \left(\sum_{j=1}^{n} \frac{1}{z - p_j} \frac{(p_j - q_1^w) \ldots (p_j - q_{n-1}^w)}{\prod_{k \neq j} (p_j - p_k)} \partial_w g_j g_j^{-1} \right),$$

$$G(z) = -\frac{(z - p_1) \ldots (z - p_n)}{(z - q_1^{\bar{w}}) \ldots (z - q_{n-1}^{\bar{w}})}$$
$$\times \left(\sum_{j=1}^{n} \frac{1}{z - p_j} \frac{(p_j - q_1^{\bar{w}}) \ldots (p_j - q_{n-1}^{\bar{w}})}{\prod_{k \neq j} (p_j - p_k)} \partial_{\bar{w}} g_j g_j^{-1} \right). \tag{2.2.22}$$

With these solutions, the 4d Chern–Simons action can be computed to give an integrable field theory that consists of coupled nonlinear sigma models, which has the explicit form

$$S_{\text{total}} = S_{\text{kin}} + S_{2-\text{form}} + S_{\text{WZ}}, \tag{2.2.23}$$

where

$$S_{\text{kin}} = \frac{k}{4\pi} \int_{\mathbb{R}^2} \left[\sum_i J_{i,\bar{w}} J_{i,w} \left(-\varphi_{i,w}(p_i) \varphi'_{i,\bar{w}}(p_i) + \varphi'_{i,w}(p_i) \varphi_{i,\bar{w}}(p_i) \right) \right.$$
$$\left. - \sum_{j \neq i} \frac{\varphi_{i,\bar{w}}(p_i) \varphi_{j,w}(p_j) - \varphi_{j,\bar{w}}(p_j) \varphi_{i,w}(p_i)}{p_i - p_j} J_{i,\bar{w}} J_{j,w} \right], \tag{2.2.24}$$

$$S_{2-\text{form}} = -\sum_{j \neq i} \frac{\varphi_{i,\bar{w}}(p_i) \varphi_{j,w}(p_j) + \varphi_{j,\bar{w}}(p_j) \varphi_{i,w}(p_i)}{p_i - p_j} J_{i,\bar{w}} J_{j,w} \tag{2.2.25}$$

and

$$S_{\text{WZ}} = -\frac{k}{2\pi} \sum_i \int_{\mathbb{R}^2 \times \mathbb{R}_{\geq 0}} \text{Tr}\left(J_{\bar{w},i} J_{w,i} J_{r,i} \right)$$
$$\times \left(\varphi_{i,w}(p_i) \varphi'_{i,\bar{w}}(p_i) + \varphi'_{i,w}(p_i) \varphi_{i,\bar{w}}(p_i) \right), \tag{2.2.26}$$

where we have used the convenient notation

$$\varphi_{i,w}(z) = \frac{\prod_k \left(z - q_k^w\right)}{\prod_{j \neq i} \left(z - p_j\right)},$$

$$\varphi_{i,\bar{w}}(z) = \frac{\prod_k \left(z - q_k^{\bar{w}}\right)}{\prod_{j \neq i} \left(z - p_j\right)}. \tag{2.2.27}$$

The subscript "2-form" in the second term of the action (2.2.23) indicates that this term arises from the integral of a two-form over $\Sigma = \mathbb{R}^2$.

2.2.2 *Unifying action for integrable field theories*

In this subsection, we shall study the interpretation of Costello and Yamazaki's derivation of integrable field theories from 4d Chern-Simons theory by Delduc, Lacroix, Magro and Vicedo [32]. In their work, a unifying action for many of the integrable field theories that can be derived from 4d Chern-Simons theory was presented. We shall review the main aspects of this derivation, and refer the readers to their original work for further details.

In this subsection, we shall mainly use the time coordinate, τ, space coordinate, σ, and the lightcone coordinates, $\sigma^\pm := \frac{1}{2}(\tau \pm \sigma)$, to parametrize Σ. We shall also use $\langle \cdot, \cdot \rangle : \mathfrak{g} \times \mathfrak{g} \to \mathbb{C}$ to denote the non-degenerate invariant symmetric bilinear form on the Lie algebra, $\mathfrak{g}$, of the gauge group, G. This was previously denoted using Tr.

We shall focus on meromorphic one-forms, ω, with at most double poles, as well as zeros.[5] Varying the 4d Chern-Simons action

$$S[A] = \frac{i}{4\pi} \int_{\Sigma \times \mathbb{C}P^1} \omega \wedge \left\langle A, \mathrm{d}A + \frac{2}{3} A \wedge A \right\rangle \tag{2.2.28}$$

with respect to the field A, we find

$$\delta S[A] = \frac{i}{2\pi} \int_{\Sigma \times \mathbb{C}P^1} \omega \wedge \langle \delta A, F(A) \rangle - \frac{i}{4\pi} \int_{\Sigma \times \mathbb{C}P^1} \mathrm{d}\omega \wedge \langle A, \delta A \rangle,$$

where $F(A) := \mathrm{d}A + A \wedge A$.[6] The equations of motion that follow are

$$\omega \wedge F(A) = 0, \tag{2.2.29a}$$

$$\mathrm{d}\omega \wedge \langle A, \delta A \rangle = 0, \tag{2.2.29b}$$

---

[5] For the generalization to the case of higher order poles, the reader is referred to the work of Benini, Schenkel and Vicedo [31].

[6] We have modified the normalization of the 4d CS action in this section to match that of [32].

where (2.2.29a) is referred to as the bulk equation of motion, while the second equation (2.2.29b) is the boundary equation of motion.

Let us express (2.2.29b) more explicitly. Let S_z denote the set of poles of ω. Set ξ_x to be a local holomorphic coordinate around an element, denoted x, of S_z. In other words, $\xi_x = z - x$ for $S_z \setminus \{\infty\}$ and $\xi_\infty = z^{-1}$ for the point at infinity. It will also be convenient to introduce the shorthand notation $f|_x := f|_{\Sigma \times \{x\}}$ for the function on Σ obtained by evaluating any function, f, on $\Sigma \times \mathbb{CP}^1$ at $x \in \mathbb{CP}^1$.

Now, using the fact that the pole part of the 1-form ω at each $x \in S_z$ can be expressed as

$$\sum_{p \geq 0} \frac{k_p^{(x)}}{\xi_x^{p+1}} \mathrm{d}\xi_x, \tag{2.2.30}$$

and that

$$\mathrm{d}\omega = -\partial_{\bar{\xi}_x} \left(\sum_{p \geq 0} \frac{k_p^{(x)}}{p!} (-1)^p \partial_{\xi_x}^p \frac{1}{\xi_x} \right) \mathrm{d}\xi_x \wedge \mathrm{d}\bar{\xi}_x$$

$$= 2\pi i \sum_{x \in S_z} \sum_{p \geq 0} \frac{k_p^{(x)}}{p!} (-1)^p \partial_{\xi_x}^p \delta(\xi_x) \mathrm{d}\xi_x \wedge \mathrm{d}\bar{\xi}_x, \tag{2.2.31}$$

the boundary equation of motion (2.2.29b) can be rewritten as

$$\sum_{x \in S_z} \sum_{p \geq 0} (\mathrm{res}_x \xi_x^p \omega) \epsilon_{ij} \frac{1}{p!} \partial_{\xi_x}^p \langle A_i, \delta A_j \rangle \big|_x = 0, \tag{2.2.32}$$

where there is an implicit sum over the repeated space-time indices $i, j = \tau, \sigma$ (this implicit sum shall be assumed hereon). When the 1-form ω has at most double poles, which is the case we shall focus on in this chapter, the boundary equation of motion (2.2.32) simply reads

$$\sum_{x \in S_z} (\mathrm{res}_x \omega) \epsilon_{ij} \langle A_i|_x, \delta A_j|_x \rangle + \sum_{x \in S_z} (\mathrm{res}_x \xi_x \omega) \epsilon_{ij} \partial_{\xi_x} \langle A_i, \delta A_j \rangle \big|_x = 0. \tag{2.2.33}$$

As in the examples considered previously, boundary conditions are imposed at the poles of ω, such that the boundary equation of motion is satisfied. As is expected for a gauge theory defined on a manifold with boundaries, the boundary conditions ought to be preserved by gauge transformations, albeit with the possibility that the gauge group is broken at the boundaries to some subgroup. However, in the present context, one also utilizes "formal" gauge transformations, which do not preserve the boundary conditions, and do not leave the 4d CS action invariant.

The utility of the formal gauge transformation is that a specific choice of formal gauge, namely the one where the $\bar{z}$-component of the gauge field vanishes, is the one wherein we may identify the gauge field with the Lax connection of an integrable field theory. The flatness of the Lax connection follows from the equation of motion involving the A_w and $A_{\bar{w}}$ components of the gauge field. This is indeed what we already observed in some examples.

The formal gauge transformation corresponds to a change of variables of the form

$$A = -\mathrm{d}\widehat{g}\widehat{g}^{-1} + \widehat{g}\mathcal{L}\widehat{g}^{-1}, \tag{2.2.34}$$

for some smooth function $\widehat{g} : \Sigma \times \mathbb{CP}^1 \to G^{\mathbb{C}}$, where $\mathcal{L} := \mathcal{L}_\sigma \mathrm{d}\sigma + \mathcal{L}_\tau \mathrm{d}\tau$ has no $\mathrm{d}\bar{z}$-component, that is, $\mathcal{L}_{\bar{z}} = 0$. In this formal gauge, the bulk equation of motion becomes

$$\omega \wedge \partial_{\bar{z}}\mathcal{L} = 0. \tag{2.2.35}$$

We immediately deduce that $\mathcal{L}$ is a meromorphic 1-form with poles at the zeros of ω, such that the order of each pole of $\mathcal{L}$ matches the order of the corresponding zero.

In explicit examples, it is often assumed that each singularity at a zero of ω is present in only one component of $\mathcal{L}$. Let S_ζ denote the set of simple zeros of the 1-form ω. We can express $\mathcal{L}$ in the following form:

$$\mathcal{L} = \sum_{y \in S_\zeta} V^y \xi_y^{-1} \mathrm{d}\sigma_y + U_\sigma \mathrm{d}\sigma + U_\tau \mathrm{d}\tau \tag{2.2.36}$$

where ξ_y is the local coordinate at $y \in S_\zeta$ and $U_\tau, U_\sigma, V^y : \Sigma \to \mathfrak{g}^{\mathbb{C}}$ are smooth functions. Here, each σ_y for $y \in S_\zeta$ is a linear combination of σ and τ.

If one is interested in deriving integrable field theories with Euclidean signature on Σ, one should consider the case where $\sigma_y = w$ for some of the zeros $y \in S_\zeta$ and $\sigma_y = \bar{w}$ for the others. If one wishes to instead derive integrable field theories with Lorentzian signature, one instead makes the choice $\sigma_y = \sigma^+$ for some of the zeros $y \in S_z$ and $\sigma_y = \sigma^-$ for the other zeros.

Now, under a formal gauge transformation as in (2.2.34), it can be shown that the Chern-Simons 3-form transforms as

$$CS(A) = \langle \mathcal{L}, \mathrm{d}\mathcal{L} \rangle + \mathrm{d}\langle \widehat{g}^{-1}\mathrm{d}\widehat{g}, \mathcal{L} \rangle + \tfrac{1}{3}\langle \widehat{g}^{-1}\mathrm{d}\widehat{g}, \widehat{g}^{-1}\mathrm{d}\widehat{g} \wedge \widehat{g}^{-1}\mathrm{d}\widehat{g} \rangle, \tag{2.2.37}$$

where we have used the fact that the 1-form $\mathcal{L}$ only has components along $\mathrm{d}\sigma$ and $\mathrm{d}\tau$, implying that $CS(\mathcal{L}) = \langle \mathcal{L}, \mathrm{d}\mathcal{L} \rangle$. Moreover, we observe that for $\mathcal{L}$ of the form (2.2.36), we have $\omega \wedge \langle \mathcal{L}, \mathrm{d}\mathcal{L} \rangle = 0$. Then, substituting (2.2.37) into the 4d CS action, we thus obtain

$$
\begin{aligned}
S[A] = &\frac{i}{12\pi} \int_{\Sigma \times \mathbb{CP}^1} \omega \wedge \langle \widehat{g}^{-1}\mathrm{d}\widehat{g}, \widehat{g}^{-1}\mathrm{d}\widehat{g} \wedge \widehat{g}^{-1}\mathrm{d}\widehat{g} \rangle \\
&+ \frac{i}{4\pi} \int_{\Sigma \times \mathbb{CP}^1} \mathrm{d}\omega \wedge \langle \widehat{g}^{-1}\mathrm{d}\widehat{g}, \mathcal{L} \rangle,
\end{aligned}
\tag{2.2.38}
$$

where Stokes' theorem has been used, and there is no boundary term since all fields are assumed to vanish at the boundaries of $\Sigma \times \mathbb{CP}^1$ (which are not poles of $\mathbb{CP}^1$). Bringing the 4d Chern-Simons action to the form given in (2.2.38) is the first main step in the derivation of the unifying action for integrable field theories.

Note that the formal gauge (2.2.34) is equivalent to saying that $A_{\bar{z}}$ is of the form

$$
A_{\bar{z}} = -\partial_{\bar{z}}\widehat{g}\widehat{g}^{-1}.
\tag{2.2.39}
$$

It is possible to put $A_{\bar{z}}$ in this form since, in defining the 4d CS theory with meromorphic 1-form ω, we consider a topologically trivial G-bundle on $\Sigma \times \mathbb{CP}^1$, and a topologically trivial complex bundle on $\mathbb{CP}^1$ is generically holomorphically trivial as well. Recall that the pure gauge form of $A_{\bar{z}}$ was also used as the starting point for the derivation of examples considered previously, such as the conformal WZW model and principal chiral model; see, e.g., (2.2.3).

Note that the smooth function $\widehat{g} : \Sigma \times \mathbb{CP}^1 \to G^{\mathbb{C}}$ is not unique. and we can multiply $\widehat{g}$ on the right by an arbitrary smooth function $h : \Sigma \to G$ since we can write

$$
A_{\bar{z}} = -\partial_{\bar{z}}(\widehat{g}h)(\widehat{g}h)^{-1},
\tag{2.2.40}
$$

which is equal to (2.2.39) since h is independent of $\bar{z}$. While this right multiplication of $\widehat{g}$ is a redundancy in expressing $A_{\bar{z}}$ in the form (2.2.39), it converts (2.2.34) to

$$
A = -\mathrm{d}\widehat{g}\widehat{g}^{-1} + \widehat{g}(-\mathrm{d}hh^{-1} + h\mathcal{L}h^{-1})\widehat{g}^{-1},
\tag{2.2.41}
$$

and thereby we observe that it corresponds to the transformation

$$\mathcal{L} \longmapsto h^{-1}\mathrm{d}h + h^{-1}\mathcal{L}h \tag{2.2.42}$$

on $\mathcal{L}$. This is the well-known two-dimensional gauge transformation of the Lax connection $\mathcal{L}$, which is always allowed in any integrable field theory, since it preserves its on-shell flatness.

On the other hand, we may also perform a gauge transformation on the connection A by a smooth function $u : \Sigma \times \mathbb{CP}^1 \to G^{\mathbb{C}}$, since the $\mathrm{d}\bar{z}$-component of the gauge transformed connection

$$A^u = -\mathrm{d}uu^{-1} + uAu^{-1} \tag{2.2.43}$$

still satisfies the form (2.2.39), i.e.,

$$A_{\bar{z}}^u = -\partial_{\bar{z}}uu^{-1} + uA_{\bar{z}}u^{-1} = -\partial_{\bar{z}}(u\widehat{g})(u\widehat{g})^{-1}. \tag{2.2.44}$$

It is worth emphasizing that the choice of u in the gauge transformation $A \mapsto A^u$ is not arbitrary in this context. The gauge transformation by u must be such that the boundary conditions imposed on A are preserved. This constraint distinguishes it as a genuine gauge transformation rather than a formal gauge transformation. It is important to note that unlike the transformation $\widehat{g} \mapsto \widehat{g}h$, the gauge transformation $\widehat{g} \mapsto u\widehat{g}$ does not modify the Lax connection, $\mathcal{L}$.

We shall now proceed to reduce the action (2.2.38) from four dimensions to two dimensions. In the work of [31], it was shown that the first term in (2.2.38) can be rewritten as

$$\frac{i}{12\pi} \int_{\Sigma \times \mathbb{CP}^1} \omega \wedge \langle \tilde{g}^{-1}\mathrm{d}\tilde{g}, \tilde{g}^{-1}\mathrm{d}\tilde{g} \wedge \tilde{g}^{-1}\mathrm{d}\tilde{g} \rangle, \tag{2.2.45}$$

where $\tilde{g} : \Sigma \times \mathbb{CP}^1 \to G^{\mathbb{C}}$ is any function that has the same restriction to the poles of ω as $\widehat{g}$. In particular, we can choose it to be of "archipelago type", meaning that it satisfies the following three archipelago conditions:

(i) The function $\tilde{g}$ satisfies $\tilde{g} = 1$ outside $\Sigma \times \bigsqcup_{x \in S_z} U_x$ for disjoint open discs U_x around the poles $x \in S_z$,

(ii) The restriction $\tilde{g}_x := \tilde{g}|_{\Sigma \times U_x}$ only depends on σ, τ as well as the radial coordinate $r_x := |\xi_x|$,

(iii) For every pole $x \in S_z$, the restriction $g_x := \tilde{g}|_{\Sigma \times \{x\}}$ to the point x only depends on σ and τ.

These archipelago conditions are visualized for a single pole, corresponding to a single "island" in an archipelago, in Figure 2.7. In Figure 2.8,

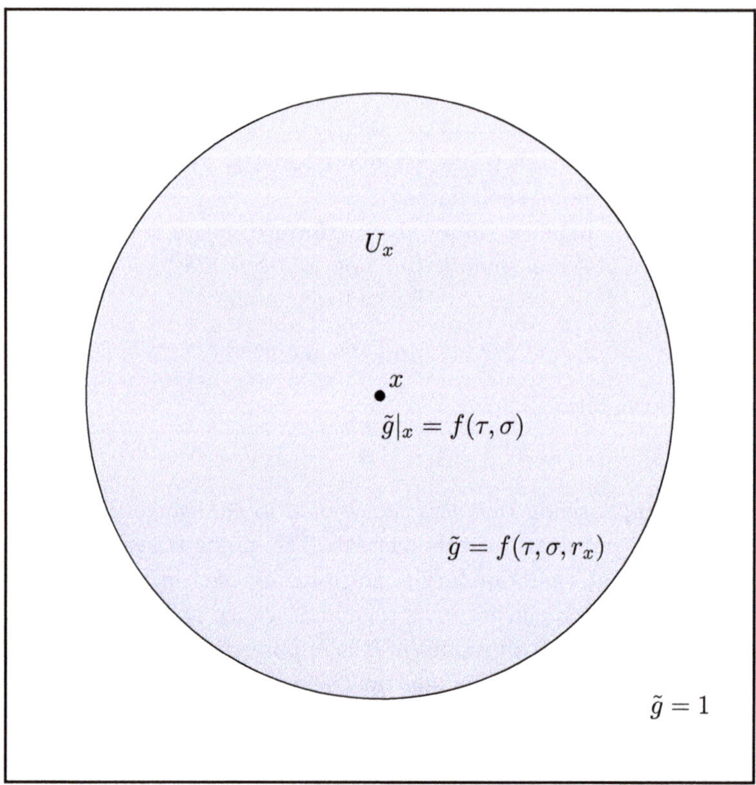

Figure 2.7.   Depiction of $\tilde{g}$ in the neighbourhood of a pole, $x \in S_z$, at a point on $\mathbb{CP}^1$.

the archipelago conditions are depicted for multiple poles on the surface $C = \mathbb{CP}^1$, giving rise to an archipelago of "islands", each of which supports a non-trivial degree of freedom g_x, sometimes known as an edge mode. These edge modes shall turn out to be fields of an integrable field theory.

For each disc, U_x, around any finite z, let us define the local polar coordinates $z = x + r_x e^{i\theta_x}$, and $z = r_\infty^{-1} e^{-i\theta_\infty}$ in U_∞ if $\infty \in S_z$. It follows that the first term in the action (2.2.38) can be expressed as

$$\frac{i}{12\pi} \sum_{x \in S_z} \int_{\Sigma \times U_x} \omega \wedge \langle \tilde{g}_x^{-1}\mathrm{d}\tilde{g}_x, \tilde{g}_x^{-1}\mathrm{d}\tilde{g}_x \wedge \tilde{g}_x^{-1}\mathrm{d}\tilde{g}_x \rangle$$

$$= -\frac{1}{12\pi} \sum_{x \in S_z \setminus \{\infty\}} \int_{\Sigma \times [0,R_x] \times [0,2\pi]} r_x e^{i\theta_x} \varphi(x + r_x e^{i\theta_x})\mathrm{d}\theta_x$$

$$\wedge \langle \tilde{g}_x^{-1}\mathrm{d}\tilde{g}_x, \tilde{g}_x^{-1}\mathrm{d}\tilde{g}_x \wedge \tilde{g}_x^{-1}\mathrm{d}\tilde{g}_x \rangle$$

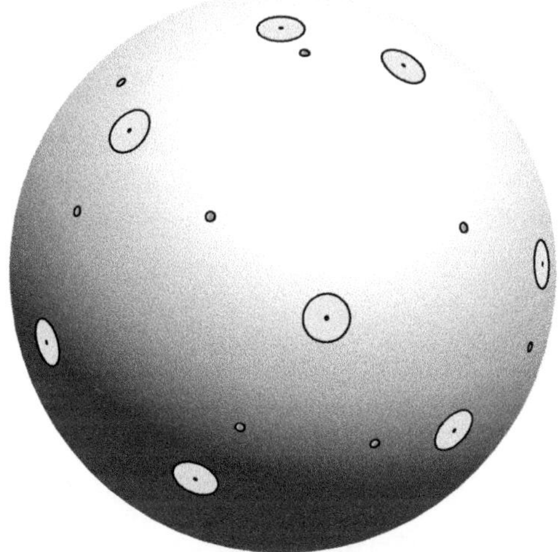

Figure 2.8.   Multiple poles of ω give rise to multiple "islands" (depicted in blue) on $C = \mathbb{CP}^1$, supporting edge modes. Here, multiple zeros of ω (depicted in red), where the gauge fields have singular behaviour (thereby realizing disorder surface defects) are also present.

$$+\frac{1}{12\pi}\sum_{x\in S_z\cap\{\infty\}}\int_{\Sigma\times[0,R_x]\times[0,2\pi]} r_x^{-1}e^{-i\theta_x}\varphi\big(r_x^{-1}e^{-i\theta_x}\big)\mathrm{d}\theta_x$$

$$\wedge\,\langle\tilde{g}_x^{-1}\mathrm{d}\tilde{g}_x,\tilde{g}_x^{-1}\mathrm{d}\tilde{g}_x\wedge\tilde{g}_x^{-1}\mathrm{d}\tilde{g}_x\rangle, \tag{2.2.46}$$

where R_x is the radius of the disc U_x around $x\in S_z$, and $\varphi(z)$ is known as the *twist function* governing the form of ω, i.e.,

$$\omega = \varphi(z)\mathrm{d}z. \tag{2.2.47}$$

To arrive at (2.2.46), we first use

$$\int\omega\wedge\langle\tilde{g}_x^{-1}\mathrm{d}\tilde{g}_x,\tilde{g}_x^{-1}\mathrm{d}\tilde{g}_x\wedge\tilde{g}_x^{-1}\mathrm{d}\tilde{g}_x\rangle$$

$$= 3\int\omega\wedge\mathrm{d}\bar{z}\wedge\mathrm{d}\tau\wedge\mathrm{d}\sigma\langle\tilde{g}_x^{-1}\partial_{\bar{z}}\tilde{g}_x,[\tilde{g}_x^{-1}\partial_\tau\tilde{g}_x,\tilde{g}_x^{-1}\partial_\sigma\tilde{g}_x]\rangle. \tag{2.2.48}$$

The first term on the RHS of (2.2.46) can be found by writing the measure on $\mathbb{CP}^1$ in terms of the polar coordinates $z = x + r_x e^{i\theta_x}$, that is, $\mathrm{d}z\wedge\mathrm{d}\bar{z} = -2ir\mathrm{d}r\wedge\mathrm{d}\theta$, and using

$$\partial_{\bar{z}}\tilde{g}_x = \frac{1}{2}e^{i\theta}\partial_r\tilde{g}_x. \tag{2.2.49}$$

Similarly, to obtain the second term on RHS of (2.2.46), corresponding to the pole at $z = \infty$, we use $\mathrm{d}\left(\frac{1}{z}\right) \wedge \mathrm{d}\left(\frac{1}{\bar{z}}\right) = -\frac{2i}{r^3}\mathrm{d}r \wedge \mathrm{d}\theta$ and $\partial_{\frac{1}{\bar{z}}}\tilde{g}_x = -\frac{r^2}{2}e^{-i\theta}\partial_r\tilde{g}_x$.

Let us choose an orientation on the discs U_x such that $r_x\mathrm{d}r_x \wedge \mathrm{d}\theta_x$ is the surface element. To compute the integral over the angular variable θ_x in the first term on the RHS of (2.2.46), we perform the Laurent expansion

$$\varphi(x + r_xe^{\theta_x}) = \sum_{p=-2}^{\infty} k_p^{(x)}(z - x)^p = \sum_{p=-2}^{\infty} k_p^{(x)}(r_xe^{i\theta_x})^p. \qquad (2.2.50)$$

The angular integral can then be computed as

$$\int_0^{2\pi} \mathrm{d}\theta_x r_x e^{i\theta_x}\varphi(x + r_xe^{\theta_x}) = \sum_{p=-2}^{\infty} k_p r_x^{1+p} \int_0^{2\pi} \mathrm{d}\theta_x e^{i(1+p)\theta} \qquad (2.2.51)$$

$$= 2\pi k_{-1} = 2\pi\mathrm{res}_x\omega.$$

A similar computation can be performed for the second term on the RHS of (2.2.46). Then, we find that the first term in the action (2.2.38) is

$$\frac{i}{12\pi} \int_{\Sigma \times \mathbb{CP}^1} \omega \wedge \langle \widehat{g}^{-1}\mathrm{d}\widehat{g}, \widehat{g}^{-1}\mathrm{d}\widehat{g} \wedge \widehat{g}^{-1}\mathrm{d}\widehat{g}\rangle = -\frac{1}{2}\sum_{x\in S_z}(\mathrm{res}_x\omega)I_{\mathrm{WZ}}[g_x],$$

where

$$I_{\mathrm{WZ}}[g_x] := -\frac{1}{3}\int_{\Sigma \times [0,R_x]} \langle \tilde{g}_x^{-1}\mathrm{d}\tilde{g}_x, \tilde{g}_x^{-1}\mathrm{d}\tilde{g}_x \wedge \tilde{g}_x^{-1}\mathrm{d}\tilde{g}_x\rangle$$

denotes a Wess-Zumino term.

We now study the second term in the action (2.2.38). Using (2.2.31), we find that it is equal to

$$-\frac{1}{2}\sum_{x\in S_z}\sum_{p\geq 0}\int_\Sigma \frac{k_p^{(x)}}{p!}\left(\partial_{\xi_x}^p \langle g_x^{-1}\mathrm{d}g_x, \mathcal{L}\rangle\right)\big|_x,$$

where $k_p^{(x)} = \mathrm{res}_x\xi_x^p\omega$ for each $p \in \mathbb{Z}_{\geq 0}$ and $x \in S_z$.

Since g_x is independent of the local coordinate ξ_x, it follows that $\langle g_x^{-1}dg_x, \mathcal{L}\rangle$ is holomorphic in a neighbourhood of the pole x of ω due to (2.2.35). Now, given a 1-form, f, that is holomorphic at x, we have

$$\mathrm{res}_x\omega \wedge f = \mathrm{res}_x\left(\sum_{p\geq 0} \frac{k_p^{(x)}}{\xi_x^{p+1}}\mathrm{d}\xi_x \wedge \sum_{q\geq 0} \frac{1}{q!}(\partial_{\xi_x}^q f)|_x\xi_x^q\right)$$

$$= \sum_{p \geq 0} \sum_{q \geq 0} \frac{k_p^{(x)}}{q!} (\partial_{\xi_x}^q f)|_x \delta_{p,q}$$

$$= \sum_{p \geq 0} \frac{k_p^{(x)}}{p!} (\partial_{\xi_x}^p f)|_x, \tag{2.2.52}$$

where we employed (2.2.30) for the pole part of ω at x, as well as the Taylor expansion of f around x. It therefore follows that

$$\frac{i}{4\pi} \int_{\Sigma \times \mathbb{CP}^1} \mathrm{d}\omega \wedge \langle \widehat{g}^{-1} \mathrm{d}\widehat{g}, \mathcal{L} \rangle = -\frac{1}{2} \sum_{x \in S_z} \int_\Sigma \mathrm{res}_x \left(\omega \wedge \langle g_x^{-1} \mathrm{d}g_x, \mathcal{L} \rangle \right)$$

$$= -\frac{1}{2} \sum_{x \in S_z} \int_\Sigma \langle g_x^{-1} \mathrm{d}g_x, \mathrm{res}_x \omega \wedge \mathcal{L} \rangle.$$

We thus arrive at the following result : the action (2.2.38) localizes to the sum of a two-dimensional term and a Wess-Zumino term for each element of S_z, namely

$$S[\{g_x\}_{x \in S_z}] = \frac{1}{2} \sum_{x \in S_z} \int_\Sigma \langle \mathrm{res}_x \omega \wedge \mathcal{L}, g_x^{-1} \mathrm{d}g_x \rangle - \frac{1}{2} \sum_{x \in S_z} (\mathrm{res}_x \omega) I_{\mathrm{WZ}}[g_x],$$

$$\tag{2.2.53}$$

with $g_x : \Sigma \to G$ denoting the restriction of $\widehat{g}$ to $\Sigma \times \{x\}$ for each $x \in S_z$. As we shall see in specific examples, the 1-form $\mathcal{L}$ can always be expressed in terms of $\{g_x\}_{x \in S_z}$, and therefore the action is only a functional of $\{g_x\}_{x \in S_z}$.

2.2.2.1 *Obtaining Real-valued Actions*

Let us now discuss the question of the reality of the action. As explained in the previous chapter, the 4d CS action is defined using complex-valued gauge fields A, and, moreover, depends on the complex-valued 1-form ω. As a result, the integrable field theories derived from 4d CS would also have complex-valued actions unless suitable reality conditions are imposed on the 4d CS action.

To this end, we define $\tau : \mathfrak{g}^{\mathbb{C}} \to \mathfrak{g}^{\mathbb{C}}$ to be an anti-linear involutive automorphism of the complex Lie algebra $\mathfrak{g}^{\mathbb{C}}$, which provides $\mathfrak{g}^{\mathbb{C}}$ with a $\mathbb{Z}_2$-action. The fixed point of the $\mathbb{Z}_2$-action is a real Lie subalgebra $\mathfrak{g}$ of $\mathfrak{g}^{\mathbb{C}}$. This anti-linear involution τ satisfies

$$\overline{\langle B, C \rangle} = \langle \tau B, \tau C \rangle \tag{2.2.54}$$

for any $B, C \in \mathfrak{g}^{\mathbb{C}}$. The notation τ shall also be used to denote the involutive automorphism on the corresponding Lie group $G^{\mathbb{C}}$.

Moreover, complex conjugation $z \mapsto \bar{z}$ on a local patch $\mathbb{C} \subset \mathbb{CP}^1$ defines an involution $\mu_t : \mathbb{CP}^1 \to \mathbb{CP}^1$, thereby furnishing $\mathbb{CP}^1$ with a $\mathbb{Z}_2$-action. The following equivariance conditions with respect to the $\mathbb{Z}_2$ action shall be imposed:

$$\bar{\omega} = \mu_t^* \omega, \qquad \tau A = \mu_t^* A. \tag{2.2.55}$$

The first condition amounts to the constraint $\overline{\varphi(z)} = \varphi(\bar{z})$ on the twist function. It can be shown that the reality conditions (2.2.55) ensure that the 4d CS action is real.

Imposing the equivariance property (2.2.55) on the 1-form A expressed as in (2.2.34) requires us to impose the equivariance properties

$$\tau \widehat{g} = \mu_t^* \widehat{g}, \qquad \tau \mathcal{L} = \mu_t^* \mathcal{L}. \tag{2.2.56}$$

In addition, note that to preserve the equivariance of $\widehat{g}$ in (2.2.56) we require that h in (2.2.40) takes values in the real subgroup $G \subset G^{\mathbb{C}}$ so that $\tau h = h$. Also, in order for A^u to be real in (2.2.43), it is necessary to impose that u is equivariant under the action of $\mathbb{Z}_2$.

Moreover, the equivariance properties (2.2.55) imply that for every $x \in S_z$, its complex conjugate is also a pole, i.e., $\bar{x} \in S_z$. Employing (2.2.56), we find that

$$\overline{\mathrm{res}_x \omega \wedge \mathcal{L}} = \mathrm{res}_{\bar{x}} \omega \wedge \mathcal{L}, \qquad \overline{\mathrm{res}_x \omega} = \mathrm{res}_{\bar{x}} \omega. \tag{2.2.57}$$

Furthermore, from (2.2.56) it follows that for any $x \in S_z$ we have $\tau(g_x) = g_{\bar{x}}$ and $\tau(\widehat{g}_x) = \widehat{g}_{\bar{x}}$. These conditions, together with the conditions given in (2.2.57), imply that the unifying action (2.2.53) is real, and thereby integrable field theory actions derived using (2.2.53) are real.

2.2.2.2 *Fixing 2d Gauge Invariance*

Before proceeding to discuss specific boundary conditions and examples of integrable field theories that can be realized from (2.2.53), let us explain a crucial property of this action.

Recall that there is a redundancy in (2.2.34) with regard to both the function $\widehat{g}$ and the 1-form $\mathcal{L}$, namely

$$\widehat{g} \longmapsto \widehat{g} h, \qquad \mathcal{L} \longmapsto h^{-1} \mathrm{d}h + h^{-1} \mathcal{L} h,$$

for an arbitrary smooth function $h : \Sigma \to G$. This transformation acts as

$$g_x \longmapsto g_x h \tag{2.2.58}$$

on the fields $\{g_x\}_{x \in S_z}$ that appear in (2.2.53), for all $x \in S_z$. It can be proved that the two-dimensional action (2.2.53) is invariant under the gauge transformation (2.2.58) for an arbitrary choice of h. In explicit examples of derivations of integrable field theories from (2.2.53), this gauge invariance is fixed by the requirement that $g_x = 1$ for some pole $x \in S_z$.

2.2.2.3 *Boundary Conditions*

As mentioned before, we restrict our attention in this subsection to the case where ω has at most double poles, in which case the boundary conditions imposed on A should ensure that (2.2.33) holds. Let us now summarize some boundary conditions that satisfy the boundary equation of motion (2.2.33).

For a double pole $x \in S_z$ of ω, with the requirement that the point x lies on the real axis in order to satisfy the reality conditions, the simplest possible choice is to impose the Dirichlet boundary condition

$$A_i|_x = 0, \qquad (2.2.59)$$

for $i = \tau, \sigma$, whereby we also have $\delta A_i|_x = 0$.

For a pair of simple poles, the boundary equation of motion is

$$(\mathrm{res}_{x_+}\omega)\epsilon_{ij}\langle A_i|_{x_+}, \delta A_j|_{x_+}\rangle + (\mathrm{res}_{x_-}\omega)\epsilon_{ij}\langle A_i|_{x_-}, \delta A_j|_{x_-}\rangle = 0. \qquad (2.2.60)$$

The two possibilities allowed by the reality conditions are those corresponding to the case where x_+ and x_- are both real and where they form a complex conjugate pair. For the first case of real simple poles, (2.2.60) can be rewritten as

$$\epsilon_{ij}\left\langle\!\!\left\langle\left(A_i|_{x_+}, A_i|_{x_-}\right), \delta\left(A_j|_{x_+}, A_j|_{x_-}\right)\right\rangle\!\!\right\rangle_{\mathfrak{d};x_\pm} = 0, \qquad (2.2.61)$$

where $\langle\!\langle\cdot,\cdot\rangle\!\rangle_{\mathfrak{d};x_\pm} : \mathfrak{d} \times \mathfrak{d} \to \mathbb{R}$ denotes the non-degenerate symmetric invariant bilinear form on $\mathfrak{d} := \mathfrak{g} \oplus \mathfrak{g}$ given by

$$\langle\!\langle(\mathsf{x},\mathsf{y}),(\mathsf{x}',\mathsf{y}')\rangle\!\rangle_{\mathfrak{d};x_\pm} := \left(\mathrm{res}_{x_+}\omega\right)\langle\mathsf{x},\mathsf{x}'\rangle + \left(\mathrm{res}_{x_-}\omega\right)\langle\mathsf{y},\mathsf{y}'\rangle, \qquad (2.2.62)$$

for arbitrary $\mathsf{x},\mathsf{y},\mathsf{x}',\mathsf{y}'$. For the second case of a complex conjugate pair of simple poles, (2.2.60) can be rewritten as

$$\epsilon_{ij}\left\langle\!\!\left\langle A_i|_{x_+}, \delta A_j|_{x_+}\right\rangle\!\!\right\rangle_{\mathfrak{g}^{\mathbb{C}};x_\pm} = 0, \qquad (2.2.63)$$

where $\langle\!\langle\cdot,\cdot\rangle\!\rangle_{\mathfrak{g}^{\mathbb{C}};x_\pm} : \mathfrak{g}^{\mathbb{C}} \times \mathfrak{g}^{\mathbb{C}} \to \mathbb{R}$ is the non-degenerate symmetric invariant bilinear form on the complexification $\mathfrak{g}^{\mathbb{C}}$, regarded as a real Lie algebra, defined by

$$\langle\!\langle\mathsf{x},\mathsf{x}'\rangle\!\rangle_{\mathfrak{g}^{\mathbb{C}};x_\pm} := 2\,\mathrm{Re}\left(\left(\mathrm{res}_{x_+}\omega\right)\langle\mathsf{x},\mathsf{x}'\rangle\right). \qquad (2.2.64)$$

for any $\mathsf{x},\mathsf{x}' \in \mathfrak{g}^{\mathbb{C}}$.

Now, for the case of real simple poles, consider a *Manin pair* $(\mathfrak{d}, \mathfrak{l})$, where $\mathfrak{d}$ is a Lie algebra and $\mathfrak{l}$ is a Lagrangian subalgebra of $\mathfrak{d}$. Here, "Lagrangian" means that $\mathfrak{l}$ is a maximal isotropic subalgebra. The boundary condition that satisfies (2.2.61) is that, for $i = \tau, \sigma$,

$$(A_i|_{x_+}, A_i|_{x_-}) \in \mathfrak{l}, \tag{2.2.65}$$

noting that we will then also have $\delta(A_i|_{x_+}, A_i|_{x_-}) \in \mathfrak{l}$.

For the second case of a complex conjugate pair of simple poles, we choose a Manin pair for the complexification of the Lie algebra, i.e., $(\mathfrak{g}^{\mathbb{C}}, \mathfrak{l})$, and the boundary condition that satisfies (2.2.63) is

$$A_i|_{x_+} \in \mathfrak{l}, \tag{2.2.66}$$

for $i = \tau, \sigma$, noting that this implies $\delta A_i|_{x_+} \in \mathfrak{l}$.

These are the simplest possible boundary conditions that can be imposed. However, (2.2.59) is a special case of a more general boundary condition that may be imposed at a double pole of ω along the real axis. Let us denote such a point as $x \in S_z$. Consider the semidirect product $\mathfrak{t} := \mathfrak{g} \ltimes \mathfrak{g}_{\mathrm{ab}}$ where $\mathfrak{g}_{\mathrm{ab}}$ is an abelian copy of $\mathfrak{g}$ on which $\mathfrak{g}$ acts by the adjoint action. In other words, $\mathfrak{t}$ is isomorphic to the direct sum $\mathfrak{g} \oplus \mathfrak{g}$ as a vector space with Lie bracket given by $[(x, y), (x', y')]_{\mathfrak{t}} = ([x, x'], [x, y'] - [x', y])$ for any $x, y, x', y' \in \mathfrak{g}$. The boundary equation of motion (2.2.33) can be rewritten in this case as

$$\epsilon_{ij} \left\langle\!\left\langle \left(A_i|_x, (\partial_{\xi_x} A_i)|_x \right), \delta\left(A_j|_x, (\partial_{\xi_x} A_j)|_x \right) \right\rangle\!\right\rangle_{\mathfrak{t};x} = 0, \tag{2.2.67}$$

where $\langle\!\langle \cdot, \cdot \rangle\!\rangle_{\mathfrak{t};x} : \mathfrak{t} \times \mathfrak{t} \to \mathbb{R}$ is the nondegenerate bilinear form on $\mathfrak{t}$ defined by

$$\langle\!\langle (\mathsf{x}, \mathsf{y}), (\mathsf{x}', \mathsf{y}') \rangle\!\rangle_{\mathfrak{t};x} := (\mathrm{res}_x\, \omega)\, \langle \mathsf{x}, \mathsf{x}' \rangle + (\mathrm{res}_x\, \xi_x \omega)\, (\langle \mathsf{x}, \mathsf{y}' \rangle + \langle \mathsf{x}', \mathsf{y} \rangle) \tag{2.2.68}$$

for every $\mathsf{x}, \mathsf{y}, \mathsf{x}', \mathsf{y}' \in \mathfrak{g}$.

Now, if we have a Manin pair $(\mathfrak{t}, \mathfrak{l})$, for a Lagrangian subalgebra $\mathfrak{l}$ of $\mathfrak{t}$, then the boundary equation of motion can be satisfied by requiring that

$$\left(A_i|_x, (\partial_{\xi_x} A_i)|_x \right) \in \mathfrak{l} \tag{2.2.69}$$

for $i = \tau, \sigma$, which further implies $\delta\left(A_i|_x, (\partial_{\xi_x} A_i)|_x \right) \in \mathfrak{l}$. We note that (2.2.59) corresponds to the special case where $\mathfrak{l} = \{0\} \ltimes \mathfrak{g}_{\mathrm{ab}}$.

2.2.2.4 *Examples of Boundary Conditions based on the Classical Yang-Baxter Equation and Manin Triples*

At both simple and double poles, we have seen that Manin pairs are useful for defining suitable boundary conditions. We shall now explore particular examples of such boundary conditions where the Manin pair can be further extended to a *Manin triple*. As we shall see, these boundary conditions allow us to derive several known integrable field theories from 4d Chern-Simons theory.

Let us first consider the case where ω has simple poles. We shall extend the Manin pairs considered previously in this context to Manin triples, i.e., we consider a Lie algebra $\tilde{\mathfrak{g}}$ (which we previously specified to be $\mathfrak{d}$ or $\mathfrak{g}_{\mathbb{C}}$) that has *two* Lagrangian subalgebras $\mathfrak{l}_1$ and $\mathfrak{l}_2$, that are moreover complementary to each other, i.e., $\tilde{\mathfrak{g}} = \mathfrak{l}_1 \dotplus \mathfrak{l}_2$.

We shall be interested in Manin triples that depend on a specific solution $R \in \mathrm{End}\,\mathfrak{g}$ of the modified classical Yang-Baxter equation

$$[R\mathsf{x}, R\mathsf{y}] - R\big([R\mathsf{x}, \mathsf{y}] + [\mathsf{x}, R\mathsf{y}]\big) = -c^2[\mathsf{x}, \mathsf{y}], \qquad (2.2.70)$$

for every $\mathsf{x}, \mathsf{y} \in \mathfrak{g}$, where either $c = 1$ or $c = i$. Moreover, we shall be interested in solutions satisfying the property of skew-symmetry, i.e.,

$$\langle R\mathsf{x}, \mathsf{y} \rangle = -\langle \mathsf{x}, R\mathsf{y} \rangle, \qquad (2.2.71)$$

for any $\mathsf{x}, \mathsf{y} \in \mathfrak{g}$.

For the case of real simple poles, we consider (2.2.70) with $c = 1$ and define the Lie algebras

$$\mathfrak{g}_R := \{((R-1)\mathsf{x}, (R+1)\mathsf{x}) \,|\, \mathsf{x} \in \mathfrak{g}\}, \qquad \mathfrak{g}^\delta := \{(\mathsf{x}, \mathsf{x}) \,|\, \mathsf{x} \in \mathfrak{g}\}, \qquad (2.2.72)$$

which are both Lie subalgebras of $\mathfrak{d}$. Pick the standard bilinear form on $\mathfrak{d}$, that is,

$$\langle\!\langle\!\langle (\mathsf{x}, \mathsf{y}), (\mathsf{x}', \mathsf{y}') \rangle\!\rangle\!\rangle_{\mathfrak{d}} := \langle \mathsf{x}, \mathsf{x}' \rangle - \langle \mathsf{y}, \mathsf{y}' \rangle$$

for any $\mathsf{x}, \mathsf{y}, \mathsf{x}', \mathsf{y}' \in \mathfrak{g}$, which, up to overall normalization, corresponds to the bilinear form (2.2.61) when $\mathrm{res}_{x_-}\omega = -\mathrm{res}_{x_+}\omega$. In this case, both $\mathfrak{g}^\delta$ and $\mathfrak{g}_R$ are isotropic, the latter being due to the skew-symmetry of R. Thus, $(\mathfrak{d}, \mathfrak{g}_R, \mathfrak{g}^\delta)$ is a Manin triple. For a pair of real simple poles x_+ and x_-, the boundary condition where the defining Lagrangian subalgebra, $\mathfrak{l}$, is $\mathfrak{g}_R$, is given explicitly by

$$A_i\big|_{z=x_\pm} = (R \mp 1)\mathsf{x}_i, \qquad (2.2.73)$$

where $\mathsf{x}_i \in \mathfrak{g}$. When the defining Lagrangian subalgebra, $\mathfrak{l}$, is $\mathfrak{g}^\delta$, the boundary condition is given explicitly by

$$A_i\big|_{z=x_\pm} = \mathsf{x}_i, \qquad (2.2.74)$$

where $\mathsf{x}_i \in \mathfrak{g}$.

For our purposes, we also need to understand how these structures lift to the Lie group level. Let the Lie algebra $\mathfrak{g}^\delta$ correspond to the Lie subgroup $G^\delta := \{(x,x) \,|\, x \in G\} \subset D$, where D is the Lie group generated by $\mathfrak{d}$. In addition, let G_R denote the Lie subgroup of D with Lie algebra $\mathfrak{g}_R$. In what follows, we shall assume that the decomposition $\mathfrak{d} = \mathfrak{g}_R \dotplus \mathfrak{g}^\delta$ lifts to the Lie group level, that is, $D = G_R G^\delta$, or at least that $G_R G^\delta$ forms a dense subset of D. A natural parametrization of the quotient $G_R \backslash D$ in the case $c = 1$ is thus furnished by elements of G^δ.

In the case of complex conjugate simple poles, we take $c = i$ and define

$$\mathfrak{g}_R := \{(R - i)\mathsf{x} \,|\, \mathsf{x} \in \mathfrak{g}\},$$

with $\mathfrak{g} \subset \mathfrak{g}^{\mathbb{C}}$ denoting the real subalgebra of $\mathfrak{g}^{\mathbb{C}}$ regarded itself as a real Lie algebra. From (2.2.70), it follows that $\mathfrak{g}_R$ is a Lie subalgebra of $\mathfrak{g}^{\mathbb{C}}$. If we assume that $\mathfrak{g}^{\mathbb{C}}$ is equipped with its standard bilinear form

$$\langle\!\langle \mathsf{x}, \mathsf{x}' \rangle\!\rangle_{\mathfrak{g}^{\mathbb{C}}} = \mathrm{Im}\langle \mathsf{x}, \mathsf{x}' \rangle$$

for any $\mathsf{x}, \mathsf{x}' \in \mathfrak{g}^{\mathbb{C}}$, that corresponds to (2.2.64) when $\mathrm{res}_{x_-}\omega = -\mathrm{res}_{x_+}\omega$, then both $\mathfrak{g}$ and $\mathfrak{g}_R$ are isotropic subalgebras. Thus, $(\mathfrak{g}^{\mathbb{C}}, \mathfrak{g}_R, \mathfrak{g})$ is a Manin triple. For a pair of complex conjugate poles x_+ and x_-, the boundary condition where the defining Lagrangian subalgebra, $\mathfrak{l}$, is $\mathfrak{g}_R$ is given explicitly by

$$A_i\big|_{z=x_\pm} = (R \mp i)\mathsf{x}_i, \tag{2.2.75}$$

where $\mathsf{x}_i \in \mathfrak{g}$.

As before, we need to understand how these structures lift to the Lie group level. To this end, let $G_R \subset G^{\mathbb{C}}$ denote the Lie subgroup with Lie algebra $\mathfrak{g}_R \subset \mathfrak{g}^{\mathbb{C}}$. In what follows, we shall assume that the decomposition $\mathfrak{g}^{\mathbb{C}} = \mathfrak{g}_R \dotplus \mathfrak{g}$ extends to the Lie group level, in particular, that $G^{\mathbb{C}} = G_R G$ or at least $G_R G$ forms a dense subset of $G^{\mathbb{C}}$. One can then furnish a natural parametrization of the quotient $G_R \backslash G^{\mathbb{C}}$ by elements of G. One example of such a parameterization is given by the Iwasawa decomposition $G^{\mathbb{C}} = ANG$, where G is the compact real form of $G^{\mathbb{C}}$ and $G_R = AN$.

Note that for each Manin triple, we obtain two Manin pairs, namely $(\mathfrak{d}, \mathfrak{l}_1)$ and $(\mathfrak{d}, \mathfrak{l}_2)$ for $(\mathfrak{d}, \mathfrak{l}_1, \mathfrak{l}_2)$, and $(\mathfrak{g}^{\mathbb{C}}, \mathfrak{l}_1)$ and $(\mathfrak{g}^{\mathbb{C}}, \mathfrak{l}_2))$ for $(\mathfrak{g}^{\mathbb{C}}, \mathfrak{l}_1, \mathfrak{l}_2)$. These models are expected to be Poisson-Lie T-dual to each other.[7] This is

---

[7] Poisson-Lie T-duality is a generalization of T-duality and non-abelian T-duality to 2d field theories without manifest global symmetries, but whose equations of motion can be expressed as a non-commutative conservation law. Since these dualities may be interpreted as canonical transformations on phase space, they preserve classical integrability, which is our primary interest in this chapter.

because, as we shall see in the next subsection, in the case of Manin triples of the form $\mathfrak{d} = \mathfrak{g}_R \dotplus \mathfrak{g}^\delta$, the two models obtained using 4d Chern-Simons theory with these boundary conditions are Poisson-Lie T-dual models, namely, the (inhomogeneous) Yang-Baxter deformation of the principal chiral model and the λ-deformation of the principal chiral model.

Now, let us consider the case of double poles. Here, the crucial point is that solutions $R \in \mathrm{End}\,\mathfrak{g}$ of the classical Yang-Baxter equation (2.2.70) with $c = 0$ furnish a class of Lie subalgebras $\mathfrak{l} \subset \mathfrak{t}$ for the boundary condition (2.2.69). Provided with such a solution, one can define a Lie subalgebra

$$\mathfrak{g}_R := \{(-R\mathsf{x}, \mathsf{x}) \mid \mathsf{x} \in \mathfrak{g}\} \tag{2.2.76}$$

of $\mathfrak{t}$. The fact that this is indeed a Lie subalgebra follows since for any $\mathsf{x}, \mathsf{y} \in \mathfrak{g}$ we have

$$\big[(-R\mathsf{x}, \mathsf{x}), (-R\mathsf{y}, \mathsf{y})\big]_\mathfrak{t} = \big([-R\mathsf{x}, -R\mathsf{y}], [-R\mathsf{x}, \mathsf{y}] - [-R\mathsf{y}, \mathsf{x}]\big) = (-R\mathsf{z}, \mathsf{z}) \in \mathfrak{g}_R, \tag{2.2.77}$$

where $\mathsf{z} = -[R\mathsf{x}, \mathsf{y}] - [\mathsf{x}, R\mathsf{y}] \in \mathfrak{g}$. Note that the Lie subalgebra $\mathfrak{g} \ltimes \{0\} \subset \mathfrak{t}$ is isotropic with respect to the bilinear form $\langle\langle \cdot, \cdot \rangle\rangle_{\mathfrak{t};x}$ defined in (2.2.68). For a solution $R \in \mathrm{End}\,\mathfrak{g}$ of (2.2.70) with $c = 0$ that is skew-symmetric, i.e.,

$$\langle R\mathsf{x}, \mathsf{y} \rangle = -\langle \mathsf{x}, R\mathsf{y} \rangle \tag{2.2.78}$$

for any $\mathsf{x}, \mathsf{y} \in \mathfrak{g}$, the subalgebra $\mathfrak{g}_R \subset \mathfrak{t}$ is also isotropic. We thus have a Manin triple $(\mathfrak{t}, \mathfrak{g}_R, \mathfrak{g} \ltimes \{0\})$. The explicit boundary condition at a double pole, denoted x, corresponding to $\mathfrak{l} = \mathfrak{g}_R$ is

$$\begin{aligned} A_i|_x &= -R\mathsf{x}_i \\ \partial_{\xi_x} A_i|_x &= \mathsf{x}_i. \end{aligned} \tag{2.2.79}$$

where $\mathsf{x}_i \in \mathfrak{g}$.

Finally, we need the aforementioned structures to lift nicely to the Lie group level. To this end, we assume that the vector space direct sum decomposition $\mathfrak{t} = (\mathfrak{g} \ltimes \{0\}) \dotplus \mathfrak{g}_R$ lifts to the Lie group level, namely that $T = G_R(G \ltimes \{0\})$, or at least that $G_R(G \ltimes \{0\})$ forms a dense subset of T.

2.2.3 *Further Examples*

In this section, we will derive the actions of various known integrable field theories from the four-dimensional Chern-Simons action by using the unifying action (2.2.53), as well as the boundary conditions studied in the previous subsection. The theories that will be derived shall have Lorentz invariance along Σ, which we parametrize by the lightcone coordinates $\sigma^\pm := \frac{1}{2}(\tau \pm \sigma)$. The reader who is unfamiliar with the integrable deformations of the principal chiral model derived in this subsection is encouraged to consult, for example, the review by Hoare [33].

2.2.3.1 *Principal chiral model with Wess-Zumino term*

We shall first rederive the principal chiral model with WZ-term, using (2.2.53). Consider the 1-form

$$\omega = K \frac{1 - z^2}{(z - k)^2} \mathrm{d}z, \qquad (2.2.80)$$

where $K, k \in \mathbb{R}$, which has a pair of double poles at k and ∞. This is equivalent to the form of ω used in the derivation of the PCM in [24], and described earlier, by a change of variable $z \mapsto z + k$.

The boundary equations of motion (2.2.33) can be satisfied by

$$A_i|_k = 0,$$
$$A_i|_\infty = 0, \qquad (2.2.81)$$

for $i = \tau, \sigma$. For the double pole at $z = k$, this follows since the corresponding boundary equation of motion is

$$-2Kk\epsilon_{ij}\langle A_i|_k, \delta A_j|_k\rangle + (1 - k^2)K\epsilon_{ij}\partial_z\langle A_i, \delta A_j\rangle\big|_k = 0, \qquad (2.2.82)$$

which is satisfied since $A|_k = 0$. In deriving (2.2.82), we have used

$$\mathrm{res}_k\omega = -2Kk. \qquad (2.2.83)$$

We shall set the restrictions of $\widehat{g}$ to the poles to be

$$g_k = g, \qquad g_\infty = 1$$

for some $g : \Sigma \to G$. The second condition fixes the gauge invariance under (2.2.58). From (2.2.34), we then find

$$A|_k = -\mathrm{d}gg^{-1} + g\mathcal{L}|_k g^{-1}, \qquad A|_\infty = \mathcal{L}|_\infty. \qquad (2.2.84)$$

The 1-form ω has simple zeros at ± 1, that is, $S_\zeta = \{1, -1\}$. Moreover, from (2.2.81) and (2.2.84), we find that $\mathcal{L}$ vanishes at infinity. It follows that $U_\sigma = U_\tau = 0$ in the general expression (2.2.36) for $\mathcal{L}$. In addition, in the notation used in (2.2.36), we choose $\sigma_{\pm 1} = \sigma^\pm$, leading to the Lax connection

$$\mathcal{L} = \frac{V^1}{z-1}\mathrm{d}\sigma^+ + \frac{V^{-1}}{z+1}\mathrm{d}\sigma^-, \tag{2.2.85}$$

for maps $V^{\pm 1} : \Sigma \to \mathfrak{g}$ that are to be determined. This can be achieved by solving $-\partial_i g g^{-1} + \mathrm{Ad}_g \mathcal{L}_i|_k = 0$ for $i = \tau, \sigma$, which follows from combining the first equations in (2.2.81) and (2.2.84), resulting in

$$-\partial_\pm g g^{-1} + \mathrm{Ad}_g \frac{V^{\pm 1}}{z \mp 1}|_k = 0, \tag{2.2.86}$$

or equivalently,

$$V^{\pm 1} = (k \mp 1)g^{-1}\partial_\pm g. \tag{2.2.87}$$

In what follows, we shall make use of the definitions

$$j_\pm := g^{-1}\partial_\pm g. \tag{2.2.88}$$

We can now use the unifying action (2.2.53) to derive an integrable field theory. Since we chose to set $g_\infty = 1$, the terms in this action corresponding to the pole $\infty \in S_z$ do not contribute. We shall need the residues

$$\mathrm{res}_k \omega \wedge \mathcal{L} = -K\big((k-1)j_+\mathrm{d}\sigma^+ + (k+1)j_-\mathrm{d}\sigma^-\big),$$

and (2.2.83). Then, the kinetic term is

$$\frac{1}{2}\sum_{x \in S_z}\int_\Sigma \langle \mathrm{res}_x \omega \wedge \mathcal{L}, g_x^{-1}\mathrm{d}g_x\rangle$$

$$= \frac{1}{2}\int_\Sigma \langle -K(k-1)g^{-1}\partial_+ g, g^{-1}\partial_- g\rangle \mathrm{d}\sigma^+ \wedge \mathrm{d}\sigma^-$$

$$ -\frac{1}{2}\int_\Sigma \langle -K(k+1)g^{-1}\partial_- g, g^{-1}\partial_+ g\rangle \mathrm{d}\sigma^+ \wedge \mathrm{d}\sigma^- \tag{2.2.89}$$

$$= K\int_\Sigma \langle g^{-1}\partial_+ g, g^{-1}\partial_- g\rangle \mathrm{d}\sigma^+ \wedge \mathrm{d}\sigma^-.$$

The Wess-Zumino term can be computed straightforwardly using (2.2.83). In summary, we obtain the 2d action

$$S[g] = \frac{K}{2}\int_\Sigma \langle j_+, j_-\rangle \mathrm{d}\sigma \wedge \mathrm{d}\tau + K k\, I_{\mathrm{WZ}}[g],$$

where we used the identity $\mathrm{d}\sigma^+ \wedge \mathrm{d}\sigma^- = \frac{1}{2}\mathrm{d}\sigma \wedge \mathrm{d}\tau$. This is the action of the principal chiral model with Wess-Zumino term.

2.2.3.2 *Homogeneous Yang-Baxter Sigma Model*

To derive the homogeneous Yang-Baxter deformation of a given integrable sigma model, we do not require modification of the underlying twist function [34]. To derive the homogeneous Yang-Baxter deformation of the PCM, we shall thus use (2.2.80) as the meromorphic one-form, where we set $k = 0$ for simplicity, that is,

$$\omega = K\frac{1 - z^2}{z^2}\mathrm{d}z. \tag{2.2.90}$$

However, the boundary condition imposed at the double pole of ω at 0 will not be the Dirichlet boundary condition. Instead, we will generalize it to a boundary condition that is associated with a choice of Lagrangian subalgebra of $\mathfrak{t} = \mathfrak{g} \ltimes \mathfrak{g}_{\mathrm{ab}}$, discussed above. In the present context, the bilinear form on $\mathfrak{t}$ is

$$\langle\!\langle(\mathsf{x}, \mathsf{y}), (\mathsf{x}', \mathsf{y}')\rangle\!\rangle_{\mathfrak{t};0} = K\big(\langle\mathsf{x}, \mathsf{y}'\rangle + \langle\mathsf{x}', \mathsf{y}\rangle\big). \tag{2.2.91}$$

As discussed above, $\mathfrak{g}_R = \{(-R\mathsf{x}, \mathsf{x}) \,|\, \mathsf{x} \in \mathfrak{g}\}$ for a skew-symmetric solution, R, of the classical Yang-Baxter equation, is a Lagrangian Lie subalgebra of $\mathfrak{t}$. The boundary equations of motion (2.2.33) are satisfied by requiring that

$$\big(A_i|_0, (\partial_z A_i)|_0\big) \in \mathfrak{g}_R, \qquad A_i|_\infty = 0, \tag{2.2.92}$$

for $i = \tau, \sigma$. Indeed, recalling that the first condition in (2.2.92) implies (2.2.79), or equivalently,

$$A_i|_0 = -R(\partial_z A_i)|_0, \tag{2.2.93}$$

we find that the boundary equation of motion for the pole at 0 is satisfied since

$$\begin{aligned}
&K\epsilon_{ij}\langle\delta A_j, \partial_z A_i\rangle|_{z=0} + K\epsilon_{ij}\langle A_i, \partial_z\delta A_j\rangle|_{z=0}\\
&=\epsilon_{ij}\langle\!\langle(A_i, \partial_z A_i), (\delta A_j, \partial_z\delta A_j)\rangle\!\rangle_{\mathfrak{t};0}
\end{aligned} \tag{2.2.94}$$

vanishes.

We shall set the restrictions of $\widehat{g}$ to the poles to be

$$g_0 = g, \qquad g_\infty = 1 \tag{2.2.95}$$

for some $g : \Sigma \to G$. The latter condition will fix the invariance of the unifying action under the gauge transformation (2.2.58). Evaluating

(2.2.34) at 0 and ∞ then leads us to

$$A|_0 = -\mathrm{d}gg^{-1} + \mathrm{Ad}_g\mathcal{L}|_0, \qquad A|_\infty = \mathcal{L}|_\infty. \tag{2.2.96}$$

In addition, taking the derivative of the gauge field with respect to z before evaluating it at 0, and since $(\partial_z\widehat{g})|_0 = 0$, one obtains

$$(\partial_z A)|_0 = \mathrm{Ad}_g(\partial_z\mathcal{L})|_0. \tag{2.2.97}$$

Given that $\mathcal{L}$ ought to be meromorphic with poles in the set $S_\zeta = \{1, -1\}$ of zeros of ω and since it vanishes at infinity due to (2.2.92) and (2.2.96), it follows from (2.2.36) that

$$\mathcal{L} = \frac{V^1}{z - 1}\mathrm{d}\sigma^+ + \frac{V^{-1}}{z + 1}\mathrm{d}\sigma^-. \tag{2.2.98}$$

Using (2.2.93), we have

$$-\partial_i gg^{-1} + g\mathcal{L}_i|_0 g^{-1} = -R\mathrm{Ad}_g\partial_z\mathcal{L}_i|_0 \tag{2.2.99}$$

which can be equivalently written as

$$-g^{-1}\partial_i g + \mathcal{L}_i|_0 = -R_g\partial_z\mathcal{L}_i|_0, \tag{2.2.100}$$

where $R_g := \mathrm{Ad}_{g^{-1}}\circ R\circ\mathrm{Ad}_g$. Substituting in the ansatz for the Lax operator (2.2.98), we find

$$\begin{aligned}
-g^{-1}\partial_+ g + \frac{V^1}{z-1}|_0 &= R_g\frac{V^1}{(z-1)^2}|_0 \\
-g^{-1}\partial_- g + \frac{V^{-1}}{z+1}|_0 &= R_g\frac{V^1}{(z+1)^2}|_0.
\end{aligned} \tag{2.2.101}$$

Therefore, we obtain

$$V^{\pm 1} = \mp\frac{1}{1 \pm R_g}j_\pm.$$

We can then proceed to derive an integrable field theory from (2.2.53). Using the facts that $\mathrm{res}_0\omega \wedge \mathcal{L} = -K(V^1\mathrm{d}\sigma^+ + V^{-1}\mathrm{d}\sigma^-)$ and $\mathrm{res}_0\omega = 0$, (2.2.53) reduces to

$$\begin{aligned}
&\frac{K}{2}\int_\Sigma\left\langle\frac{1}{1+R_g}j_+, j_-\right\rangle\mathrm{d}\sigma^+ \wedge \mathrm{d}\sigma^- - \frac{K}{2}\int_\Sigma\left\langle j_+, \frac{1}{1-R_g}j_-\right\rangle\mathrm{d}\sigma^- \wedge \mathrm{d}\sigma^+ \\
&= \frac{K}{2}\int_\Sigma\left\langle j_+, \frac{1}{1-R_g}j_-\right\rangle\mathrm{d}\sigma^+ \wedge \mathrm{d}\sigma^- + \frac{K}{2}\int_\Sigma\left\langle j_+, \frac{1}{1-R_g}j_-\right\rangle\mathrm{d}\sigma^+ \wedge \mathrm{d}\sigma^-
\end{aligned} \tag{2.2.102}$$

where we have used $\langle A, \left(\frac{1}{1+R_g}\right)^T B\rangle = \langle A, \frac{1}{1-R_g}B\rangle$ which follows from $\langle R_g A, B\rangle = -\langle A, R_g B\rangle$. In this way, we arrive at the action

$$S[g] = \frac{K}{2}\int_\Sigma \left\langle j_+, \frac{1}{1-R_g}j_-\right\rangle \mathrm{d}\sigma \wedge \mathrm{d}\tau, \tag{2.2.103}$$

which is the homogeneous Yang-Baxter deformation of the principal chiral model.

2.2.3.3   *Inhomogeneous Yang-Baxter Sigma Model*

To derive the Yang-Baxter sigma model associated with solutions of the modified classical Yang-Baxter equation (2.2.70) where $c = 1$ or $c = i$, we shall begin with 4d Chern-Simons theory with the meromorphic 1-form

$$\omega = \frac{K}{1-c^2\eta^2}\frac{1-z^2}{z^2-c^2\eta^2}\mathrm{d}z, \tag{2.2.104}$$

where $K, \eta \in \mathbb{R}$.

Pick a skew-symmetric solution $R \in \mathrm{End}\,\mathfrak{g}$ of the modified classical Yang-Baxter equation (2.2.70). Since $\mathrm{res}_{c\eta}\omega = -\mathrm{res}_{-c\eta}\omega$, we know that $\mathfrak{g}_R$ is a Lagrangian subalgebra of $\mathfrak{d}\,(\mathfrak{g}^{\mathbb{C}})$ when $c = 1$ ($c = i$). We can then satisfy the boundary equations of motion (2.2.33) using

$$(A_i|_\eta, A_i|_{-\eta}) \in \mathfrak{g}_R, \qquad A_i|_\infty = 0, \tag{2.2.105a}$$

for $i = \tau, \sigma$, for $c = 1$, or

$$A_i|_{i\eta} \in \mathfrak{g}_R, \qquad A_i|_\infty = 0, \tag{2.2.105b}$$

for $i = \tau, \sigma$, for $c = i$. Indeed, from (2.2.73) and (2.2.75), we know that these boundary conditions can be restated as

$$(R+c)A_i|_{c\eta} = (R-c)A_i|_{-c\eta}, \qquad A_i|_\infty = 0. \tag{2.2.106}$$

Noting that $\mathrm{res}_{c\eta}\,\omega = -\,\mathrm{res}_{-c\eta}\,\omega$, the boundary equation of motion is solved since

$$\epsilon^{ij}(\langle A_i|_{c\eta}, \delta A_j|_{c\eta}\rangle - \langle A_i|_{-c\eta}, \delta A_j|_{-c\eta}\rangle)$$

$$= \epsilon^{ij}\left(\langle \frac{R-c}{R+c}A_i|_{-c\eta}, \frac{R-c}{R+c}\delta A_j|_{-c\eta}\rangle - \langle A_i|_{-c\eta}, \delta A_j|_{-c\eta}\rangle\right)$$

$$= \epsilon^{ij}\left(\langle \frac{R+c}{R-c}\frac{R-c}{R+c}A_i|_{-c\eta}, \delta A_j|_{-c\eta}\rangle - \langle A_i|_{-c\eta}, \delta A_j|_{-c\eta}\rangle\right) \tag{2.2.107}$$

$$= \epsilon^{ij}(\langle A_i|_{-c\eta}, \delta A_j|_{-c\eta}\rangle - \langle A_i|_{-c\eta}, \delta A_j|_{-c\eta}\rangle)$$

$$= 0,$$

where we have used $\langle A, RB\rangle = -\langle RA, B\rangle$.

We can now choose the values of $\widehat{g}$ at the poles to be

$$g_{\pm c\eta} = g, \qquad g_{\infty} = 1 \tag{2.2.108}$$

for some $g : \Sigma \to G$.

For the case of $c = 1$, this follows since there is gauge invariance of the boundary condition (2.2.106) with respect to elements of G_R, meaning that the value of $\widehat{g}$ at the pair of points $\pm\eta$ defines a field on Σ valued in $G_R \backslash D$, after gauge-fixing. We know that this can be parametrized by the diagonal subgroup G^δ, allowing us to pick $\widehat{g}$ such that $(\widehat{g}|_\eta, \widehat{g}|_{-\eta}) = (g, g)$.

Moreover, for the case of $c = i$, due to the gauge invariance of (2.2.106) with respect to elements of G_R, the value of $\widehat{g}$ at the point $i\eta$ defines a field on Σ valued in $G_R \backslash G^{\mathbb{C}}$, after gauge-fixing. This can be parametrized by the real subgroup G, allowing us to pick $\widehat{g}$ such that $\widehat{g}|_{i\eta} = g$. We can then use the property that $\widehat{g}$ is equivariant to obtain $g_{-i\eta} = \tau(g_{i\eta}) = g$.

Then, evaluating (2.2.34) at the poles of ω we obtain

$$A|_{\pm c\eta} = -dgg^{-1} + \mathrm{Ad}_g \mathcal{L}|_{\pm c\eta}, \qquad A|_\infty = \mathcal{L}|_\infty. \tag{2.2.109}$$

Since the simple zeros of ω are at ± 1, we have $S_\zeta = \{1, -1\}$. Furthermore, from (2.2.105) and (2.2.109) we find that $\mathcal{L}$ should vanish at infinity. The Lax connection must thus be of the form

$$\mathcal{L} = \frac{V^1}{z-1} d\sigma^+ + \frac{V^{-1}}{z+1} d\sigma^-, \tag{2.2.110}$$

for $\mathfrak{g}$-valued maps $V^{\pm 1} : \Sigma \to \mathfrak{g}$.

Using (2.2.106) together with (2.2.109) and (2.2.110) we find

$$-(R+c)dgg^{-1} + (R+c)\mathrm{Ad}_g\left(\frac{1}{c\eta - 1}V^1 d\sigma^+ + \frac{1}{c\eta + 1}V^{-1} d\sigma^-\right)$$

$$= -(R-c)dgg^{-1} - (R-c)\mathrm{Ad}_g\left(\frac{1}{c\eta + 1}V^1 d\sigma^+ + \frac{1}{c\eta - 1}V^{-1} d\sigma^-\right),$$

which provides two equations for the two unknowns $V^{\pm 1}$, which are

$$-(R+c)\partial_\pm gg^{-1} + (R+c)\mathrm{Ad}_g\frac{1}{c\eta \mp 1}V^{\pm 1}$$

$$= -(R-c)\partial_\pm gg^{-1} - (R-c)\mathrm{Ad}_g\frac{1}{c\eta \pm 1}V^{\pm 1}, \tag{2.2.111}$$

or equivalently,

$$-2c\partial_\pm gg^{-1} + (2Rc\eta \pm 2c)\frac{\mathrm{Ad}_g V^{\pm 1}}{c^2\eta^2 - 1} = 0. \tag{2.2.112}$$

Solving these equations gives

$$V^{\pm 1} = \pm \frac{c^2\eta^2 - 1}{1 \pm \eta R_g} j_\pm$$

where $R_g = \mathrm{Ad}_{g^{-1}} \circ R \circ \mathrm{Ad}_g$ and $j_\pm = g^{-1}\partial_\pm g$.

There are no Wess-Zumino terms, for the following reasons. Firstly, since $\mathrm{res}_{\pm c\eta}\omega = \pm K/2c\eta$ and $g_{c\eta} = g_{-c\eta}$, the Wess-Zumino terms in (2.2.53) associated with the simple poles $\pm c\eta$ cancel each other. Secondly, since $g_\infty = 1$, there is no Wess-Zumino term corresponding to the double pole at ∞.

To obtain the explicit form of the remaining term in the unifying action (2.2.53), we compute

$$\mathrm{res}_{\pm c\eta}\omega \wedge \mathcal{L} = (\mathrm{res}_{\pm c\eta}\omega)\mathcal{L}|_{\pm c\eta} = \pm \frac{K}{2c\eta}\left(\frac{c\eta + 1}{1 \pm \eta R_g} j_\pm \mathrm{d}\sigma^\pm - \frac{c\eta - 1}{1 \mp \eta R_g} j_\mp \mathrm{d}\sigma^\mp \right). \tag{2.2.113}$$

The kinetic term in (2.2.53) is then

$$\frac{1}{2}\int_\Sigma \left\langle \frac{K}{2c\eta}\left(\frac{c\eta + 1}{1 + \eta R_g} j_+ \mathrm{d}\sigma^+ - \frac{c\eta - 1}{1 - \eta R_g} j_- \mathrm{d}\sigma^- \right), g^{-1}\mathrm{d}g \right\rangle$$

$$- \frac{1}{2}\int_\Sigma \left\langle \frac{K}{2c\eta}\left(\frac{c\eta + 1}{1 - \eta R_g} j_- \mathrm{d}\sigma^- - \frac{c\eta - 1}{1 + \eta R_g} j_+ \mathrm{d}\sigma^+ \right), g^{-1}\mathrm{d}g \right\rangle \tag{2.2.114}$$

$$= \frac{K}{2}\int_\Sigma \left\langle \frac{1}{1 + \eta R_g} j_+, j_- \right\rangle \mathrm{d}\sigma^+ \wedge \mathrm{d}\sigma^-$$

$$- \frac{K}{2}\int_\Sigma \left\langle \frac{1}{1 - \eta R_g} j_-, j_+ \right\rangle \mathrm{d}\sigma^- \wedge \mathrm{d}\sigma^+.$$

The final step is to use $\left\langle A, \left(\frac{1}{1+\eta R_g}\right)^T B \right\rangle = \left\langle A, \frac{1}{1-\eta R_g} B \right\rangle$ which follows from $\langle R_g A, B \rangle = -\langle A, R_g B \rangle$. Then, we obtain the Yang-Baxter sigma model action, which takes the form

$$S[g] = \frac{K}{2}\int_\Sigma \left\langle j_+, \frac{1}{1 - \eta R_g} j_- \right\rangle \mathrm{d}\sigma \wedge \mathrm{d}\tau. \tag{2.2.115}$$

2.2.3.4   *λ-deformation of the principal chiral model*

To derive the λ-deformation of the principal chiral model, we shall use 4d Chern-Simons theory with the meromorphic 1-form

$$\omega = \frac{K}{1 - \alpha^2}\frac{1 - z^2}{z^2 - \alpha^2}\mathrm{d}z, \qquad \lambda = \frac{1 + \alpha}{1 - \alpha}, \tag{2.2.116}$$

with $K, \alpha \in \mathbb{R}$.

Now, given that $\mathrm{res}_\alpha \omega = -\mathrm{res}_{-\alpha}\omega$, we know that $\mathfrak{g}^\delta$ is a Lagrangian subalgebra of $\mathfrak{d}$. The boundary equation of motion can thus be solved by imposing

$$(A_i|_\alpha, A_i|_{-\alpha}) \in \mathfrak{g}^\delta, \qquad A_i|_\infty = 0 \tag{2.2.117}$$

for $i = \tau, \sigma$, which, by (2.2.74), is equivalent to $A_i|_\alpha = A_i|_{-\alpha}$ and $A_i|_\infty = 0$. Indeed, this follows since

$$\begin{aligned}
\epsilon^{ij}(&\langle A_i|_\alpha, \delta A_j|_\alpha\rangle - \langle A_i|_{-\alpha}, \delta A_j|_{-\alpha}\rangle) \\
&= \epsilon^{ij}(\langle A_i|_{-\alpha}, \delta A_j|_{-\alpha}\rangle - \langle A_i|_{-\alpha}, \delta A_j|_{-\alpha}\rangle) \\
&= 0.
\end{aligned} \tag{2.2.118}$$

Due to gauge invariance of the boundary condition $A_i|_\alpha = A_i|_{-\alpha}$ with respect to elements of G^δ, $(\widehat{g}|_\alpha, \widehat{g}|_{-\alpha})$ defines a field on Σ valued in $G^\delta \backslash D$, after gauge-fixing. We can parametrize this quotient by elements of the form $(h, 1)$ for $h \in G$. The field $\widehat{g}$ can thus be chosen such that

$$g_\alpha = g, \qquad g_{-\alpha} = 1, \qquad g_\infty = 1$$

for some $g : \Sigma \to G$. Furthermore, from (2.2.34), we obtain

$$A|_\alpha = -dgg^{-1} + \mathrm{Ad}_g\mathcal{L}|_\alpha, \qquad A|_{-\alpha} = \mathcal{L}|_{-\alpha}, \qquad A|_\infty = \mathcal{L}|_\infty. \tag{2.2.119}$$

As before, we can deduce that $\mathcal{L}|_\infty = 0$. This leads to $\mathcal{L}$ being of the form

$$\mathcal{L} = \frac{\alpha + 1}{z - 1}U_+\mathrm{d}\sigma^+ + \frac{\alpha + 1}{z + 1}U_-\mathrm{d}\sigma^- \tag{2.2.120}$$

for a pair of $\mathfrak{g}$-valued fields $U_\pm = (\alpha + 1)^{-1}V^{\pm 1}$ on Σ.

Now, computing the value of $\mathcal{L}$ at $\pm\alpha$ gives us

$$\mathcal{L}|_\alpha = -\lambda U_+\mathrm{d}\sigma^+ + U_-\mathrm{d}\sigma^-, \qquad \mathcal{L}|_{-\alpha} = -U_+\mathrm{d}\sigma^+ + \lambda U_-\mathrm{d}\sigma^-. \tag{2.2.121}$$

Using the boundary conditions at $\pm\alpha$ and the first two equations in (2.2.119) gives us

$$-dgg^{-1} - \lambda\mathrm{Ad}_g U_+\mathrm{d}\sigma^+ + \mathrm{Ad}_g U_-\mathrm{d}\sigma^- = -U_+\mathrm{d}\sigma^+ + \lambda U_-\mathrm{d}\sigma^-.$$

We can thus equate the coefficients of $\mathrm{d}\sigma^\pm$ on both sides to give us two equations in the two unknowns U_+ and U_-, which are

$$\begin{aligned}
-\partial_+gg^{-1} - \lambda\mathrm{Ad}_g U_+ &= -U_+ \\
-\partial_-gg^{-1} + \mathrm{Ad}_g U_- &= \lambda U_-.
\end{aligned} \tag{2.2.122}$$

Solving these equations gives us

$$U_+ = \frac{\partial_+ g g^{-1}}{1 - \lambda \mathrm{Ad}_g},$$

$$U_- = \frac{-\partial_- g g^{-1}}{\lambda - \mathrm{Ad}_g}. \tag{2.2.123}$$

Substituting the solutions back into (2.2.121) leads us to

$$\mathcal{L}|_\alpha = -\frac{\lambda \mathrm{Ad}_g}{1 - \lambda \mathrm{Ad}_g} j_+ \mathrm{d}\sigma^+ + \frac{\mathrm{Ad}_g}{\mathrm{Ad}_g - \lambda} j_- \mathrm{d}\sigma^-,$$

where $j_\pm := g^{-1} \partial_\pm g$. Thus, we obtain

$$\mathrm{res}_\alpha \omega \wedge \mathcal{L} = (\mathrm{res}_\alpha \omega) \mathcal{L}|_\alpha = 2k \frac{\lambda \mathrm{Ad}_g}{1 - \lambda \mathrm{Ad}_g} j_+ \mathrm{d}\sigma^+ - 2k \frac{\mathrm{Ad}_g}{\mathrm{Ad}_g - \lambda} j_- \mathrm{d}\sigma^-$$

since $\mathrm{res}_\alpha \omega = -2k$, where $k = -K/4\alpha$.

The unifying action (2.2.53) can then be computed. The kinetic term is

$$\frac{1}{2} \sum_{x \in S_z} \int_\Sigma \langle \mathrm{res}_x \omega \wedge \mathcal{L}, g_x^{-1} \mathrm{d}g_x \rangle$$

$$= \frac{1}{2} \int_\Sigma \langle 2k \frac{\lambda \mathrm{Ad}_g}{1 - \lambda \mathrm{Ad}_g} j_+, j_- \rangle \mathrm{d}\sigma^+ \wedge \mathrm{d}\sigma^- - \frac{1}{2} \int_\Sigma \langle 2k \frac{\mathrm{Ad}_g}{\mathrm{Ad}_g - \lambda} j_-, j_+ \rangle \mathrm{d}\sigma^- \wedge \mathrm{d}\sigma^+$$

$$= \frac{1}{2} \int_\Sigma \langle 2k \frac{\lambda \mathrm{Ad}_g}{1 - \lambda \mathrm{Ad}_g} j_+, j_- \rangle \mathrm{d}\sigma^+ \wedge \mathrm{d}\sigma^- - \frac{1}{2} \int_\Sigma \langle 2k \frac{\lambda}{\mathrm{Ad}_g - \lambda} j_-, j_+ \rangle \mathrm{d}\sigma^- \wedge \mathrm{d}\sigma^+$$

$$\quad - \frac{1}{2} \int_\Sigma \langle 2k j_-, j_+ \rangle \mathrm{d}\sigma^- \wedge \mathrm{d}\sigma^+$$

$$= \frac{1}{2} \int_\Sigma \langle 2k \frac{\lambda \mathrm{Ad}_g}{1 - \lambda \mathrm{Ad}_g} j_+, j_- \rangle \mathrm{d}\sigma^+ \wedge \mathrm{d}\sigma^- - k \int_\Sigma \langle j_-, \frac{\lambda}{\mathrm{Ad}_g^{-1} - \lambda} j_+ \rangle \mathrm{d}\sigma^- \wedge \mathrm{d}\sigma^+$$

$$\quad - \frac{1}{2} \int_\Sigma \langle 2k j_-, j_+ \rangle \mathrm{d}\sigma^- \wedge \mathrm{d}\sigma^+$$

$$= \int_\Sigma \langle 2k \frac{\lambda \mathrm{Ad}_g}{1 - \lambda \mathrm{Ad}_g} j_+, j_- \rangle \mathrm{d}\sigma^+ \wedge \mathrm{d}\sigma^- - \frac{1}{2} \int_\Sigma \langle 2k j_-, j_+ \rangle \mathrm{d}\sigma^- \wedge \mathrm{d}\sigma^+. \tag{2.2.124}$$

where we have used $\langle A, \frac{1}{\mathrm{Ad}_g - \lambda} B \rangle = \langle \frac{1}{\mathrm{Ad}_g^{-1} - \lambda} A, B \rangle$, which is a consequence of $\langle A, \mathrm{Ad}_g B \rangle = \langle \mathrm{Ad}_g^{-1} A, B \rangle$. The Wess-Zumino term is straightforward to compute, using $\mathrm{res}_\alpha \omega = -2k$.

Thus, we find that the action (2.2.53) reduces to

$$S[g] = \frac{k}{2} \int_\Sigma \langle g^{-1}\partial_+ g, g^{-1}\partial_- g\rangle \mathrm{d}\sigma \wedge \mathrm{d}\tau$$

$$+ k \int_\Sigma \left\langle \frac{1}{\lambda^{-1} - \mathrm{Ad}_g} \partial_+ g g^{-1}, g^{-1}\partial_- g \right\rangle \mathrm{d}\sigma \wedge \mathrm{d}\tau \qquad (2.2.125)$$

$$+ k\, I_{\mathrm{WZ}}[g],$$

which is the action of the λ-deformation of the principal chiral model.

2.3 Maillet Bracket from 4d Chern–Simons Dirac Bracket

We shall now briefly describe an important result derived by Vicedo [35], who showed that one can derive the known Poisson bracket algebra of the Lax operators of an integrable field theory from 4d Chern–Simons theory. This is achieved by performing a Hamiltonian analysis of 4d Chern–Simons theory with meromorphic 1-form $\omega = \varphi(z)\mathrm{d}z$ defined on $\mathbb{CP}^1$ for generic $\varphi(z)$, with one of the Σ directions as time.

The analysis involves considering the first-class constraints arising from the gauge symmetry of the theory and the second-class constraints arising from the linearity of the Lagrangian with respect to the time derivative. The Dirac bracket of the coordinate on the reduced phase space, which is the spatial component of the gauge field along Σ in a natural gauge, can then be computed to be

$$\{A_{\sigma 1}(z,\sigma), A_{\sigma 2}(z',\sigma')\}^* = [\mathcal{R}_{12}(z,z'), A_{\sigma 1}(z,\sigma)]\,\delta_{\sigma\sigma'}$$

$$- [\mathcal{R}_{21}(z',z), A_{\sigma 2}(z',\sigma)]\,\delta_{\sigma\sigma'}$$

$$- (\mathcal{R}_{12}(z,z') + \mathcal{R}_{21}(z',z))\,\delta'_{\sigma\sigma'}, \qquad (2.3.1)$$

where $A_{\sigma 1} = A_\sigma \otimes \mathbb{1}$, $A_{\sigma 2} = \mathbb{1} \otimes A_\sigma$,

$$\mathcal{R}_{12}(z,z') := 2\pi \frac{C_{12}}{z'-z} \varphi(z')^{-1}, \qquad (2.3.2)$$

C_{12} is the tensor Casimir $T_1^a \otimes T_{a2}$, where σ indicates the spatial direction on Σ and where $\delta_{\sigma\sigma'} = \delta(\sigma - \sigma')$.

The expression (2.3.1) is simply the Poisson bracket structure for the Lax connection of an integrable field theory with twist function $\varphi(z)$. This Poisson bracket structure was first described by Maillet [36], who further derived the related Poisson bracket algebra of monodromy matrices after introducing a suitable regularization procedure, from where one can show the existence of an infinite number of Poisson-commuting conserved charges. Finally, we note that when the term in (2.3.1) which includes the derivative of a delta function is non-zero, the corresponding integrable field theory is known as being of non-ultralocal type, and the non-ultralocality in the Poisson bracket is an obstacle towards discretization and integrable quantization of such theories.

2.4   Integrable Field Theories from Chern–Simons Theory on Twistor Space

In the final section of this chapter, we describe the relationship of 4d Chern–Simons theory to a higher-dimensional Chern–Simons theory, namely 6d Chern–Simons theory on twistor space.

This relationship is particularly important, since another well-known unifying approach to two-dimensional integrable systems is that of the anti-self-dual Yang–Mills (ASDYM) equation in four dimensions, and its reduction to various two-dimensional integrable systems via symmetry reductions.[8] The anti-self-dual Yang–Mills equation itself can be realized as the equation of motion of a higher-dimensional analogue of the WZW model, that is known as the 4d WZW model.

A natural question is then to ask whether the 4d WZW model can be obtained from a six-dimensional Chern–Simons theory using techniques analogous to those of Costello and Yamazaki [24], described in this chapter. This was answered positively by Bittleston and Skinner [37] and Penna [142]. Furthermore, the former showed that one can derive various 4d integrable systems realizing the anti-self-dual Yang–Mills equation from Chern–Simons theory on twistor space, and moreover, elucidated the relationship to 4d Chern–Simons theory and two-dimensional integrable field theories. Their results can be encapsulated in the diamond-like diagram shown in Figure 2.9.

The basic example of the relationship indicated in Figure 2.9 is the case of 6d Chern–Simons theory realizing the 4d WZW model, which can be dimensionally reduced to the 2d principal chiral model with Wess–Zumino term. The 6d Chern–Simons theory setup itself can be dimensionally

---

[8]A pedagogical review of the relationship between the anti-self-dual Yang–Mills equation and two-dimensional integrable systems can be found in the book by Mason and Woodhouse [143].

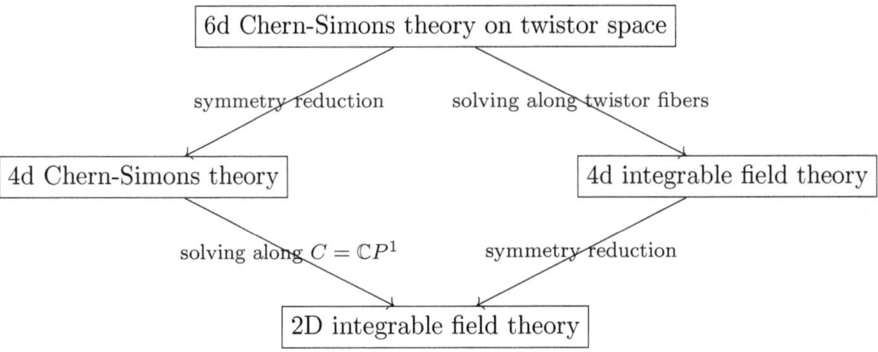

Figure 2.9.   Relationship between 4d Chern–Simons theory and 6d Chern–Simons theory on twistor space.

reduced to 4d Chern–Simons theory with disorder defects that give rise to the principal chiral model with Wess–Zumino term.

We shall review the derivation of the 4d WZW model from 6d Chern–Simons theory on twistor space in what follows.[9]

Let us quickly recall some facts on twistor space. Firstly, recall that the twistor space, $\mathbb{PT}$, of complexified Minkowski space, $\mathbb{CM}^4$, is the total space of the holomorphic vector bundle

$$\mathcal{O}(1) \oplus \mathcal{O}(1) \to \mathbb{C}P^1, \tag{2.4.1}$$

which can be endowed with homogeneous coordinates $Z^\alpha = \left(\omega^A, \pi_{A'}\right)$ defined with respect to the equivalence relation $Z^\alpha \sim t Z^\alpha$ for $t \in \mathbb{C}^*$. Here, the unprimed (primed) indices $A, B, C, \ldots$ $(A', B', C', \ldots)$ label elements of the right-handed (left-handed) spin bundle on $\mathbb{CM}^4$. More precisely, we have $\omega^A = \left(\omega^0, \omega^1\right)$ and $\pi_{A'} = (\pi_{0'}, \pi_{1'})$, that are, respectively, coordinates on the base and fiber of (2.4.1). The points $x^{AA'} \in \mathbb{CM}^4$ are in bijection with holomorphic lines, that is,

$$\iota_x : \mathbb{C}P^1_x \hookrightarrow \mathbb{PT}, \quad \pi_{A'} \mapsto \left(\omega^A, \pi_{A'}\right) = \left(x^{AB'}\pi_{B'}, \pi_{A'}\right). \tag{2.4.2}$$

To derive the 4d WZW model, one starts from Chern–Simons theory on (Euclidean) twistor space (where we have imposed the reality structure

---

[9]The reader who is unfamiliar with twistor notation is encouraged to refer to, e.g., the review by Adamo [38].

associated with 4d Euclidean space, $\mathbb{R}^4$, on $\mathbb{CM}^4)$,[10] with the action

$$S_\Omega[\overline{\mathcal{A}}] = \frac{1}{2\pi i}\int_{\mathbb{PT}} \Omega \wedge \mathrm{HCS}(\overline{\mathcal{A}}), \qquad (2.4.4)$$

with

$$\mathrm{HCS}(\bar{\mathcal{A}}) = \mathrm{Tr}\left(\bar{\mathcal{A}} \wedge \bar{\partial}\bar{\mathcal{A}} + \frac{2}{3}\bar{\mathcal{A}} \wedge \bar{\mathcal{A}} \wedge \bar{\mathcal{A}}\right) \qquad (2.4.5)$$

and

$$\Omega = \mathrm{D}^3 Z \otimes \Phi, \qquad (2.4.6)$$

where

$$\mathrm{D}^3 Z = \frac{\langle \mathrm{d}\pi\pi\rangle \wedge \mathrm{d}^2 x^{A'B'}\,\pi_{A'}\pi_{B'}}{2} \qquad (2.4.7)$$

and

$$\Phi = (\langle\pi\alpha\rangle\langle\pi\beta\rangle)^{-2}, \qquad (2.4.8)$$

where we use the notation $\langle\alpha\gamma\rangle = \alpha^{A'}\gamma^{B'}\varepsilon_{A'B'} = \alpha^{A'}\gamma_{A'}$, and where A' and B' are now indices that label elements of the left-handed spin bundle over $\mathbb{R}^4$. Similarly, unprimed indices now label elements of the right-handed spin bundle over $\mathbb{R}^4$. We have also used the expression $\mathrm{d}^2 x^{A'B'} := \mathrm{d}x^{AA'} \wedge \mathrm{d}x_A^{B'}$. Here, Φ is a meromorphic section of $\mathcal{O}(-4) \to \mathbb{PT}$. In addition, Φ has been restricted to only depend on the fiber direction of twistor space $\mathbb{PT} \to \mathbb{CP}^1$, i.e., Φ only depends on the coordinates $\pi^{A'}$.

To ensure that the equations of motion are local, we ought to impose the boundary condition that $\overline{\mathcal{A}}$ is divisible by $\langle\pi\alpha\rangle\langle\pi\beta\rangle$. The equations of motion are then of the expected form

$$\overline{\mathcal{F}}(\overline{\mathcal{A}}) = \bar{\partial}\overline{\mathcal{A}} + \overline{\mathcal{A}} \wedge \overline{\mathcal{A}} = 0. \qquad (2.4.9)$$

---

[10]Recall that in Euclidean signature, primed (unprimed) spinors are mapped to primed (unprimed) spinors, that is, for $\omega^A = (\omega^0, \omega^1)$ and $\pi_{A'} = (\pi_0, \pi_1)$, we have

$$\omega^A \to \hat{\omega}^A = \left(-\overline{\omega^1}, \overline{\omega^0}\right) \quad \text{and} \quad \pi_{A'} \to \hat{\pi}_{A'} = (-\overline{\pi_{1'}}, \overline{\pi_{0'}}). \qquad (2.4.3)$$

The components of the gauge fields can be further decomposed along the base and fibers of $\mathbb{PT} = \left(\mathcal{O}(1) \oplus \mathcal{O}(1) \to \mathbb{CP}^1 \right)$:

$$\overline{\mathcal{A}} = \overline{\mathcal{A}}_0 \bar{e}^0 + \widehat{\mathcal{A}}_A \widehat{e}^A, \tag{2.4.10}$$

such that the equations of motion are

$$\overline{\mathcal{F}}_{0A}(\overline{\mathcal{A}}) = \bar{\partial}_0 \widehat{\mathcal{A}}_A - \widehat{\partial}_A \overline{\mathcal{A}}_0 + \left[\overline{\mathcal{A}}_0, \widehat{\mathcal{A}}_A \right] = 0$$

$$\overline{\mathcal{F}}_{AB}(\overline{\mathcal{A}}) = \widehat{\partial}_A \widehat{\mathcal{A}}_B - \widehat{\partial}_B \widehat{\mathcal{A}}_B + \left[\widehat{\mathcal{A}}_A, \widehat{\mathcal{A}}_B \right] = 0. \tag{2.4.11}$$

Here, we utilize a frame of holomorphic $(0,1)$-forms adapted to the non-holomorphic coordinates $\left(x^{AA'}, \pi_{A'} \right)$ on $\mathbb{PT}$:

$$\bar{e}^0 = \frac{\langle \mathrm{d}\widehat{\pi}\widehat{\pi} \rangle}{\|\pi\|^4} \in \Omega^{0,1}(\mathbb{PT}, \mathcal{O}(-2)), \quad \widehat{e}^A = \frac{\mathrm{d}x^{AA'}\widehat{\pi}_{A'}}{\|\pi\|^2} \in \Omega^{0,1}(\mathbb{PT}, \mathcal{O}(-1)), \tag{2.4.12}$$

originally introduced in Ref. [39], where $\|\pi\|^2 = \pi_{A'}\widehat{\pi}^{A'}$. The dual frame of $(0,1)$-vectors is given by

$$\bar{\partial}_0 = \|\pi\|^2 \pi^{A'} \frac{\partial}{\partial \widehat{\pi}^{A'}}, \quad \widehat{\partial}_A = \pi^{A'} \frac{\partial}{\partial x^{AA'}} = \pi^{A'} \partial_{AA'}. \tag{2.4.13}$$

We can now perform a derivation analogous to that of Costello–Yamazaki, and solve the first of the equations in (2.4.11) (we do not solve the second one; it will eventually be equivalent to the equations of motion of the effective 4d theory). As in the derivation of 2d integrable field theories from 4d CS, we start by setting the pullback of the gauge field to the $\mathbb{CP}^1$ fiber to be pure gauge, i.e.,

$$\overline{\mathcal{A}}_x = \widehat{\sigma}^{-1}\bar{\partial}_{\mathbb{CP}^1_x}\widehat{\sigma}, \tag{2.4.14}$$

for a frame field $\widehat{\sigma} : \mathbb{PT} \to G$, where $\mathbb{CP}^1_x$ is the fiber over $x \in \mathbb{R}^4$. There are gauge redundancies in this definition, which we fix by picking a specific $\widehat{\sigma}$ which equals a map $\sigma : \mathbb{R}^4 \to G$ in a neighborhood of α and is the identity in a neighborhood of β. We shall also require that $\widehat{\sigma}$ is invariant under the circle action on $\mathbb{CP}^1_x$ which preserves α and β.

The equation of motion of interest reduces to

$$\overline{\mathcal{F}}_{0A}(\overline{\mathcal{A}}) = \widehat{\sigma}^{-1}\overline{\mathcal{F}}_{0A}\left(\overline{\mathcal{A}}'\right)\widehat{\sigma} = \widehat{\sigma}^{-1}\left(\bar{\partial}_0\widehat{\mathcal{A}}'_A\right)\widehat{\sigma} = 0, \tag{2.4.15}$$

where we define

$$\bar{\partial} + \overline{\mathcal{A}} = \widehat{\sigma}^{-1}\left(\bar{\partial} + \overline{\mathcal{A}}'\right)\widehat{\sigma}, \tag{2.4.16}$$

where $\overline{\mathcal{A}}'$ can be thought of as $\overline{\mathcal{A}}$ in the gauge where $\overline{\mathcal{A}}_x$ vanishes (this is analogous to the formal gauge transformation between the 4d CS gauge field and the Lax connection, studied earlier).

The equation of motion (2.4.15) can be solved while satisfying the boundary conditions by setting

$$\widehat{\mathcal{A}}_A = \widehat{\sigma}^{-1}\widehat{\partial}_A\widehat{\sigma} + \widehat{\sigma}^{-1}\widehat{\mathcal{A}}'_A\widehat{\sigma} = \widehat{\sigma}^{-1}\pi^{B'}\partial_{AB'}\widehat{\sigma} + \widehat{\sigma}^{-1}\widehat{\mathcal{A}}'_A\widehat{\sigma} \tag{2.4.17}$$

where

$$\widehat{\mathcal{A}}'_A = -\langle\pi\beta\rangle\alpha^{B'}\partial_{AB'}\sigma\sigma^{-1}, \tag{2.4.18}$$

where we have chosen α and β such that $\langle\alpha\beta\rangle = 1$. Indeed, one can check directly that (2.4.17) satisfies Dirichlet boundary conditions at both second-order poles of Ω.

Let us show that the twistor holomorphic Chern–Simons action itself reduces to the 4d WZW model whose equations of motion are classically equivalent to the ASDYM equations. This will follow from substituting the solution (2.4.17) into the holomorphic Chern–Simons action. First, observe that

$$\mathrm{HCS}(X + Y) = \mathrm{HCS}(X) + 2\,\mathrm{Tr}(\overline{\mathcal{F}}(X)Y) - \bar{\partial}\,\mathrm{Tr}(XY)$$

$$+ 2\,\mathrm{Tr}\left(XY^2\right) + \mathrm{HCS}(Y). \tag{2.4.19}$$

Setting $\overline{\mathcal{J}} = -\bar{\partial}\widehat{\sigma}\widehat{\sigma}^{-1}$, and defining $X = \widehat{\sigma}^{-1}\bar{\partial}\widehat{\sigma} = -\widehat{\sigma}^{-1}\overline{\mathcal{J}}\widehat{\sigma}$ and $Y = \widehat{\sigma}^{-1}\widehat{\mathcal{A}}'\widehat{\sigma}$, we have

$$\mathrm{HCS}(\widehat{\mathcal{A}}) = \frac{1}{3}\,\mathrm{Tr}\left(\overline{\mathcal{J}}^3\right) + \bar{\partial}\,\mathrm{Tr}\left(\overline{\mathcal{J}}\widehat{\mathcal{A}}'\right) - 2\,\mathrm{Tr}\left(\overline{\mathcal{J}}\widehat{\mathcal{A}}'^2\right)$$

$$+ \mathrm{Tr}\left(\widehat{\sigma}^{-1}\widehat{\mathcal{A}}'\widehat{\sigma}\bar{\partial}\left(\widehat{\sigma}^{-1}\widehat{\mathcal{A}}'\widehat{\sigma}\right)\right). \tag{2.4.20}$$

In order for the final term to contribute to the action, $\bar{\partial}$ must act as $\bar{e}^0\bar{\partial}_0$, since $\widehat{\mathcal{A}}'$ has no $\bar{e}^0$ component. Furthermore, the equations of motion imply that $\bar{\partial}_0\widehat{\mathcal{A}}'_A = 0$, meaning that

$$\mathrm{Tr}\left(\widehat{\sigma}^{-1}\widehat{\mathcal{A}}'\widehat{\sigma}\bar{\partial}\left(\widehat{\sigma}^{-1}\widehat{\mathcal{A}}'\widehat{\sigma}\right)\right) = -2\,\mathrm{Tr}\left(\bar{\partial}\widehat{\sigma}\widehat{\sigma}^{-1}\widehat{\mathcal{A}}'^2\right) = 2\,\mathrm{Tr}\left(\overline{\mathcal{J}}\widehat{\mathcal{A}}'^2\right), \tag{2.4.21}$$

which cancels the penultimate term. We thus arrive at

$$\mathrm{HCS}(\widehat{\mathcal{A}}) = \frac{1}{3}\mathrm{Tr}\left(\overline{\mathcal{J}}^3\right) + \bar{\partial}\,\mathrm{Tr}\left(\overline{\mathcal{J}}\widehat{\mathcal{A}}'\right). \tag{2.4.22}$$

Now, consider the second of these terms. Its contribution to the action is

$$\frac{1}{2\pi i}\int_{\mathrm{PT}}\Omega \wedge \bar{\partial}\,\mathrm{Tr}\left(\overline{\mathcal{J}}\wedge\widehat{\mathcal{A}}'\right) = \frac{1}{2\pi i}\int_{\mathrm{PT}}\bar{\partial}\Omega \wedge \mathrm{Tr}\left(\overline{\mathcal{J}}\wedge\widehat{\mathcal{A}}'\right)$$

$$= \frac{1}{4\pi i}\int_{\mathbb{CP}^1}\bar{\partial}_{\mathbb{CP}^1}\left(\frac{\langle\mathrm{d}\pi\pi\rangle}{\langle\pi\alpha\rangle^2\langle\pi\beta\rangle^2}\right)\int_{\mathbb{R}^4}\mathrm{d}^2x^{A'B'}\pi_{A'}\pi_{B'}\wedge\mathrm{Tr}\left(\overline{\mathcal{J}}\wedge\widehat{\mathcal{A}}'\right).$$

To evaluate this action, we can employ an inhomogeneous coordinate, denoted z, on the twistor sphere, which is related to π^A as

$$\pi^A \sim \alpha^A - z\beta^A \tag{2.4.23}$$

(the symbol $\sim$ here denotes equivalence up to an overall scale factor, which is equal to $\langle\pi\beta\rangle$), where $z = \frac{\langle\pi\alpha\rangle}{\langle\pi\beta\rangle}$, with

$$\mathrm{d}z = \frac{e^0}{\langle\pi\beta\rangle^2}, \quad \frac{\partial}{\partial z} = \langle\pi\beta\rangle^2\partial_0. \tag{2.4.24}$$

We shall also use

$$e^0 = \langle\mathrm{d}\pi\pi\rangle, \tag{2.4.25}$$

and

$$\bar{e}^0 = \frac{1}{\langle\pi\beta\rangle^2}\frac{\mathrm{d}\bar{z}}{(1+|z|^2)^2}, \quad \bar{\partial}_0 = \langle\pi\beta\rangle^2(1+|z|^2)^2\frac{\partial}{\partial\bar{z}}. \tag{2.4.26}$$

We then find that

$$\frac{1}{4\pi i}\int_{\mathbb{CP}^1}\bar{\partial}_{\mathbb{CP}^1}\left(\frac{\langle\,\mathrm{d}\pi\pi\rangle}{\langle\pi\alpha\rangle^2\langle\pi\beta\rangle^2}\right)\int_{\mathbb{R}^4}\mathrm{d}^2x^{\tilde{A}\tilde{B}}\pi_{\tilde{A}}\pi_{\tilde{B}}\wedge\mathrm{Tr}\left(\overline{\mathcal{J}}\wedge\widehat{\mathcal{A}}'\right)$$

$$= \frac{1}{2\pi i}\int_{\mathbb{CP}^1}\left(-\partial_{\bar{z}}\left(\frac{\mathrm{d}z}{z^2}\right)\wedge\mathrm{d}\bar{z}\right)\int_{\mathbb{R}^4}\varepsilon^{AB}\mathrm{vol}_4\wedge\mathrm{Tr}\left(\overline{\mathcal{J}}_B\wedge\widehat{\mathcal{A}}'_A\right)\frac{1}{\langle\pi\beta\rangle^2}$$

$$= \frac{1}{2\pi i}\int_{\mathbb{CP}^1}(-2\pi i\partial_z\delta^{(2)}(z))\int_{\mathbb{R}^4}\varepsilon^{AB}\mathrm{vol}_4$$

$$\wedge\mathrm{Tr}\left((-\pi^{B'}\partial_{BB'}\widehat{\sigma}\widehat{\sigma}^{-1})\wedge(-1)\langle\pi\beta\rangle\alpha^{A'}\partial_{AA'}\sigma\sigma^{-1}\right)\frac{1}{\langle\pi\beta\rangle^2}$$

$$= \frac{1}{2\pi i}\int_{\mathbb{CP}^1}(-2\pi i\partial_z\delta^{(2)}(z))\int_{\mathbb{R}^4}\varepsilon^{AB}\mathrm{vol}_4$$

$$\wedge\mathrm{Tr}\left((\alpha^{B'} - z\beta^{B'})\partial_{BB'}\widehat{\sigma}\widehat{\sigma}^{-1}\wedge\alpha^{A'}\partial_{AA'}\sigma\sigma^{-1}\right). \tag{2.4.27}$$

Here, we have used the identity $\mathrm{d}^2 x^{A'B'} \pi_{A'} \pi_{B'} \wedge \widehat{e}^A \wedge \widehat{e}^B = -2\varepsilon^{AB} \,\mathrm{vol}_4$, $\langle \pi\alpha \rangle^2 = \langle \pi\beta \rangle^2 z^2$, and the complex analysis lemma $\partial_{\bar{z}} \frac{1}{z^2} \mathrm{d}z \wedge \mathrm{d}\bar{z} = 2\pi i \partial_z \delta^{(2)}(z) \mathrm{d}z \wedge \mathrm{d}\bar{z}$, where $\delta^{(2)}(z)$ is a delta function on $\mathbb{CP}^1$.

Using the antisymmetry of the wedge form, this term can be written as

$$
\frac{1}{4\pi i} \int_{\mathbb{CP}^1} (-2\pi i \partial_z \delta^{(2)}(z)) \int_{\mathbb{R}^4} \varepsilon^{AB} \,\mathrm{vol}_4
$$
$$
\wedge \operatorname{Tr}\left((\alpha^{B'} - z\beta^{B'}) \partial_{BB'} \widehat{\sigma} \widehat{\sigma}^{-1} \wedge \alpha^{A'} \partial_{AA'} \sigma \sigma^{-1} \right. \tag{2.4.28}
$$
$$
\left. - (\alpha^{A'} - z\beta^{A'}) \partial_{AA'} \widehat{\sigma} \widehat{\sigma}^{-1} \wedge \alpha^{B'} \partial_{BB'} \sigma \sigma^{-1} \right).
$$

Integrating by parts, we find the kinetic term

$$
-\frac{1}{2} \int_{\mathbb{R}^4} \mathrm{vol}_4 \, \varepsilon^{AB} (\alpha^{A'} \beta^{B'} - \beta^{A'} \alpha^{B'}) \operatorname{Tr}\left(\partial_{AA'} \sigma \sigma^{-1} \partial_{BB'} \sigma \sigma^{-1} \right)
$$
$$
= -\frac{1}{2} \int_{\mathbb{R}^4} \mathrm{vol}_4 \, \varepsilon^{AB} \varepsilon^{A'B'} \operatorname{Tr}\left(\partial_{AA'} \sigma \sigma^{-1} \partial_{BB'} \sigma \sigma^{-1} \right), \tag{2.4.29}
$$

since we have fixed the normalization

$$
\alpha^{A'} \beta^{B'} - \beta^{A'} \alpha^{B'} = \varepsilon^{A'B'}. \tag{2.4.30}
$$

The remaining term has a contribution to the action of the form

$$
\frac{1}{12\pi i} \int_{\mathbb{PT}} \Omega \wedge \operatorname{Tr}\left(\overline{\mathcal{J}}^3 \right), \tag{2.4.31}
$$

which is equal to

$$
\frac{1}{6\pi i} \int_{\mathbb{PT}} \frac{\langle \mathrm{d}\pi\pi \rangle \wedge \mathrm{d}^2 x^{A'B'} \pi_{A'} \pi_{B'}}{2(\langle \pi\alpha \rangle \langle \pi\beta \rangle)^2} \wedge \operatorname{Tr}\left(\overline{\mathcal{J}}^3 \right)
$$
$$
= \frac{1}{6\pi i} \int_{\mathbb{PT}} \frac{\mathrm{d}z \wedge \mathrm{d}^2 x^{A'B'} (\alpha_{A'} - z\beta_{A'})(\alpha_{B'} - z\beta_{B'})}{2z^2} \wedge \operatorname{Tr}\left(\overline{\mathcal{J}}^3 \right). \tag{2.4.32}
$$

Since $\widehat{\sigma}$ is gauge fixed to not depend on the S^1 fiber direction of $\mathbb{CP}^1$ (when viewed as an S^1 fibration over the interval), we know that the $\mathrm{d}z$ integral involves only the angular coordinate parametrizing the S^1. That is, only the second term in $\mathrm{d}z = e^{i\theta} \mathrm{d}r + i e^{i\theta} r \mathrm{d}\theta$ contributes, where we defined polar coordinates via $z = r e^{i\theta}$.

Then, only the terms linear in z in $(\alpha_{A'} - z\beta_{A'})(\alpha_{B'} - z\beta_{B'}) = \alpha_{A'}\alpha_{B'} - z\alpha_{A'}\beta_{B'} - z\beta_{A'}\alpha_{B'} + z^2 \beta_{A'}\beta_{B'}$ will contribute non-zero values when integrated over θ (since the remaining terms involve some power of $e^{i\theta}$, whose integral is zero), leading to

$$\frac{1}{6}\int_{\mathbb{R}^4\times I}\mathrm{d}^2x^{A'B'}(-\beta_{A'}\alpha_{B'}-\alpha_{A'}\beta_{B'})\wedge\mathrm{Tr}(\tilde{J}^3),\qquad(2.4.33)$$

where $\tilde{J}=-\mathrm{d}\tilde{\sigma}\sigma^{-1}$, with $\tilde{\sigma}$ being a smooth homotopy from σ to the identity, 1. Using the fact that $\mathrm{d}^2x^{A'B'}=\mathrm{d}x^{AA'}\wedge\mathrm{d}x_A^{B'}=\mathrm{d}x^{AB'}\wedge\mathrm{d}x_A^{A'}$, we arrive at the Wess–Zumino term

$$-\frac{1}{3}\int_{\mathbb{R}^4\times I}\mathrm{d}^2x^{A'B'}\alpha_{A'}\beta_{B'}\wedge\mathrm{Tr}(\tilde{J}^3).\qquad(2.4.34)$$

In this way, we obtain the complete 4d WZW action

$$-\frac{1}{2}\int_{\mathbb{R}^4}\mathrm{vol}_4\,\varepsilon^{AB}\varepsilon^{A'B'}\,\mathrm{Tr}\left(\partial_{AA'}\sigma\sigma^{-1}\partial_{BB'}\sigma\sigma^{-1}\right)-\frac{1}{3}\int_{\mathbb{R}^4\times[0,1]}\mu_{\alpha,\beta}\wedge\mathrm{Tr}\left(\tilde{J}^3\right)$$
$$(2.4.35)$$

where $\mu_{\alpha,\beta}=\mathrm{d}^2x^{A'B'}\alpha_{A'}\beta_{B'}$.

Next, we shall review the relationship between 6d holomorphic Chern–Simons theory on twistor space and 4d Chern–Simons theory. Let H denote a 2-dimensional group of translations along $\mathbb{R}^4$. The aim is to perform a symmetry reduction of the 6d holomorphic Chern–Simons action with respect to this group. To this end, we define the generators of this group to be translations along the complex null vectors

$$\chi=\widehat{\kappa}^A\widehat{\mu}^{A'}\partial_{AA'}=\partial_z,\quad\bar{\chi}=\kappa^A\mu^{A'}\partial_{AA'}=\partial_{\bar{z}},\qquad(2.4.36)$$

where we employ double-null coordinates on $\mathbb{R}^4$ given by

$$z=x^{AA'}\kappa_A\mu_{A'},\quad\bar{z}=x^{AA'}\widehat{\kappa}_A\widehat{\mu}_{A'},\quad w=-x^{AA'}\widehat{\kappa}_A\mu_{A'},\quad\bar{w}=x^{AA'}\kappa_A\widehat{\mu}_{A'}.$$
$$(2.4.37)$$

We shall also make use of the vectors

$$\omega=\kappa^A\widehat{\mu}^{A'}\partial_{AA'}=\partial_w,\quad\bar{\omega}=-\widehat{\kappa}^A\mu^{A'}\partial_{AA'}=\partial_{\bar{w}}.\qquad(2.4.38)$$

The reduction from a six-dimensional theory to a four-dimensional theory is performed by imposing invariance of the six-dimensional gauge field with respect to χ and $\bar{\chi}$, that is, we require

$$\mathcal{L}_\chi\overline{\mathcal{A}}=\mathcal{L}_{\bar{\chi}}\overline{\mathcal{A}}=0,\qquad(2.4.39)$$

where $\mathcal{L}$ is the Lie derivative acting on differential forms. The reduction is carried out by contracting the bivector $\chi\wedge\bar{\chi}$ with the Lagrangian density of 6d twistor space Chern–Simons theory. Using

$$\iota_{\chi\wedge\bar\chi}\Omega = \frac{\langle \mathrm{d}\pi\pi\rangle\langle\pi\mu\rangle\langle\pi\widehat\mu\rangle}{\langle\pi\alpha\rangle^2\langle\pi\beta\rangle^2},\tag{2.4.40}$$

it can be shown that

$$\iota_{\chi\wedge\bar\chi}(\Omega\wedge\mathrm{hCS}(\overline{\mathcal{A}})) = \frac{\langle \mathrm{d}\pi\pi\rangle\langle\pi\mu\rangle\langle\pi\widehat\mu\rangle}{\langle\pi\alpha\rangle^2\langle\pi\beta\rangle^2}\,\mathrm{phCS}(\bar A),\tag{2.4.41}$$

where

$$\bar A = \bar e^0\overline{\mathcal{A}}_0 + \left(\iota_\omega\overline{\mathcal{A}} - \frac{\langle\pi\widehat\mu\rangle}{\langle\pi\mu\rangle}\iota_{\bar\chi}\overline{\mathcal{A}}\right)\mathrm{d}w + \left(\iota_{\bar\omega}\overline{\mathcal{A}} + \frac{\langle\pi\mu\rangle}{\langle\pi\widehat\mu\rangle}\iota_\chi\overline{\mathcal{A}}\right)\mathrm{d}\bar w\tag{2.4.42}$$

satisfies $\iota_\chi\bar A = \iota_{\bar\chi}\bar A = 0$, and

$$\mathrm{phCS}(\bar A) = \mathrm{Tr}\left(\bar A\,\mathrm{d}'\bar A + \frac{2}{3}\bar A\wedge\bar A\wedge\bar A\right)\tag{2.4.43}$$

with $\mathrm{d}' = \bar e^0\bar\partial_0 + \mathrm{d}w\partial_w + \mathrm{d}\bar w\partial_{\bar w}$. In particular, the w and $\bar w$ components of the 4d gauge field are related to the 6d gauge field via

$$\bar A_w = \iota_\omega\overline{\mathcal{A}} - \frac{\langle\pi\widehat\mu\rangle}{\langle\pi\mu\rangle}\iota_{\bar\chi}\overline{\mathcal{A}}$$

$$\bar A_{\bar w} = \iota_{\bar\omega}\overline{\mathcal{A}} + \frac{\langle\pi\mu\rangle}{\langle\pi\widehat\mu\rangle}\iota_\chi\overline{\mathcal{A}}.\tag{2.4.44}$$

These relations allow us to map boundary conditions obeyed by the 6d gauge field to boundary conditions obeyed by the 4d gauge field.

The result of the symmetry reduction is thus 4d Chern–Simons theory with the Lagrangian density (2.4.41). Translating to inhomogeneous coordinates on $\mathbb{CP}^1$ using (2.4.23), we find that this is precisely the 4d Chern–Simons theory setup that localizes to the principal chiral model with Wess–Zumino term on $\mathbb{R}^2$ (with complex coordinates w and $\bar w$)! Moreover, the symmetry reduction procedure can also be applied to the 4d WZW model action given in (2.4.35), whereby one arrives at the same 2d principal chiral model with Wess–Zumino term. To perform the symmetry reduction, the identities

$$X_{A'} = \langle X\widehat\gamma\rangle\gamma_{A'} - \langle X\gamma\rangle\widehat\gamma_{A'}\tag{2.4.45}$$

and

$$X^A = [X\widehat\mu]\mu^A - [X\mu]\widehat\mu^A\tag{2.4.46}$$

for unit norm spinors $\gamma_{A'}$ and μ^A, where $[\mu\nu] = \mu^A \nu^B \varepsilon_{AB} = \mu^A \nu_A$, prove useful. In this way, we arrive at an example of the relationship depicted in Figure 2.9.

Finally, we note that the work of Bittleston and Skinner has been generalized to the case of deformations of the 2d principal chiral model [40] (see also [144] for prior work in this direction), as well as 2d integrable coset models [41].

Chapter 3

4d Chern–Simons Theory from Supersymmetric Gauge Theory and String Theory

In this chapter, we shall study various embeddings of 4d Chern–Simons theory in string theory and supersymmetric gauge theory. We shall first understand how 4d Chern–Simons theory is T-dual, at the level of quantum field theory, to 3d Chern–Simons theory, based on the work of Yamazaki [42]. We also review the topological string theory realization of 4d Chern–Simons theory, whereby the aforementioned T-duality to 3d Chern–Simons theory is further elucidated, as shown by Yamazaki [42]. Subsequently, we review two embeddings of 4d Chern–Simons theory in superstring theory and supersymmetric gauge theory, with the first, due to Costello and Yagi [46], involving an Ω-deformed, topological-holomorphic twist of 6d maximally supersymmetric Yang–Mills (SYM) theory, and the second, due to Ashwinkumar and Tan [53], involving a topological-holomorphic twist of 5d maximally supersymmetric Yang–Mills theory with a boundary.

3.1   T-Duality between Chern–Simons Theories

To gain some intuition for string-theoretric embeddings of 4d Chern–Simons theory, we shall first attempt to understand how it is related via dimensional reduction to 3d Chern–Simons theory, and moreover, study how 4d Chern–Simons theory enjoys a duality (known as T-duality), to Chern–Simons theories in other dimensions, following the work of Yamazaki [42]. This is helpful, since 3d Chern–Simons theory has various embeddings in string theory that are well understood [44, 55, 145].

103

Recall that 3d Chern–Simons theory on $\mathbb{R}^3$, parametrized by the coordinates x, y, and t, is defined via the action functional

$$
\begin{aligned}
S_{3d} &= \frac{1}{\hbar_{3d}} \int_{\mathbb{R}^3} \mathrm{Tr}(A \wedge \mathrm{d}A + \frac{2}{3} A \wedge A \wedge A) \\
&= \frac{2}{\hbar_{3d}} \int_{\mathbb{R}^3} \mathrm{d}^3x\, \mathrm{Tr}\left(-A_x \partial_t A_y + A_t \left(\partial_x A_y - \partial_y A_x\right) + [A_x, A_y]\, A_t\right).
\end{aligned}
$$

$$(3.1.1)$$

Here, and in what follows, we shall consider all fields to be analytically-continued to be complex valued, so there is no quantization condition on $\hbar_{3d}$. We shall pick the gauge group to be $G = GL(N, \mathbb{C})$. The first observation of relevance is that the action (3.1.1) can be obtained from the 4d Chern–Simons theory action expressed in the form

$$
\begin{aligned}
S_{4d} &= \frac{1}{\hbar_{4d}} \int_{\mathbb{R}^2 \times \mathbb{R}_t \times S^1} \mathrm{d}\tilde{z} \wedge \mathrm{Tr}(A \wedge \mathrm{d}A + \frac{2}{3} A \wedge A \wedge A) \\
&= \frac{2}{\hbar_{4d}} \int_{\mathbb{R}^2 \times \mathbb{R}_t \times S^1} \mathrm{d}^4x\, \mathrm{Tr}\left(-A_x \partial_{\tilde{z}} A_y + A_{\tilde{z}} \left(\partial_x A_y - \partial_y A_x\right) + [A_x, A_y]\, A_{\tilde{z}}\right),
\end{aligned}
$$

$$(3.1.2)$$

by dimensional reduction along S^1. Here, $\tilde{z} = t + i\theta$ is a complex coordinate on $\mathbb{R}_t \times S^1$ (we shall reserve the symbol z for another complex coordinate), where t is the coordinate along $\mathbb{R}_t$ and θ is the coordinate parametrizing S^1. The dimensional reduction procedure involves demanding that fields are independent of θ, whereby the $\partial_{\tilde{z}}$ derivative can be replaced by $\frac{1}{2}\partial_t$. Then, identifying $A_{\tilde{z}}$ with $\frac{1}{2}A_t$, one obtains (3.1.1), upon appropriately defining $\hbar_{3d}$.

Next, observe that one can define an analogue of 3d Chern–Simons theory on an infinite-dimensional covering space of $\mathbb{R}^3$, with the action functional

$$
\begin{aligned}
S = \frac{2}{\hbar_{3d}} \int_{\mathbb{R}^3} \mathrm{Tr}\,\big(&A_x^{m,n} \partial_t A_y^{n,m} + A_t^{m,n} \left(\partial_x A_y^{n,m} - \partial_y A_x^{n,m}\right) + A_x^{m,p} A_y^{p,q} A_t^{q,m} \\
&- A_y^{m,p} A_x^{p,q} A_t^{q,m} \big),
\end{aligned}
$$

$$(3.1.3)$$

where the Einstein summation convention is assumed for repeated indices, with each index taking values in $\mathbb{Z}$. The intuition employed for the theory on the covering space is that we are T-dualizing along a circle $S^1 = \mathbb{R}/2\pi\mathbb{Z}$, where the quotient by $2\pi\mathbb{Z}$ is responsible for creating an infinite number of mirror images of $\mathbb{R}^3$. In addition, $A_i^{m,n}$ denotes the winding mode relating the m-th mirror image to the n-th mirror image, with winding number

$m - n$. Moreover, since there is gauge symmetry given by G at each sheet, the entire theory has an infinite-dimensional gauge group. Gauge invariance constrains the form of the action, and this means that terms of the form $\mathrm{Tr}\left(A_x^{m,p} A_y^{p,q} A_t^{q,n}\right)$ with $m \neq n$ do not appear in the action.

The next step in the T-duality procedure is to impose the periodicity relations

$$
\begin{aligned}
A_i^{m,n} &= A_i^{m-1,n-1} && (i = x, y), \\
A_t^{m,n} &= A_t^{m-1,n-1} && (m \neq n), \\
A_t^{n,n} &= (2\pi R')\,\mathrm{Id}_{N \times N} + A_t^{n-1,n-1},
\end{aligned}
\tag{3.1.4}
$$

where A_t has a diagonal component which is absent for A_i. The parameter R' will play the role of the radius of the extra circular dimension. Now, using the periodicity relations, with the definition

$$
A_i^m := A_i^{m,0} = A_i^{0,-m},
\tag{3.1.5}
$$

(the equality between $A_i^{m,0}$ and $A_i^{0,-m}$ follows from (3.1.4)) we find

$$
\begin{aligned}
S &= \frac{2}{\hbar_{3d}} \int_{\mathbb{R}^3} \mathrm{d}^3x\, \mathrm{Tr}\left(- A_x^{m,0}\partial_t A_y^{0,m} + A_t^{m,0}\left(\partial_x A_y^{0,m} - \partial_y A_x^{0,m}\right)\right. \\
&\qquad\left. + A_x^{m,0} A_y^{0,-n} A_t^{-n,m} - A_y^{m,0} A_x^{0,-n} A_t^{-n,m}\right) \\
&= \frac{2}{\hbar_{3d}} \int_{\mathbb{R}^3} \mathrm{d}^3x\, \mathrm{Tr}\left(- A_x^m \left(\partial_t - 4m\pi R'\right) A_y^{-m} + A_t^m\left(\partial_x A_y^{-m} - \partial_y A_x^{-m}\right)\right. \\
&\qquad\left. + A_x^m A_y^n A_t^{-m-n} - A_y^m A_x^n A_t^{-m-n}\right),
\end{aligned}
\tag{3.1.6}
$$

where we used

$$
\begin{aligned}
&\sum_{m,n\in\mathbb{Z}} \mathrm{Tr}(A_x^{m,0} A_y^{0,-n} A_t^{-n,m}) \\
&= \sum_{m,n\in\mathbb{Z}, m\neq -n} \mathrm{Tr}(A_x^{m,0} A_y^{0,-n} A_t^{-n,m}) + \sum_{m,n\in\mathbb{Z}, m=-n} \mathrm{Tr}(A_x^{m,0} A_y^{0,-n} A_t^{-n,m}) \\
&= \sum_{m,n\in\mathbb{Z}, m\neq -n} \mathrm{Tr}(A_x^{m,0} A_y^{0,-n} A_t^{-m-n,0}) \\
&\quad + \sum_{m,n\in\mathbb{Z}, m=-n} \mathrm{Tr}(A_x^{m,0} A_y^{0,-n} A_t^{-m-n,0}) + \sum_m m(2\pi R')\mathrm{Tr}(A_x^m A_y^{-m}),
\end{aligned}
\tag{3.1.7}
$$

and the analogous identity

$$- \sum_{m,n\in\mathbb{Z}} \mathrm{Tr}(A_x^{0,-n} A_t^{-n,m} A_y^{m,0})$$

$$= - \sum_{m,n\in\mathbb{Z}} \mathrm{Tr}(A_x^{0,-n} A_t^{-n-m,0} A_y^{m,0}) + \sum_{m\in\mathbb{Z}} m(2\pi R')\mathrm{Tr}(A_x^m A_y^{-m}).$$

$$(3.1.8)$$

We shall now compare the action (3.1.6) to an action obtained from 4d Chern–Simons theory on $\Sigma \times C$, where $\Sigma = \mathbb{R}^2$ and $C = \mathbb{R}_t \times S^1 \cong \mathbb{C}^\times$. As explained in Chapter 1, the latter takes the form

$$S = \frac{1}{\hbar} \int_{\mathbb{R}^2 \times \mathbb{R}_t \times S^1} \frac{\mathrm{d}z}{z} \wedge \mathrm{CS}(A), \qquad (3.1.9)$$

where $z = e^{(t+i\theta)/R}$, with t and θ denoting the cylindrical coordinates on $\mathbb{R}_t \times S^1$, and where θ has period $2\pi R$. The partial connection which is the gauge field in this theory is defined as

$$A = A_x \, \mathrm{d}x + A_y \, \mathrm{d}y + A_{\bar{z}} \left(R \frac{\mathrm{d}\bar{z}}{\bar{z}} \right), \qquad (3.1.10)$$

where R is a parameter with dimensions of length. Its inclusion is to keep track of dimensions, since $R/\bar{z}$ is a dimensionless quantity. For the same reason, we shall also multiply dz/z by R. The 4d CS action can be written as

$$S = \frac{2}{\hbar} \int_{\mathbb{R}^2 \times \mathbb{R}_t \times S^1} \mathrm{d}x \wedge \mathrm{d}y \wedge R \frac{\mathrm{d}z}{z}$$

$$\wedge R \frac{\mathrm{d}\bar{z}}{\bar{z}} \mathrm{Tr} \left(-A_x \bar{z} \partial_{\bar{z}} A_y + A_{\bar{z}} (\partial_x A_y - \partial_y A_x) + [A_x, A_y] A_{\bar{z}} \right)$$

$$(3.1.11)$$

$$= -\frac{4i}{\hbar} \int_{\mathbb{R}^2 \times \mathbb{R}_t \times S^1} \mathrm{d}x \wedge \mathrm{d}y \wedge \mathrm{d}t$$

$$\wedge \mathrm{d}\theta \, \mathrm{Tr} \left(-A_x \bar{z} \partial_{\bar{z}} A_y + A_{\bar{z}} (\partial_x A_y - \partial_y A_x) + [A_x, A_y] A_{\bar{z}} \right),$$

where we have used $\frac{R}{z} dz \wedge \frac{R}{\bar{z}} d\bar{z} = -i2dt \wedge d\theta$. Performing a Fourier expansion of the gauge field components in the θ direction, i.e.,

$$A_x = \sum_{n\in\mathbb{Z}} A_x^n(x,y,t) e^{\frac{in\theta}{R}}, \qquad A_y = \sum_{n\in\mathbb{Z}} A_y^n(x,y,t) e^{\frac{in\theta}{R}},$$

$$(3.1.12)$$

$$A_{\bar{z}} = \frac{R}{2} \sum_{n\in\mathbb{Z}} A_t^n(x,y,t) e^{\frac{in\theta}{R}}$$

we obtain

$$S = -\frac{4\pi i R}{\hbar} \left[-\sum_{n \in \mathbb{Z}} \int_{\mathbb{R}^2 \times \mathbb{R}_t} \mathrm{Tr}\left(A_x^{-n} \left(\partial_t A_y^n - \frac{n}{R} A_y^n \right) + A_t^{-n} \left(\partial_x A_y^n - \partial_y A_x^n \right) \right) \right.$$

$$\left. + \sum_{m,n \in \mathbb{Z}} \int_{\mathbb{R}^2 \times \mathbb{R}_t} \mathrm{Tr}\left(\left[A_x^m, A_y^n \right] A_t^{-m-n} \right) \right], \tag{3.1.13}$$

using $\int_0^{2\pi} d\theta \, e^{\frac{in\theta}{R}} e^{\frac{im\theta}{R}} = 2\pi \delta_{m,-n}$, where the identity $\bar{z} \partial_{\bar{z}} = \frac{R}{2} \left(\frac{\partial}{\partial t} + i \frac{\partial}{\partial \theta} \right)$ has been utilized.

Thus, there is agreement between the 3d Chern–Simons theory (3.1.6) defined on an infinite-dimensional covering space of $\mathbb{R}^3$ and 4d Chern–Simons theory (3.1.13) on $\mathbb{R}^2 \times \mathbb{R}_t \times S^1$, as long as the following identifications are made:

$$4\pi R' \longleftrightarrow \frac{1}{R}.$$

$$-\frac{\hbar}{2\pi i R} \longleftrightarrow \hbar_{3d}. \tag{3.1.14}$$

The first of these relations is reminiscent of the relation between radii encountered in the study of T-duality in string theory, and can thus be understood as an incarnation of T-duality in quantum field theory.

One can repeat this procedure starting from Chern–Simons theories in dimensions higher than three. For instance, defining 4d CS on an infinite dimensional covering space of $\mathbb{R}^2 \times \mathbb{C}^\times$ leads to T-duality with 5d Chern–Simons theory which has the action

$$S = \frac{1}{\hbar} \int_{\mathbb{R} \times \mathbb{C}^\times \times \mathbb{C}^\times} \frac{dz_1}{z_1} \wedge \frac{dz_2}{z_2} \wedge \mathrm{CS}(A). \tag{3.1.15}$$

A non-commutative generalization of this theory has been studied in relation to Ω-deformed M-theory and the *affine* Yangian by Costello [43].

Likewise, 5d Chern–Simons theory can be shown to be T-dual to 6d holomorphic Chern–Simons theory, with the action

$$S = \frac{1}{\hbar} \int_{\mathbb{C}^\times \times \mathbb{C}^\times \times \mathbb{C}^\times} \frac{dz_1}{z_1} \wedge \frac{dz_2}{z_2} \wedge \frac{dz_3}{z_3} \wedge \mathrm{CS}(A). \tag{3.1.16}$$

3.1.1 *Topological string realization of 4d CS*

Having gained some intuition into how Chern–Simons theories in various dimensions are related via T-duality, let us use it to understand how 4d Chern–Simons theory can be embedded into topological string theory.

It is well known from the work of Witten [44] that three-dimensional Chern–Simons theory defined on $\mathbb{R}^3$, with gauge group $SU(N)$, can be understood as a theory of N D-branes wrapping the base $\mathbb{R}^3$ of the cotangent bundle of $\mathbb{R}^3$, i.e., $T^*\mathbb{R}^3 \simeq \mathbb{C}^3$, in A-type topological string string theory on $T^*\mathbb{R}^3$. Let us denote the coordinates of the base as x, y, t, and the coordinates of the fiber as p_x, p_y, p_t. Defining the symplectic form on $T^*\mathbb{R}^3$ to be

$$\omega = \mathrm{d}x \wedge \mathrm{d}p_x + \mathrm{d}y \wedge \mathrm{d}p_y + \mathrm{d}t \wedge \mathrm{d}p_t, \tag{3.1.17}$$

the base $\mathbb{R}^3$ is a Lagrangian submanifold, which we denote by L_{base} (recall that the simplest branes in A-type topological string theory are those that wrap Lagrangian submanifolds).

In this setup, it is also possible to realize Wilson lines that give rise to knots/links in 3d Chern–Simons theory. From the work of Ooguri and Vafa [45], these Wilson lines can be identified with a different set of D-branes that wrap Lagrangian submanifolds, where the intersection of these D-branes with the original N D-branes provides the knots/links in the base manifold.

Let us consider Wilson lines along the x and y directions, realized by Lagrangian submanifolds L_x and L_y, supported along the x and y directions, respectively. In addition, we can include an order *surface* defect realized by a Lagrangian submanifold, L_s, that is supported along both the x and y directions. The resulting configuration that incorporates both line operators and a surface operator is shown in the following table:

	x	y	t	p_x	p_y	p_t
L_{base}	×	×	×			
L_x	×				×	×
L_y		×		×		×
L_s	×	×				×

If we compactify one of the non-compact directions of $T^*\mathbb{R}^3$ to be a circle, we can T-dualize this brane configuration. Picking this direction to be p_t, T-duality gives rise to the following brane configuration:

	x	y	t	p_x	p_y	p_t
L'_{base}	×	×	×			×
L'_x	×				×	
L'_y		×		×		
L'_s	×	×				

$$\tag{3.1.18}$$

We find that T-duality results in the (t, p_t) direction becoming "holomorphic", in the sense that it is described by the B-type topological string (the

B-model), while the (x, y, p_x, p_y) directions are still described by the A-type topological string (the A-model). Thus, we have arrived at a mixture of an A-model and B-model.

This is the brane configuration that realizes 4d Chern–Simons theory, with Wilson line and surface operator insertions necessary to realize integrable models. This can be observed by restricting to the (x, y, t, p_t) directions:

$$
\begin{array}{c|cc|cc}
 & x & y & t & p_t \\
\hline
\text{4d CS} & \times & \times & \times & \times \\
\hline
\text{Wilson}_x & \times & & & \\
\hline
\text{Wilson}_y & & \times & & \\
\hline
\text{Surface} & \times & \times & &
\end{array}
\tag{3.1.19}
$$

Here, Wilson lines along the x and y directions intersect to realize R-matrices, and multiple such Wilson line intersections would realize an integrable lattice model, as explained in Chapter 1. Multiple order surface defects along the $x - y$-plane realize integrable field theories, as explained in Chapter 2.

Further T-dualities will take us to brane configurations that realize 5d and 6d Chern–Simons theories. The former is also realized by the mixture of an A-model and a B-model, while the latter is realized by a pure B-model. This realization of 6d Chern–Simons theory on a Calabi–Yau 3-fold has been known for some time from the work of Witten [44].

3.2 Embedding of 4d CS in Ω-Deformed 6d $\mathcal{N} = (1, 1)$ Gauge Theory and Type IIB String Theory

The first embedding of 4d Chern–Simons theory in superstring theory that we shall study is its embedding in type IIB string theory by Costello and Yagi [46]. This formulation involves a topological-holomorphic twist of 6d $\mathcal{N} = (1, 1)$ supersymmetric Yang–Mills theory that describes a stack of D5-branes,[1] as well as a continuous deformation of the theory known as the Ω-deformation.[2]

---

[1] The reader unfamiliar with topological twisting of quantum field theories will benefit from perusing the seminal work on this topic by Witten [47].

[2] The details of the derivations that follow can be compared with a similar embedding of 3d analytically-continued Chern–Simons theory in an Ω-deformation of a fully topological twist of 5d $\mathcal{N} = 2$ super Yang–Mills theory [145].

3.2.1   *6d $\mathcal{N} = (1,1)$ supersymmetric Yang–Mills theory*

Let us first briefly recall some details about 6d $\mathcal{N} = (1,1)$ supersymmetric Yang–Mills theory.

A convenient description of this theory is as a dimensional reduction of ten-dimensional super Yang–Mills theory with gauge group G, which also has 16 supersymmetries. The fields of the ten-dimensional theory are the gauge field A and a fermionic gaugino field Ψ, which is a positive chirality Majorana–Weyl spinor valued in the adjoint representation of the gauge group.

Let us use $I, J, K, \ldots = 0, \ldots, 9$ to denote 10d Lorentz indices and η to denote the Minkowski metric in ten dimensions with signature $(-, +, +, \ldots, +)$. We shall utilize gamma matrices that are denoted Γ_I, and obey the Clifford algebra

$$\{\Gamma_I, \Gamma_J\} = 2\eta_{IJ}. \tag{3.2.1}$$

The gamma matrices are chosen to be real 32×32 matrices, and we shall denote as $\Gamma_{I_1 \ldots I_k}$ the matrix that equals $\Gamma_{I_1} \cdots \Gamma_{I_k}$ if $I_1, \ldots, I_k$ are all different and vanishes otherwise. In addition, the conventions chosen are such that the gamma matrix Γ_0 is antihermitian, while each Γ_I for $I \neq 0$ is hermitian. We can define the chirality and charge conjugation matrices as

$$\widehat{\Gamma}_{10d} = \Gamma_{0123456789}, \tag{3.2.2}$$

and

$$C_{10d} = -\Gamma_{03579}, \tag{3.2.3}$$

respectively. These matrices satisfy $\widehat{\Gamma}_{10d}^2 = 1$, $C_{10d}^2 = -1$ and $(C_{10d})^T = -C_{10d}$. Moreover, the gamma matrices satisfy $C_{10d}\Gamma_I C_{10d}^{-1} = -\Gamma_I^T$.[3]

In the notation defined above, the positive chirality condition on the gaugino field is $\widehat{\Gamma}_{10d}\Psi = \Psi$ and the Majorana–Weyl condition is

$$\Psi = C_{10d}\bar{\Psi}^T, \tag{3.2.4}$$

or equivalently, $\bar{\Psi} = \Psi^T C_{10d}$, where $\bar{\Psi} = \Psi^\dagger \Gamma^0$.

---

[3]Recall that there are two possible charge conjugation matrices in 10d Minkowski spacetime, satisfying $C\Gamma_I C^{-1} = \Gamma_I^T$ or $C\Gamma_I C^{-1} = -\Gamma_I^T$, and we can choose to work with either one of these matrices.

The ten-dimensional super Yang–Mills theory has the following classical action

$$-\frac{1}{e^2}\int d^{10}x\,\mathrm{Tr}\left(\frac{1}{2}F^{IJ}F_{IJ}-i\bar{\Psi}\Gamma^I D_I\Psi\right),\qquad(3.2.5)$$

Here, e is the gauge coupling, F_{IJ} is the field strength of the gauge field, $D_I=\partial_I+[A_I,\cdot]$ is the covariant derivative, and Tr denotes a negative-definite symmetric bilinear form on the Lie algebra, $\mathfrak{g}$, of the compact Lie group, G.

The supersymmetry variations

$$\delta_\epsilon A_I=i\bar{\epsilon}\Gamma_I\Psi,\quad \delta_\epsilon\Psi=\frac{1}{2}F_{IJ}\Gamma^{IJ}\epsilon\qquad(3.2.6)$$

leave the action invariant, where ϵ is a constant spinor of positive chirality (having eigenvalue equal to 1 when acted on by $\widehat{\Gamma}_{10d}$). These supersymmetry transformations are generated by the supercharge Q_ϵ, which satisfies the supersymmetry algebra

$$\{Q_\epsilon,Q_{\epsilon'}\}=\bar{\epsilon}\Gamma^I\epsilon'P_I\qquad(3.2.7)$$

on-shell. Here, P_I denotes the momentum in the x^I-direction.

To dimensionally reduce the theory to six dimensions, we require that the fields have translation invariance along the directions parametrized by x^6,x^7,x^8,x^9. As we review below, the result is 6d $\mathcal{N}=(1,1)$ super Yang–Mills theory. A consequence of the dimensional reduction is that the ten-dimensional Lorentz group $\mathrm{Spin}(9,1)$ decomposes into the product $\mathrm{Spin}(5,1)\times\mathrm{Spin}(4)_R$, where the $\mathrm{Spin}(4)_R$ denotes the R-symmetry group of the dimensionally-reduced theory.

The dimensional reduction involves expressing the 10d gamma matrices in terms of 8×8 6d gamma matrices, denoted using $\tilde{\Gamma}$, as

$$\Gamma_I=\tilde{\Gamma}_I\otimes\mathbb{1},\quad \Gamma_{\mu+6}=\widehat{\Gamma}_{6d}\otimes\gamma_\mu,\qquad(3.2.8)$$

where $I=0,1,2,3,4,5$ and $\mu=0,1,2,3$, where $\widehat{\Gamma}_{6d}=\tilde{\Gamma}_0\tilde{\Gamma}_1\tilde{\Gamma}_2\tilde{\Gamma}_3\tilde{\Gamma}_4\tilde{\Gamma}_5$ is the 6d chirality matrix and

$$\gamma_\mu=\begin{pmatrix}0 & \sigma^\dagger_{\mu+1}\\ \sigma_{\mu+1} & 0\end{pmatrix},\qquad(3.2.9)$$

with σ_i for $i=1,2,3$ denoting the Pauli matrices, and $\sigma_4=i\mathbb{1}$. The 10d spinor Ψ is then expressed in terms of 6d symplectic Majorana–Weyl spinors with opposite chirality:

$$\Psi = \begin{pmatrix} \lambda_+^A \\ \lambda_-^A \end{pmatrix}, \quad \widehat{\Gamma}_{6d}\lambda_\pm^A = \pm\lambda_\pm^A, \quad \lambda_\pm^A = \pm\varepsilon^{AB}C_{6d}\overline{\lambda_\pm^B}^T \tag{3.2.10}$$

where $A, B = 1, 2$, ϵ^{AB} is the two index Levi–Civita symbol, and $C_{6d} = \tilde{\Gamma}_0\tilde{\Gamma}_3\tilde{\Gamma}_5$ is the 6d charge conjugation matrix. Each of these spinors can be shown to be doublets with respect to the different $SU(2)$ factors of the R-symmetry group $\mathrm{Spin}(4)_R$. These statements are also true for the supersymmetry variation parameters, which implies that the resulting six-dimensional gauge theory has $\mathcal{N} = (1,1)$ supersymmetry. The six-dimensional action obtained via dimensional reduction can be shown to be

$$-\frac{1}{e^2}\int d^6x\,\mathrm{Tr}\left(\frac{1}{2}\sum_{I,J=0}^{5} F^{IJ}F_{IJ} + \sum_{I=0}^{5}\sum_{\mu=0}^{3} D^I\phi^\mu D_I\phi_\mu\right.$$

$$+ \frac{1}{2}\sum_{\mu=0}^{3}[\phi^\mu,\phi^\nu][\phi_\mu,\phi_\nu] - i\sum_{A=1}^{2}\sum_{I=0}^{5}\overline{\lambda_+^A}\tilde{\Gamma}^I D_I\lambda_+^A - i\sum_{A=1}^{2}\sum_{I=0}^{5}\overline{\lambda_-^A}\tilde{\Gamma}^I D_I\lambda_-^A$$

$$\left. - i\sum_{A,B=1}^{2}\sum_{\mu=0}^{3}\left(\overline{\lambda_+^A}(\bar{\sigma}_\mu)^A{}_B\left[i\phi^\mu, -\lambda_-^B\right] + \overline{\lambda_-^A}(\sigma_\mu)^A{}_B\left[i\phi^\mu, \lambda_+^B\right]\right)\right),$$

$$\tag{3.2.11}$$

where ϕ_μ, $\mu = 0,\ldots,3$, are scalar fields that transform in the vector representation of the R-symmetry group, and that arise from the gauge field components in the dimensionally-reduced directions. The supersymmetry transformations that leave (3.2.11) invariant can also be derived from (3.2.6) using dimensional reduction.

3.2.2   *Topological-holomorphic twist*

The topological-holomorphic twist of 6d $\mathcal{N} = (1,1)$ super Yang–Mills theory can be performed when the theory is defined on a product of a four-manifold, M, and a Riemann surface, C, where the latter corresponds to the holomorphic plane of 4d Chern–Simons theory. Here, we are working with a metric with Euclidean signature, denoted g.[4]

The topological twist is implemented by noting that the structure group of the spinor bundle on $M \times C$ is $\mathrm{Spin}(4)_M \times \mathrm{Spin}(2)_C$, and replacing

---

[4]The Euclidean theory is obtained from the Lorentzian theory via Wick rotation, $x^0 \mapsto -ix^0$.

$\mathrm{Spin}(4)_M$ with the diagonal subgroup

$$\mathrm{Spin}(4)'_M \subset \mathrm{Spin}(4)_M \times \mathrm{Spin}(4)_R. \tag{3.2.12}$$

As a result, the fields that were scalars prior to twisting now transform as the components of a one-form

$$\phi = \phi_\mu \mathrm{d}x^\mu \tag{3.2.13}$$

with respect to $\mathrm{Spin}(4)'_M$. Moreover, it can be shown that the fermions transform under $\mathrm{Spin}(4)'_M \times \mathrm{Spin}(2)_C$ as

$$2(\mathbf{1},\mathbf{1})^{-1} \oplus (\mathbf{1},\mathbf{3})^{-1} \oplus (\mathbf{3},\mathbf{1})^{-1} \oplus 2(\mathbf{2},\mathbf{2})^{1}. \tag{3.2.14}$$

The first three summands correspond to two scalars on M, denoted ξ and ξ', and a two-form on M, denoted χ, which are all negative chirality spinors on C (denoted by "-1" in the superscript), while the last summand corresponds to two one-forms on M, denoted ψ and ψ', which are positive chirality spinors on C.

The topological-holomorphic theory is defined with respect to super-charges that are scalar along M. The corresponding supersymmetry parameters are annihilated by the generators of $\mathrm{Spin}(4)'_M$, i.e.,

$$(\Gamma_{\mu\nu} + \Gamma_{\mu+6,\nu+6})\,\epsilon = 0, \quad \mu,\nu = 0,\dots,3. \tag{3.2.15}$$

These equations can be reexpressed as

$$\epsilon = \Gamma_{\mu\nu\mu+6,\nu+6}\epsilon \tag{3.2.16}$$

and they have two independent solutions. This follows since (3.2.16) are three independent constraints on ϵ, implying that there are $16 \times (1/2)^3 = 2$ preserved supercharges. We can identify a specific supercharge by further demanding

$$\epsilon = -i\Gamma_{\mu,\mu+6}\epsilon. \tag{3.2.17}$$

This equation is compatible with (3.2.16), which can be seen by multiplying both sides by $\Gamma_{\nu,\nu+6}$. In addition, (3.2.17) is compatible with the chirality condition

$$i\widehat{\Gamma}_{10d}\epsilon = \epsilon \tag{3.2.18}$$

in Euclidean signature, which can be seen by multiplying both sides of (3.2.17) by $\widehat{\Gamma}_{10d}$.

The condition (3.2.17) results in a unique solution to (3.2.16), up to rescaling, and we refer to the corresponding supercharge as Q. Analogously, imposing the condition

$$\epsilon = i\Gamma_{\mu,\mu+6}\epsilon$$

furnishes another supercharge, Q'. Both supercharges can be shown to square to zero:

$$Q^2 = \left(Q'\right)^2 = 0. \tag{3.2.19}$$

Moreover, (3.2.16) and (3.2.18) together imply

$$\epsilon = i\Gamma_{45}\epsilon. \tag{3.2.20}$$

This follows since

$$\begin{aligned}
\epsilon &= i\widehat{\Gamma}_{10d}\epsilon \\
&= i\Gamma_{0167}\Gamma_{45}\Gamma_{2389}\epsilon \\
&= i\Gamma_{0167}\Gamma_{45}\epsilon \\
&= i\Gamma_{45}\epsilon.
\end{aligned} \tag{3.2.21}$$

The condition (3.2.20) leads to the following the supersymmetry algebra for the scalar supercharges:

$$\{Q, Q'\} \propto P_4 - iP_5. \tag{3.2.22}$$

Introducing the complex coordinate

$$z = \frac{1}{2}\left(x^4 - ix^5\right), \tag{3.2.23}$$

the supercharges are normalized in such a way that

$$\{Q, Q'\} = P_{\bar{z}}. \tag{3.2.24}$$

This expression suggests that "translations" in the $\bar{z}$ direction are Q or Q'-exact, and that we are dealing with a theory with holomorphic dependence on z.

Now, substituting (3.2.20) expressed as $\left(\Gamma_4 - i\Gamma_5\right)\epsilon = 0$ into the super-symmetry variations (3.2.6), we see that $A_{\bar{z}}$ is invariant under the action of Q and Q'. Moreover, the condition (3.2.17) implies $\left(\Gamma_\mu + i\Gamma_{\mu+6}\right)\epsilon = 0$. This is equivalent to $\bar{\epsilon}(\Gamma_\mu + i\Gamma_{\mu+6}) = 0$, meaning that $A_\mu + iA_{\mu+6}$ is annihilated by Q. The twist thus results in the Q-invariant partial connection

$$\mathcal{A} = \mathcal{A}_\mu dx^\mu + A_{\bar{z}}\, d\bar{z}, \tag{3.2.25}$$

where $\mathcal{A}_\mu = A_\mu + i\phi_\mu$. Analogously, if we define

$$\overline{\mathcal{A}} = \left(A_\mu - i\phi_\mu\right)dx^\mu + A_z\, dz, \tag{3.2.26}$$

then $\overline{\mathcal{A}}_\mu dx^\mu + A_{\bar{z}}\, d\bar{z}$ is Q'-invariant. We shall employ $\mathcal{D}$ and $\mathcal{F}$ for the covariant derivative and curvature of $\mathcal{A} + A_z\, dz$, and $\overline{\mathcal{D}}$ and $\overline{\mathcal{F}}$ for those of $\overline{\mathcal{A}} + A_{\bar{z}}\, d\bar{z}$.

We can now deduce the action of Q on the rest of the fields, which are

$$
\begin{aligned}
\delta \mathcal{A}_\mu &= 0, & \delta \overline{\mathcal{A}}_\mu &= \psi_\mu, \\
\delta A_z &= \xi, & \delta A_{\bar z} &= 0, \\
\delta \xi &= 0, & \delta \xi' &= \mathrm{P}, \\
\delta \psi_\mu &= 0, & \delta \psi'_\mu &= \mathcal{F}_{\mu \bar z}, \\
\delta \mathrm{P} &= 0, & \delta \chi_{\mu\nu} &= \mathcal{F}_{\mu\nu}.
\end{aligned}
\tag{3.2.27}
$$

Here, we have introduced an auxiliary bosonic scalar P such that the supersymmetry algebra holds off-shell. The logic behind the derivation of these supersymmetry transformations is as follows. Firstly, we know that $\mathcal{A}_\mu$ and $A_{\bar z}$ are Q-invariant, and moreover that $Q^2 = 0$. We also know the transformation properties of the fields under $\mathrm{Spin}(4)'_M \times \mathrm{Spin}(2)_C$, and that the supersymmetry variation of a fermion is a linear combination of the field strength F_{IJ} in ten dimensions (see equations (3.2.6)). The transformations of $\overline{\mathcal{A}}_\mu$ and A_z are nonzero, but from the nilpotency of Q, it must be the case that their fermionic superpartners, ξ and ψ_μ, are Q-invariant. Similarly, the Q-transformation of ξ' ought to be a bosonic field, which is itself Q-invariant. Finally, the variations of the remaining fermions ought to be field strengths that arise from $\mathcal{A}_\mu$ and $A_{\bar z}$, in order to ensure the nilpotency of Q.

The variation of the fields under the action of Q' can be identified with the help of relation (3.2.24). With some work, the following formulas for the supersymmetry transformations generated by Q' can be deduced:

$$
\begin{aligned}
\delta' \mathcal{A}_\mu &= -\psi'_\mu, & \delta' \overline{\mathcal{A}}_\mu &= 0, \\
\delta' A_z &= -\xi', & \delta' A_{\bar z} &= 0, \\
\delta' \xi &= \mathrm{P} - F_{z\bar z}, & \delta' \xi' &= 0, \\
\delta' \psi_\mu &= -\overline{\mathcal{F}}_{\mu \bar z}, & \delta' \psi'_\mu &= 0, \\
\delta' \mathrm{P} &= D_{\bar z} \xi', & \delta' \chi_{\mu\nu} &= -\left(\star_M \overline{\mathcal{F}} \right)_{\mu\nu}.
\end{aligned}
\tag{3.2.28}
$$

The action we would like to study can be written as the sum of two parts, S_1 and S_2, where S_1 is

$$
S_1 = \frac{1}{e^2} \int_{M \times C} \mathrm{d}^2 z \, \mathrm{Tr} \left(\delta'(\mathcal{F} \wedge \chi) + \chi \wedge \star_M \overline{\mathcal{D}} \psi + \frac{1}{2} \chi \wedge D_{\bar z} \chi \right),
\tag{3.2.29}
$$

and S_2 is

$$
\begin{aligned}
S_2 &= \frac{1}{e^2} \int_{M \times C} \sqrt{g}\ \mathrm{d}^6 x\, \delta\delta'\, \mathrm{Tr}\left(-\xi\xi' + 2iF_{\mu z}\phi^\mu\right)\\[4pt]
&= \frac{1}{e^2} \int_{M \times C} \sqrt{g}\ \mathrm{d}^6 x\, \delta\, \mathrm{Tr}\left(\left(-\mathrm{P} + F_{z\bar z} + 2iD^\mu\phi_\mu\right)\xi' - \overline{\mathcal{F}}_{\mu z}\psi'^\mu\right)\\[4pt]
&= \frac{1}{e^2} \int_{M \times C} \sqrt{g}\ \mathrm{d}^6 x\, \mathrm{Tr}\left(-\mathrm{P}\left(\mathrm{P} - F_{z\bar z} - 2iD^\mu\phi_\mu\right) - 2\overline{\mathcal{F}}^{\mu\bar z}\mathcal{F}_{\mu\bar z}\right.\\[4pt]
&\qquad\left. + \xi' D_{\bar z}\xi + \xi'\mathcal{D}_\mu\psi^\mu + \psi'_\mu \overline{\mathcal{D}}^\mu \xi + D_z\psi^\mu\psi'_\mu\right).
\end{aligned}
\tag{3.2.30}
$$

Here, we have dropped boundary terms, which can be justified by the imposition of suitable boundary conditions. In order to define the volume form $\sqrt{g}\ \mathrm{d}^6 x$, we have endowed C with the metric $g_C = \left(\mathrm{d}x^4\right)^2 + \left(\mathrm{d}x^5\right)^2$. The full metric on $M \times C$ is the direct sum $g = g_M \oplus g_C$, with $g_{z\bar z} = g_{\bar z z} = 2$. The action $S^1 + S^2$ is both Q and Q'-invariant.

The bosonic part of the complete action $S_1 + S_2$ has the form

$$
\begin{aligned}
&\frac{1}{e^2} \int_{M \times C} \sqrt{g}\ \mathrm{d}^6 x\, \mathrm{Tr}\left(-\frac{1}{2}\overline{\mathcal{F}}^{\mu\nu}\mathcal{F}_{\mu\nu} - \mathrm{P}\left(\mathrm{P} - F_{z\bar z} - 2iD^\mu\phi_\mu\right) - 2\overline{\mathcal{F}}^{\mu\bar z}\mathcal{F}_{\mu\bar z}\right)\\[6pt]
&= \frac{1}{e^2} \int_{M \times C} \sqrt{g}\ \mathrm{d}^6 x\, \mathrm{Tr}\left(-\left(\mathrm{P} - \frac{1}{2}F_{z\bar z} - iD^\mu\phi_\mu\right)^2 - \frac{1}{2}F^{\mu\nu}F_{\mu\nu} - 2F^{\mu z}F_{\mu z}\right.\\[6pt]
&\qquad\left. - F^{z\bar z}F_{z\bar z} - D^\mu\phi^\nu D_\mu\phi_\nu - 2D^z\phi^\mu D_z\phi_\mu - \frac{1}{2}[\phi^\mu,\phi^\nu][\phi_\mu,\phi_\nu] - R_{\mu\nu}\phi^\mu\phi^\nu\right),
\end{aligned}
$$

where $R_{\mu\nu}$ is the Ricci curvature of g_M. Setting $M = \mathbb{R}^4$, this reproduces the bosonic sector of the action of $\mathcal{N} = (1,1)$ super Yang–Mills theory given in (3.2.11).

Having constructed a twist with two supercharges that square to zero, we shall pick one of them, say Q, and consider the Q-invariant sector of the theory. Since the dependence of the action on the metric of M is contained in the Q-exact part, the theory is topological along M (once we restrict it to the Q-invariant sector). However, the Q-invariant sector of the theory depends on the complex structure of C, but not its metric. Since $P_{\bar z}$ is Q-exact, correlation functions of Q-closed operators supported at points on C depend holomorphically on C. Thus, the twisted theory is a topological-holomorphic theory on $M \times C$.

3.2.2.1 *B-twisted 2d gauge theory*

We would like to eventually show that a deformed version of the topological-holomorphic theory on $M \times C$ is equivalent to 4d Chern–Simons theory. To this end, a useful fact is that the six-dimensional topological-holomorphic theory can be written as a B-twisted $\mathcal{N} = (2,2)$ gauged sigma model in two dimensions, with a space of maps as target space. Let us review such sigma models before proceeding.

We shall write $\mathcal{G}$ for the gauge group of a B-twisted gauge theory. The spacetime of the theory is a surface, denoted D, and the gauge bundle is a principal $\mathcal{G}$-bundle $\mathcal{P} \to D$.

The B-twisted gauge theory consists of vector multiplets and chiral multiplets. A vector multiplet comprises of a connection, A, on $\mathcal{P}$, bosonic fields

$$\sigma \in \Gamma\left(\Lambda_D^1 \otimes \mathrm{ad}(\mathcal{P})\right), \quad \mathrm{D} \in \Gamma\left(\Lambda_D^0 \otimes \mathrm{ad}(\mathcal{P})\right), \tag{3.2.31}$$

and fermionic fields

$$\alpha \in \Gamma\left(\Lambda_D^0 \otimes \mathrm{ad}(\mathcal{P})\right), \quad \lambda \in \Gamma\left(\Lambda_D^1 \otimes \mathrm{ad}(\mathcal{P})\right), \quad \zeta \in \Gamma\left(\Lambda_D^2 \otimes \mathrm{ad}(\mathcal{P})\right). \tag{3.2.32}$$

A chiral multiplet is valued in a unitary representation R of $\mathcal{G}$. It comprises the bosonic fields

$$\varphi \in \Gamma\left(\Lambda_D^0 \otimes R(\mathcal{P})\right), \quad \mathrm{F} \in \Gamma\left(\Lambda_D^2 \otimes R(\mathcal{P})\right), \tag{3.2.33}$$

and fermionic fields

$$\bar{\eta} \in \Gamma\left(\Lambda_D^0 \otimes \bar{R}(\mathcal{P})\right), \quad \rho \in \Gamma\left(\Lambda_D^1 \otimes R(\mathcal{P})\right), \quad \bar{\mu} \in \Gamma\left(\Lambda_D^2 \otimes \bar{R}(\mathcal{P})\right), \tag{3.2.34}$$

where $R(\mathcal{P})$ denotes the vector bundle associated to $\mathcal{P}$ constructed from R, and $\bar{R}$ is the complex conjugate of R.

The B-twist of an $\mathcal{N} = (2,2)$ supersymmetric theory has two scalar supercharges, Q and $\widetilde{Q}$. With respect to Q, the vector multiplet transforms as

$$\begin{aligned} \delta A &= 0, \ \delta \bar{A} = \lambda, \\ \delta \lambda &= 0, \ \delta \alpha = \mathrm{D}, \\ \delta \mathrm{D} &= 0, \ \delta \zeta = \mathcal{F}, \end{aligned} \tag{3.2.35}$$

while the chiral multiplet transforms as

$$\begin{aligned}
\delta\varphi &= 0, & \delta\bar{\varphi} &= \bar{\eta}, \\
\delta\rho &= \mathcal{D}\varphi, & \delta\bar{\eta} &= 0, \\
\delta\mathrm{F} &= \mathcal{D}\rho - \zeta\varphi, & \delta\overline{\mathrm{F}} &= 0, \\
& & \delta\bar{\mu} &= \overline{\mathrm{F}}.
\end{aligned} \tag{3.2.36}$$

Here, we have defined

$$\mathcal{A} = A + i\sigma, \quad \overline{\mathcal{A}} = A - i\sigma. \tag{3.2.37}$$

Moreover, the fields $\bar{\varphi}$ and $\overline{\mathrm{F}}$ are the hermitian conjugates of φ and F. The other supercharge $\widetilde{Q}$ acts on the vector multiplet as

$$\begin{aligned}
\tilde{\delta}\mathcal{A} &= \star\lambda, & \tilde{\delta}\overline{\mathcal{A}} &= 0, \\
\tilde{\delta}\lambda &= 0, & \tilde{\delta}\alpha &= \star\overline{\mathcal{F}}, \\
\tilde{\delta}\mathrm{D} &= -\star\overline{\mathcal{D}}\lambda, & \tilde{\delta}\zeta &= -\star\mathrm{D} + 2iD\star\sigma,
\end{aligned} \tag{3.2.38}$$

where $\star$ is the Hodge star on D. In addition, $\widetilde{Q}$ acts on the chiral multiplet as

$$\begin{aligned}
\tilde{\delta}\varphi &= 0, & \tilde{\delta}\bar{\varphi} &= -\star\bar{\mu} \\
\tilde{\delta}\rho &= -\star\overline{\mathcal{D}}\varphi, & \tilde{\delta}\bar{\eta} &= \star\overline{\mathrm{F}}, \\
\tilde{\delta}\mathrm{F} &= \overline{\mathcal{D}}\star\rho - \star\alpha\varphi, & \tilde{\delta}\overline{\mathrm{F}} &= 0 \\
& & \tilde{\delta}\bar{\mu} &= 0.
\end{aligned} \tag{3.2.39}$$

The action for the vector multiplet is

$$\begin{aligned}
S_{\mathrm{V}} &= \int_D \delta\tilde{\delta}\,\mathrm{Tr}(\zeta\alpha) \\
&= \int_D \delta\,\mathrm{Tr}((-\star\mathrm{D} + 2iD\star\sigma)\alpha - \zeta\star\overline{\mathcal{F}}) \\
&= \int_D \mathrm{Tr}(-\overline{\mathcal{F}}\star\mathcal{F} - \mathrm{D}\star(\mathrm{D} - 2i\star D\star\sigma) + \alpha\mathcal{D}\star\lambda + \zeta\star\overline{\mathcal{D}}\lambda),
\end{aligned} \tag{3.2.40}$$

while the chiral multiplet action is

$$\begin{aligned}
S_{\mathrm{C}} &= \int_D \delta\tilde{\delta}(-\bar{\varphi}\mathrm{F}) \\
&= \int_D \delta(-\bar{\varphi}(\overline{\mathcal{D}}\star\rho - \star\alpha\varphi) + \bar{\mu}\star\mathrm{F}) \\
&= \int_D (-\bar{\varphi}\overline{\mathcal{D}}\star\mathcal{D}\varphi + \star\bar{\varphi}\mathrm{D}\varphi + \overline{\mathrm{F}}\star\mathrm{F} \\
&\qquad - \bar{\eta}\overline{\mathcal{D}}\star\rho - \bar{\mu}\star\mathcal{D}\rho + \star\bar{\eta}\alpha\varphi - \bar{\varphi}\lambda\wedge\star\rho + \bar{\mu}\star\zeta\varphi).
\end{aligned} \tag{3.2.41}$$

Furthermore, we include a superpotential, W, in the action. Here, W is a gauge invariant, holomorphic function of the chiral multiplet scalars. It gives rise to the interaction terms

$$
\begin{aligned}
S_W &= \int_D \left(\mathrm{F}\frac{\partial W}{\partial \varphi} + \frac{1}{2}\rho \wedge \rho \frac{\partial^2 W}{\partial\varphi\partial\varphi} + \star\delta\tilde{\delta}\bar{W} \right)\\
&= \int_D \left(\mathrm{F}\frac{\partial W}{\partial \varphi} + \frac{1}{2}\rho \wedge \rho \frac{\partial^2 W}{\partial\varphi\partial\varphi} - \bar{\mathrm{F}}\frac{\partial \bar{W}}{\partial \bar\varphi} - \bar\eta\bar\mu\frac{\partial^2 \bar{W}}{\partial\bar\varphi\partial\bar\varphi} \right).
\end{aligned}
\tag{3.2.42}
$$

Unlike S_V and S_C, this expression is neither Q-exact nor $\widetilde{Q}$-exact. Moreover, it is not invariant under Q or $\widetilde{Q}$ if D has a boundary. In fact, Q-invariance demands that

$$
\int_{\partial D} \rho\frac{\partial W}{\partial \varphi} = 0
\tag{3.2.43}
$$

holds at the boundary of D, while for the $\widetilde{Q}$-invariance we need

$$
\int_{\partial D} \star\rho\frac{\partial W}{\partial \varphi} = 0
\tag{3.2.44}
$$

to hold at the boundary of D. For these to be satisfied, suitable supersymmetric boundary conditions must be imposed.

3.2.2.2 *6d Topological-holomorphic theory as a B-twisted gauge theory*

With $M = D \times \Sigma$, we shall now describe the six-dimensional topological-holomorphic theory on $D \times \Sigma \times C$ as a B-twisted gauge theory on D. We use $i, j, \ldots$ for vector indices on D and $m, n, \ldots$ for vector indices on Σ. We shall assume that appropriate supersymmetric boundary conditions are imposed at the boundaries, such that there are no non-vanishing boundary terms that arise from integration by parts. A choice of such boundary conditions will be discussed below after introducing a necessary deformation of the 6d theory.

The aim is to organize the fields of the six-dimensional theory into supermultiplets of B-twisted gauge theory. The B-twisted gauge theory has a single vector multiplet whose gauge field $A_i\mathrm{d}x^i$ is part of the six-dimensional gauge field. By comparing the supersymmetry transformations of the B-twisted gauge theory and the 6d topological-holomorphic theory, we identify the remaining fields in the 2d vector multiplet as

$$
\sigma_i = \phi_i, \quad \mathrm{D} = \mathrm{P}, \quad \alpha = \xi', \quad \lambda_i = \psi_i, \quad \zeta_{ij} = \chi_{ij}.
\tag{3.2.45}
$$

In particular, the combinations $\mathcal{A}_i = A_i + i\sigma_i$ and $\overline{\mathcal{A}}_i = A_i - i\sigma_i$ in 2d lift to $\mathcal{A}_i = A_i + i\phi_i$ and $\overline{\mathcal{A}}_i = A_i - i\phi_i$ in 6d.

The vector multiplet action (3.2.40) can be lifted to six dimensions by identifying the bilinear form on the Lie algebra of $\mathcal{G}$ (the gauge group of the B-twisted gauge theory) as

$$\frac{1}{e^2} \int_{\Sigma \times C} \star_{\Sigma \times C} \operatorname{Tr}, \tag{3.2.46}$$

where Tr in the integrand stands for the bilinear form on $\mathfrak{g}$ (the Lie algebra of the gauge group of the 6d theory). This furnishes the action

$$S_{\mathrm{V}} = \frac{1}{e^2} \int_{D \times \Sigma \times C} \sqrt{g} \ \mathrm{d}^6 x \delta \operatorname{Tr}\left(\left(-\mathrm{P} + 2i D^i \phi_i\right) \xi' - \frac{1}{2}\chi^{ij}\overline{\mathcal{F}}_{ij} \right). \tag{3.2.47}$$

Next, we need to address the fact that in six dimensions, the gauge field has six components, while in two dimensions, it only has two components. What is the B-model interpretation of the remaining four components of the 6d gauge field? The answer is that these fields are the lowest components of *chiral* multiplets from the B-model perspective.

Indeed, the fields $\mathcal{A}_m$ and $A_{\bar{z}}$ are annihilated by Q, and this matches the behaviour of the field φ of the 2d B-model. Hence, we conclude that we have three chiral multiplets in the adjoint representation, whose lowest components are $\mathcal{A}_m$ and $A_{\bar{z}}$. Let us call the fields of these multiplets $(\mathcal{A}_m,\ \mathrm{F}_m, \bar{\eta}_m, \rho_m, \bar{\mu}_m)$ and $(A_{\bar{z}},\ \mathrm{F}_{\bar{z}}, \bar{\eta}_z, \rho_{\bar{z}}, \bar{\mu}_z)$.

Now, when we lift the chiral multiplet supersymmetry transformations (3.2.36) to six dimensions, it is important to note that we have to replace the B-model scalar fields with covariant derivatives appropriately, such that when we perform a dimensional reduction along $\Sigma \times C$, we obtain the original transformations. This leads us to the Q-variations

$$
\begin{aligned}
\delta \mathcal{A}_m &= 0, & \delta \overline{\mathcal{A}}_m &= \bar{\eta}_m, \\
\delta \rho_{mi} &= \mathcal{F}_{im}, & \delta \bar{\eta}_m &= 0, \\
\delta \mathrm{F}_{mij} &= \mathcal{D}_i \rho_{mj} - \mathcal{D}_j \rho_{mi} + \mathcal{D}_m \zeta_{ij}, & \delta \overline{\mathrm{F}}_{mij} &= 0, \\
\delta \bar{\mu}_{mij} &= \overline{\mathrm{F}}_{mij} &&
\end{aligned}
\tag{3.2.48}
$$

and

$$
\begin{aligned}
\delta A_{\bar{z}} &= 0, & \delta A_z &= \bar{\eta}_z, \\
\delta \rho_{\bar{z}i} &= \mathcal{F}_{i\bar{z}}, & \delta \bar{\eta}_z &= 0, \\
\delta \mathrm{F}_{\bar{z}ij} &= \mathcal{D}_i \rho_{\bar{z}j} - \mathcal{D}_j \rho_{\bar{z}i} + \mathcal{D}_{\bar{z}} \zeta_{ij}, & \delta \overline{\mathrm{F}}_{zij} &= 0, \\
\delta \bar{\mu}_{zij} &= \overline{\mathrm{F}}_{zij}. &&
\end{aligned}
\tag{3.2.49}
$$

Note that the lift from scalar fields to covariant derivatives has replaced $\mathcal{D}\phi$ in the 2d B-model transformations with $[\mathcal{D}_i, \mathcal{D}_m] = \mathcal{F}_{im}$ and $[\mathcal{D}_i, \mathcal{D}_{\bar{z}}] = \mathcal{F}_{i\bar{z}}$. From the Q-variations that involve the gauge field in (3.2.27), we are able to make the identifications

$$\bar{\eta}_m = \psi_m, \quad \rho_{mi} = \chi_{im}, \quad \bar{\eta}_z = \xi, \quad \rho_{\bar{z}i} = \psi_i'. \tag{3.2.50}$$

Via these identifications, we have

$$\begin{aligned} \delta\mathrm{F}_{mij} &= (\mathcal{D}\chi)_{mij}, \\ \delta\mathrm{F}_{\bar{z}ij} &= (\mathcal{D}\psi')_{ij} + D_{\bar{z}}\chi_{ij}. \end{aligned} \tag{3.2.51}$$

Moreover, from (3.2.30), we can derive the equations of motion

$$\begin{aligned} \mathcal{D}_M \chi &= -\star_M D_z\psi + \star_M \mathcal{D}_M \xi = -\star_M \iota_{\partial_z}\delta\overline{\mathcal{F}}, \\ \mathcal{D}_M \psi' + D_{\bar{z}}\chi &= -\star_M \overline{\mathcal{D}}_M\psi = -\star_M \delta\overline{\mathcal{F}}, \end{aligned} \tag{3.2.52}$$

where $\mathcal{D}_M = \mathcal{D}_\mu \mathrm{d}x^\mu$. Comparing (3.2.51) and (3.2.52), we may obtain the on-shell relations

$$\begin{aligned} \mathrm{F}_{mij} &= -\left(\star_M \iota_{\partial_z}\overline{\mathcal{F}}\right)_{mij}, & \mathrm{F}_{\bar{z}ij} &= -\left(\star_M \overline{\mathcal{F}}\right)_{ij}, \\ \overline{\mathrm{F}}_{mij} &= -\left(\star_M \iota_{\partial_{\bar{z}}}\mathcal{F}\right)_{mij}, & \overline{\mathrm{F}}_{zij} &= -\left(\star_M \mathcal{F}\right)_{ij}. \end{aligned} \tag{3.2.53}$$

In addition, we obtain the relations

$$\bar{\mu}_{mij} = \left(\star_M \psi'\right)_{mij}, \quad \bar{\mu}_{zij} = -\left(\star_M \chi\right)_{ij}. \tag{3.2.54}$$

We can then lift the chiral multiplet action to six dimensions, obtaining the expression

$$\begin{aligned} S_{\mathrm{C}} = \frac{1}{e^2} \int_{D\times\Sigma\times C} &\sqrt{g}\, \mathrm{d}^6x\, \mathrm{Tr}\left(\delta\left(-\overline{\mathcal{F}}^{im}\chi_{im} - \overline{\mathcal{F}}_{iz}\psi^i + (2iD^m\phi_m + F_{z\bar{z}})\xi'\right)\right. \\ &\left. + \frac{1}{2}\overline{\mathrm{F}}^{mij}\,\mathrm{F}_{mij} + \frac{1}{2}\overline{\mathrm{F}}_z^{ij}\,\mathrm{F}_{\bar{z}ij}\right) \\ + \frac{1}{e^2} \int_{D\times\Sigma\times C} &\mathrm{d}^2z\, \mathrm{Tr}\left(-\mathcal{D}\chi \wedge \psi_\Sigma' + \chi_\Sigma \wedge (\mathcal{D}\psi' + D_{\bar{z}}\chi)\right), \end{aligned} \tag{3.2.55}$$

where we have defined

$$\chi_\Sigma = \frac{1}{2}\chi_{mn}\,\mathrm{d}x^m \wedge \mathrm{d}x^n, \quad \psi_\Sigma' = \psi_m'\mathrm{d}x^m. \tag{3.2.56}$$

It is now crucial to determine the superpotential. To this end, recall that the equation of motion for the auxiliary field F in the 2d B-twisted gauge theory is

$$\mathrm{F} = \frac{\partial \bar{W}}{\partial \bar{\varphi}}. \tag{3.2.57}$$

In order to reproduce the equations of motion (3.2.53), the superpotential must be the 4d CS action functional

$$W = -\frac{i}{e^2} \int_{\Sigma \times C} \mathrm{d}z \wedge \mathrm{Tr}\left(\mathcal{A} \wedge \mathrm{d}\mathcal{A} + \frac{2}{3}\mathcal{A} \wedge \mathcal{A} \wedge \mathcal{A}\right). \tag{3.2.58}$$

The superpotential terms in the action are, therefore,

$$\begin{aligned}
S_W = &\frac{1}{e^2} \int_{D \times \Sigma \times C} \sqrt{g}\, \mathrm{d}^6 x\, \mathrm{Tr}\left(\frac{1}{2}\,\mathrm{F}^{mij}\left(\star_M \iota_{\partial_{\bar{z}}}\mathcal{F}\right)_{mij} + \frac{1}{2}\,\mathrm{F}^{ij}_{\bar{z}}\left(\star_M \mathcal{F}\right)_{ij}\right. \\
&\left. + \frac{1}{2}\overline{\mathrm{F}}^{mij}\left(\star_M \iota_{\partial_z}\overline{\mathcal{F}}\right)_{mij} + \frac{1}{2}\overline{\mathrm{F}}^{ij}_z\left(\star_M \overline{\mathcal{F}}\right)_{ij} - \delta\overline{\mathcal{F}}_{mz}\psi'^m + \frac{1}{2}\chi^{mn}\delta\overline{\mathcal{F}}_{mn}\right) \\
&+ \frac{1}{e^2}\int_{D \times \Sigma \times C} \mathrm{d}^2 z\, \mathrm{Tr}\left(\frac{1}{2}\chi_{D|\Sigma} \wedge D_{\bar{z}}\chi + \chi_{D|\Sigma} \wedge \mathcal{D}\psi'_D\right),
\end{aligned}$$

$$\tag{3.2.59}$$

where

$$\chi_{D|\Sigma} = \chi_{im}\, \mathrm{d}x^i \wedge \mathrm{d}x^m, \quad \psi'_D = \psi'_i \mathrm{d}x^i. \tag{3.2.60}$$

The remaining terms necessary to find agreement with the six-dimensional action, i.e.,

$$\frac{1}{e^2}\int_{D \times \Sigma \times C} \sqrt{g}\, \mathrm{d}^6 x\, \mathrm{Tr}\left(-\frac{1}{2}\overline{\mathcal{F}}^{mn}\mathcal{F}_{mn} - 2\overline{\mathcal{F}}^{m\bar{z}}\mathcal{F}_{m\bar{z}}\right), \tag{3.2.61}$$

are furnished when the auxiliary fields are integrated out.

We have thus identified the supersymmetry transformations generated by the supercharge Q and the action of the topological-holomorphic twist of 6d $\mathcal{N} = (1,1)$ super Yang–Mills theory with that of the 2d B-twisted supersymmetric gauge theory.

In particular, since the chiral multiplet scalars of the B-model correspond to the partial connection $\mathcal{A} = \mathcal{A}_m \mathrm{d}x^m + A_{\bar{z}}\mathrm{d}\bar{z}$ on $\Sigma \times C$, the target space of the B-model is the space of such connections, i.e., it is a space of maps. The gauge group, $\mathcal{G}$, of the B-model is the group of maps from $\Sigma \times C$ to G, the gauge group of the 6d theory. In addition, we can deduce from the action that the target space of the B-model is endowed with the $\mathcal{G}$-invariant Kähler metric

$$
g_X = -\frac{1}{2e^2} \int_{\Sigma \times C} \sqrt{g_\Sigma}\, \mathrm{d}^2 x \; \mathrm{d}^2 z \, \mathrm{Tr} \left(\delta \mathcal{A}^m \otimes \delta \overline{\mathcal{A}}_m + \delta \overline{\mathcal{A}}^m \otimes \delta \mathcal{A}_m \right.
$$
$$
\left. + \, \delta A_{\bar{z}} \otimes \delta A_z + \delta A_z \otimes \delta A_{\bar{z}} \right). \tag{3.2.62}
$$

3.2.3 Ω-deformed B-model

We would now like to perform an Ω-deformation of the B-model in two dimensions, with the aim of Ω-deforming the 6d topological-holomorphic theory. The Ω-deformation is a continuous deformation of certain twisted supersymmetric field theories, and in the present context of the B-model, it is defined with respect to a Killing vector field, V, that generates a $U(1)$ isometry of the disk that the B-model is defined on. A crucial difference from the undeformed topological theory is that the supercharge Q is no longer nilpotent, but instead squares to a Lie derivative associated with the vector field, V.

Let us review the supersymmetry transformations of the Ω-deformed B-model generated by the deformation of the supercharge Q, denoted Q_V. The action of the deformed supercharge, Q_V, is given by

$$
\delta_V \mathcal{A} = \iota_V \zeta, \qquad\qquad \delta_V \overline{\mathcal{A}} = \lambda - \iota_V \zeta,
$$
$$
\delta_V \lambda = 2\iota_V F - 2i D \iota_V \sigma, \quad \delta_V \zeta = \mathcal{F}, \tag{3.2.63}
$$
$$
\delta_V \alpha = \mathrm{D}, \qquad\qquad \delta_V \mathrm{D} = \iota_V \mathcal{D} \alpha
$$

for the vector multiplet and

$$
\delta_V \varphi = \iota_V \rho, \quad \delta_V \bar{\varphi} = \bar{\eta},
$$
$$
\delta_V \rho = \mathcal{D}\varphi + \iota_V \, \mathrm{F}, \quad \delta_V \bar{\eta} = \iota_V \mathcal{D}\bar{\varphi},
$$
$$
\delta_V \, \mathrm{F} = \mathcal{D}\rho - \zeta \varphi, \quad \delta_V \overline{\mathrm{F}} = \mathcal{D}\iota_V \bar{\mu}, \tag{3.2.64}
$$
$$
\delta_V \bar{\mu} = \overline{\mathrm{F}}
$$

for the chiral multiplet. One can check that Q_V is not nilpotent, but instead squares to the covariant Lie derivative defined with respect to the complexification of the gauge field, i.e., $\mathcal{A} = A + i\sigma$:

$$
\delta_V^2 = \mathrm{d}_{\mathcal{A}} \iota_V + \iota_V \, \mathrm{d}_{\mathcal{A}}. \tag{3.2.65}
$$

It can also be checked that for $V = 0$, the supersymmetry transformations given here reduce to those given in (3.2.35) and (3.2.36).

The Ω-deformed B-twisted gauge theory action can be written as $S_V + S_C + S_W$, where

$$S_\mathrm{V} = \delta_V \int_D \mathrm{Tr}((-\star \mathrm{D} + 2iD \star \sigma)\alpha - \zeta \star \overline{\mathcal{F}}), \tag{3.2.66}$$

$$S_\mathrm{C} = \delta_V \int_D \left((\overline{\mathcal{D}}\bar\varphi + \iota_{\bar V}\overline{\mathrm{F}}) \wedge \star\rho + \star\bar\varphi\alpha\varphi + \bar\mu \star \mathrm{F} \right), \tag{3.2.67}$$

$$S_W = \int_D \left(\mathrm{F}\frac{\partial W}{\partial\varphi} + \frac{1}{2}\rho \wedge \rho\frac{\partial^2 W}{\partial\varphi\partial\varphi} - \delta_V \left(\bar\mu\frac{\partial\bar W}{\partial\bar\varphi}\right) \right) - \int_{\partial D} W\frac{\mathrm{d}\theta}{V^\theta}. \tag{3.2.68}$$

Notably, the boundary term in S_W is needed for unbroken Q_V-invariance at the boundary.

The Ω-deformed B-model with the space of partial connections of the form $\mathcal{A} = \mathcal{A}_m \mathrm{d}x^m + A_{\bar z}\mathrm{d}\bar z$ on $\Sigma \times C$ as target space, with the metric (3.2.62) and superpotential (3.2.58), would thus be an Ω-deformation of the 6d topological-holomorphic gauge theory that we have studied thus far. As we shall review below, this theory can be shown to be equivalent to 4d Chern–Simons theory. To this end, we shall discuss suitable boundary conditions for the Ω-deformed B-model.

3.2.3.1 *Boundary Conditions*

Since we are studying Ω-deformed B-models defined on the disk, D, we ought to impose boundary conditions at its boundary, ∂D. These boundary conditions must ensure that no boundary terms arise when performing a variation of the action, while also being Q_V-invariant. In what follows, we shall utilize polar coordinates r and θ on D, and further endow it with the metric of a hemisphere, which has the form

$$g(r,\theta) = g_{rr}(r)\mathrm{d}r^2 + g_{\theta\theta}(r)\mathrm{d}\theta^2. \tag{3.2.69}$$

In addition, the hatted indices $\widehat{r}$ and $\widehat{\theta}$ are used to represent components of tensors with respect to the orthonormal vectors $\partial_{\widehat\theta} = \sqrt{g^{\theta\theta}}\partial_\theta$, $\partial_{\widehat r} = \sqrt{g^{rr}}\partial_r$ and the dual one-forms $\mathrm{d}\widehat\theta = \sqrt{g_{\theta\theta}}\mathrm{d}\theta$ and $\mathrm{d}\widehat r = \sqrt{g_{rr}}\mathrm{d}r$. In these coordinates, the Killing vector field takes the form

$$V = \epsilon\partial_\theta \tag{3.2.70}$$

for $\epsilon \in \mathbb{C}$, a parameter that is often referred to as the Ω-deformation parameter.

We shall first consider the vector multiplet. The choice of boundary condition on the gauge field that preserves gauge symmetry at the boundary is the Neumann boundary condition

$$F_{r\theta}|_{\partial D} = 0. \qquad (3.2.71)$$

It shall also be useful to pick the gauge whereby $A_r = 0$ at the boundary, whereby the boundary condition becomes

$$\partial_r A_\theta|_{\partial D} = 0. \qquad (3.2.72)$$

Since we encounter the combination $\mathcal{A} = A + i\sigma$ repeatedly in our analysis, it shall be natural to impose the Neumann boundary condition on the field σ_θ and the Dirichlet boundary condtion on σ_r, that is, we impose

$$\sigma_r|_{\partial D} = \partial_r \sigma_\theta|_{\partial D} = 0. \qquad (3.2.73)$$

To ensure Q_V-invariance, we further impose

$$\zeta_{r\theta}|_{\partial D} = \lambda_r|_{\partial D} = \partial_r \lambda_\theta|_{\partial D} = \partial_r \alpha|_{\partial D} = \partial_r \mathrm{D}|_{\partial D} = 0. \qquad (3.2.74)$$

Next, let us consider the chiral multiplet. Denoting the target space of the B-model by X, we shall pick a boundary condition for the scalar field such that it is valued in a submanifold, γ, of X, that is,

$$\varphi|_{\partial D} \in \gamma. \qquad (3.2.75)$$

Since we require that gauge symmetry is unbroken at the boundary, the submanifold γ shall be required to have a G-isometry. Moreover, one assumes that $\mathrm{Re}(W/\epsilon)$ is bounded above along γ, such that the boundary term in (3.2.68) does not result in the path integral becoming divergent. Q_V-invariance of the boundary conditions then requires

$$(\iota_V \rho, \bar{\eta})\,|_{\partial D} \in T_\varphi \gamma \otimes \mathbb{C}. \qquad (3.2.76)$$

In addition, we should also impose the Q_V-invariant boundary conditions

$$\left(\rho_{\hat{r}}, -\|V\|^{-2}(\iota_{\bar{V}}\bar{\mu})_{\hat{r}}\right)|_{\partial D} \in N_\varphi \gamma \otimes \mathbb{C} \qquad (3.2.77)$$

and

$$\left(\partial_{\hat{r}}\varphi + (\iota_V \mathrm{F})_{\hat{r}}, -\|V\|^{-2}\left(\iota_{\bar{V}}\overline{\mathrm{F}}\right)_{\hat{r}}\right)|_{\partial D} \in N_\varphi \gamma \otimes \mathbb{C}, \qquad (3.2.78)$$

where $N\gamma$ is the normal bundle of γ with respect to the Kähler metric on the target space X, and where $\|V\|$ is the norm of V. We have now fully specified the boundary conditions of the Ω-deformed B-model.

3.2.3.2 *Localization of the Ω-deformed B-model*

We would like to perform the evaluation of the path integral of the 6d topological-holomorphic field theory on $D \times \Sigma \times C$, with Ω-deformation with respect to D. As is the case for Ω-deformed quantum field theories, the path integral localizes to fixed points of the action of the Killing vector field, V, thereby producing a quantum field theory on $\Sigma \times C$.

Let us first explain localization for general Ω-deformed B-models on D, with the aforementioned boundary conditions imposed. We shall outline the main features of the localization computation; further details can be found in the original reference [46].

The fact that Q_V-exact terms in the action of the Ω-deformed B-model can be multiplied by a large factor without changing the value of the path integral implies that the path integral localizes to certain Q_V-invariant field configurations. The evaluation of the path integral then involves expanding each field around Q_V-invariant field configurations, and computing the bosonic and fermionic one-loop determinants, that can be shown to cancel. The Q_V-invariant configurations are such that the non-Q_V-exact terms other than the boundary term also vanish. Since we are working with a gauge theory, it is also important to implement BRST gauge-fixing to remove redundant degrees of freedom from the path integral.

The end result of the localization is a zero-dimensional theory with action proportional to the superpotential W, and integration cycle specified by the Q_V-invariant configurations and boundary conditions at the boundary of the disk. This zero-dimensional theory governs maps from a point to the submanifold, γ, of the target space (that was used to define the boundary conditions of the chiral multiplet). Schematically, the result of the localization computation is of the form

$$\int d\varphi_0 \exp\left(\frac{2\pi}{\epsilon} W(\varphi_0)\right), \tag{3.2.79}$$

with additional terms responsible for gauge-fixing suppressed. Here, φ_0 is a "zero-mode", i.e., a scalar field configuration that is Q_V-invariant. The factor of 2π multiplying $W(\varphi_0)$ arises because φ_0 is independent of the angular coordinate parametrizing ∂D, and integration over this coordinate produces 2π. This property of φ_0 follows since two of the Q_V-invariant configurations are

$$F_{ij} = 0, \quad D_i\varphi = 0. \tag{3.2.80}$$

Since the disk, D, is simply connected, the first of these conditions implies that the gauge field components on D are pure gauge, and we can pick the gauge where these components vanish (recall that gauge symmetry is preserved along the boundary). Then, the second condition becomes $\partial_i \varphi = 0$, meaning that the zero-mode of φ is a constant, in particular, along ∂D.

In the original work by Costello and Yagi [46], the Ω-deformed B-model was studied on a cigar, and it was shown that the path integral localized to configurations that could be identified with gradient flow equations for the superpotential, W. These equations specify the integration cycle for the path integral, and ensure its convergence.

Finally, the aforementioned localization procedure works straightforwardly in the specific case of the Ω-deformed B-model on the space of partial connections of the form $A_m \mathrm{d}x^m + A_{\bar{z}} \mathrm{d}\bar{z}$ on $\Sigma \times C$, and leads to the path integral of 4d Chern–Simons theory

$$\int_\gamma \mathcal{D}\mathcal{A} \exp\left(\frac{i}{\pi\hbar} \int_{\Sigma \times C} \mathrm{d}z \wedge \mathrm{CS}(\mathcal{A}) \right), \tag{3.2.81}$$

with an integration cycle, γ, determined by gradient flow equations, that ensure the convergence of the path integral, and where the coupling

$$\hbar = -\frac{\epsilon e^2}{2\pi^2}, \tag{3.2.82}$$

is proportional to the Ω-deformation parameter.

3.2.4 *Type IIB string theory embedding*

The above construction can be embedded in type IIB string theory, employing the fact that Ω-deformation can be equivalently viewed as a cigar geometry [48], and can therefore be realized via the fluxbrane background studied in [49].

The 10d (Euclidean) spacetime is specified to take the form

$$\mathrm{d}s^2 = \mathrm{d}s_{T^*\Sigma}^2 + \mathrm{d}s_C^2 + \mathrm{d}s_{\mathrm{TN}}^2, \tag{3.2.83}$$

where $\mathrm{d}s_{\mathrm{TN}}^2$ refers to the following Taub-NUT background,

$$\mathrm{d}s_{\mathrm{TN}}^2 = U\, \mathrm{d}\vec{x} \cdot \mathrm{d}\vec{x} + \frac{1}{U}(\, \mathrm{d}\theta + \vec{\omega} \cdot \mathrm{d}\vec{x})^2. \tag{3.2.84}$$

Here, $\vec{x}$ is a coordinate of $\mathbb{R}^3$ and θ parametrizes a circle of radius r, which is referred to as the Taub-NUT circle. In addition, $U = \frac{1}{r} + \frac{1}{\lambda^2}$, while $\vec{\omega}$ is a vector on $\mathbb{R}^3$ that satisfies $\mathrm{d}U = \star_{\mathbb{R}^3}(\vec{\omega} \cdot \mathrm{d}\vec{x})$. Parametrizing $\mathbb{R}^3$ by a radial coordinate, $\rho = \sqrt{\vec{x} \cdot \vec{x}}$, and two angular coordinates, one finds that a 2d surface at fixed values of these angular coordinates is a cigar with coordinates r and θ.

The supergravity background of interest preserves a supercharge, Q, that induces an Ω-deformation of the worldvolume theory of any D-brane that wraps a cigar in the Taub-NUT geometry, such that the theory can be described as an Ω-deformed B-model whose target space is a space of maps. This supercharge squares to a Lie derivative generating a rotation of the Taub-NUT circle. In addition, the type IIB string theory background of interest includes a nontrivial dilaton and RR 2-form.

To realize the $U(N)$ Ω-deformed 6d topological-holomorphic theory (and thereby, the $GL(N,\mathbb{C})$ 4d CS theory), one needs to place a stack of N D5-branes on a product of C, a cigar in the Taub-NUT geometry, and $\Sigma \subset T^*\Sigma$.[5] Crucially, terms in the action that involve the superpotential of 4d CS form can be understood to arise from a Wess-Zumino term in the D-brane action associated with the aforementioned RR 2-form, C_2, which is proportional to

$$\int_{D\times\Sigma\times C} C_2 \wedge \mathrm{Tr}(F \wedge F). \tag{3.2.85}$$

Here, the restriction of the RR 2-form to the D-brane worldvolume has the form

$$C_2 \propto \frac{r^2}{(1 + |\epsilon|^2 r^2)^2}\, \mathrm{d}\theta \wedge \mathrm{Re}(\bar\epsilon\, \mathrm{d}z). \tag{3.2.86}$$

Indeed, we find that (3.2.85) contains a term proportional to

$$2i\,\mathrm{Im} \int_D r\mathrm{d}r \wedge \mathrm{d}\theta \wedge \frac{\bar\epsilon r}{(1 + |\epsilon|^2 r^2)^2} \partial_r \left(-\frac{i}{e^2} \int_{\Sigma\times C} \mathrm{d}z \wedge \mathrm{CS}(A) \right), \tag{3.2.87}$$

which can be identified with one of the terms in the six-dimensional action, which is of the form $2i\,\mathrm{Im}\left(\bar{V}^{\hat\theta}\partial_{\hat r}W\right)/\left(1 + \|V\|^2\right)$.

The type IIB configuration described above can be encapsulated in the following table denoting the stack of N D5-branes, where we have denoted the Taub-NUT by $\mathbb{R}^2_\epsilon \times \mathbb{R}^2_{-\epsilon}$ for convenience, where the opposite signs of the deformation parameters indicate that the supercharge Q squares to a

---

[5]As described in [50], the twist of the normal bundle to $\Sigma \subset T^*\Sigma$ realizes the twist that ensures topological invariance along Σ.

symmetry that rotates the two cigars in opposite directions:

	$\overbrace{\hphantom{xx}}^{\Sigma}$		$\overbrace{\hphantom{xx}}^{\mathbb{R}^2_\epsilon}$		$\overbrace{\hphantom{xx}}^{C}$		$\overbrace{\hphantom{xx}}^{N\Sigma\subset T^*\Sigma}$		$\overbrace{\hphantom{xx}}^{\mathbb{R}^2_{-\epsilon}}$	
	0	**1**	**2**	**3**	**4**	**5**	**6**	**7**	**8**	**9**
D5	×	×	×	×	×	×				

One of the main advantages of this string theory embedding is that it can be easily shown to be related via string dualities to other realizations of integrable models in terms of supersymmetric field theories, such as the Bethe/Gauge correspondence of Nekrasov and Shatashvili [51]. Details of this relationship can be found in the original work of Costello and Yagi [46].

3.2.4.1 *Wilson Lines from D3-branes*

Now, we would like to understand how Wilson lines can be realized using D-branes in this setup, allowing us to interpret R-matrices and lattice models in terms of string theory. This can be accomplished by using D3-branes, which, depending on the directions of spacetime they wrap, give rise to Wilson lines in different representations of the gauge group. Consider the following configuration of D5- and D3-branes:

	$\overbrace{\hphantom{xx}}^{\Sigma}$		$\overbrace{\hphantom{xx}}^{\mathbb{R}^2_\epsilon}$		$\overbrace{\hphantom{xx}}^{C}$		$\overbrace{\hphantom{xx}}^{N\Sigma\subset T^*\Sigma}$		$\overbrace{\hphantom{xx}}^{\mathbb{R}^2_{-\epsilon}}$	
	0	**1**	**2**	**3**	**4**	**5**	**6**	**7**	**8**	**9**
D5	×	×	×	×	×	×				
$\mathbf{D3}^+_b$	×		×	×				×		
$\mathbf{D3}^-_b$		×	×	×			×			
$\mathbf{D3}^+_f$	×							×	×	×
$\mathbf{D3}^-_f$		×					×		×	×

$$(3.2.88)$$

Here, we have included D3-branes that spread along x^0 or x^1 and wrap $\mathbb{R}^2_\epsilon$, denoted $\mathrm{D3}^\pm_b$, and D3-branes that spread along x^0 or x^1 and wrap $\mathbb{R}^2_{-\epsilon}$, denoted $\mathrm{D3}^\pm_f$. As we shall explain, $\mathrm{D3}^\pm_b$-branes give rise to line operators supporting symmetric representations, while $\mathrm{D3}^\pm_f$-branes give rise to line operators supporting antisymmetric representations. A configuration with multiple $\mathrm{D3}^+_b$ and $\mathrm{D3}^-_b$ branes, forming a lattice, is depicted in Figure 3.1.

Let us first understand why the $\mathrm{D3}^\pm_b$-branes preserve the topological-holomorphic supercharge, Q. Recall that type IIB supersymmetry involves two 10d Weyl spinors of identical chirality, a linear combination of which

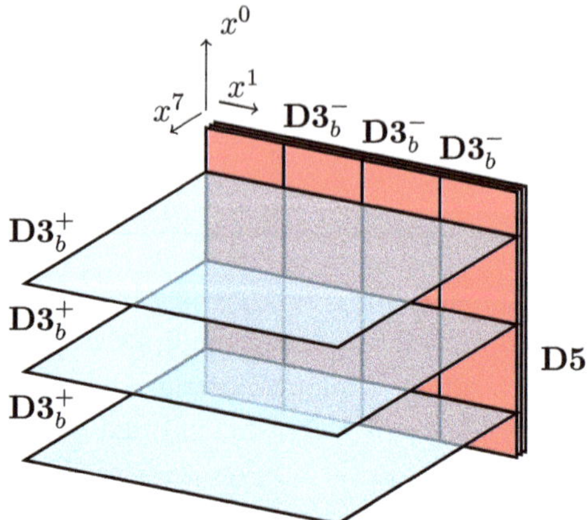

Figure 3.1. Depiction of the type IIB superstring theory realization of integrable lattice models. D3$_b^+$- and D3$_b^-$-branes intersect and form a lattice along the worldvolume of a stack of D5-branes subject to the topological-holomorphic twist with Ω-deformation. Here, we have only depicted the x^0, x^1, and x^7 directions, while the remaining directions are suppressed.

can be written as $\epsilon_L Q_L + \epsilon_R Q_R$. The chirality constraints are $i\Gamma_{0\ldots9}\epsilon_L = \epsilon_L$, $i\Gamma_{0\ldots9}\epsilon_R = \epsilon_R$. The D5-branes constrain the type IIB supersymmetries by:

$$\epsilon_R = i\Gamma_{012345}\epsilon_L \qquad (3.2.89)$$

which halves the number of supercharges to 16 . Since ϵ_L is determined entirely in terms of ϵ_R, we can use ϵ_L to refer to the preserved supercharge. For convenience, we shall denote ϵ_L as ϵ.

Recall that the topological-holomorphic twist of the D5-brane worldvolume theory was implemented by imposing the three constraints (3.2.15), or equivalently, (3.2.16), and that the supercharge, Q, was singled out by imposing the additional constraint (3.2.17).

Now, we consider a D3$_b^+$ brane along the 0237 directions. We shall actually require this brane to be an *anti*-D3-brane to preserve supersymmetry, which implies the constraint

$$\epsilon_R = -i\Gamma_{0237}\epsilon_L \, . \qquad (3.2.90)$$

Then, using (3.2.89), we have

$$i\Gamma_{012345}\epsilon = -i\Gamma_{0237}\epsilon \, . \qquad (3.2.91)$$

Using the constraint (3.2.20), we then find via the Clifford algebra that

$$\epsilon = -i\Gamma_{17}\epsilon \, , \qquad (3.2.92)$$

which is one of the relations of (3.2.17). Hence, the D3$_b^+$-brane preserves the supercharge Q. Next, let us consider the D3$_b^-$-brane along the 1236 directions. In order to preserve the supercharge, Q, this should be a D3-brane, as opposed to an anti-D3-brane. The corresponding constraint is

$$\epsilon_R = i\Gamma_{1236}\epsilon_L \,, \tag{3.2.93}$$

which, using (3.2.89), implies

$$i\Gamma_{012345}\epsilon = i\Gamma_{1236}\epsilon \,. \tag{3.2.94}$$

Using the Clifford algebra and (3.2.20), we arrive at

$$\epsilon = -i\Gamma_{06}\epsilon \,, \tag{3.2.95}$$

which is one of the relations of (3.2.17). Thus, the topological-holomorphic supercharge, Q, corresponding to the condition (3.2.17), is preserved in the D5-D3-anti-D3-brane configuration described above, when considering D3$_b$-branes. Analogous arguments can be used to show that a configuration involving D5-branes and intersecting D3$_f^+$- and D3$_f^-$-branes preserves the supercharge Q, using one of the constraints given in (3.2.15), $(\Gamma_{23} + \Gamma_{89})\epsilon = 0$.

Let us now try to understand why the D3$_b$-brane realizes a Wilson line in 4d Chern–Simons theory. The intersecting D3$_b$-brane and D5-brane share a 3d worldvolume with topology $\mathbb{R} \times \mathbb{R}_\epsilon^2$, where strings stretched between them give rise to a 3d $\mathcal{N} = 4$ hypermultiplet, which further localizes to a 1d theory of bosons [65], with an action of the form

$$\frac{1}{\hbar_{1d}} \int_{\mathbb{R}} \mathrm{Tr}_{\mathbb{C}^N} \left(\varphi \mathrm{d}_A \tilde{\varphi}\right) , \tag{3.2.96}$$

where φ and $\tilde{\varphi}$ are bosonic fields in the fundamental representation of $GL(N, \mathbb{C})$, d^A is an appropriately defined covariant derivative which couples to the 4d Chern–Simons gauge field, and $\hbar_{1d}$ is a coupling constant. As explained in Appendix C, the path integral of this 1d quantum mechanical system can be shown to be equivalent to a Wilson line in the direct sum of all symmetric representations of $GL(N, \mathbb{C})$.

Note that if we were to consider a stack of k D3$_b$ branes instead, we would obtain bosonic fields in the bifundamental representation of $GL(N, \mathbb{C}) \times GL(k, \mathbb{C})$ (we shall study such configurations in detail below). In general, the quantum mechanical system would have a coupling to the $GL(k, \mathbb{C})$ gauge field on the worldvolume of the D3$_b$-branes, and this gauge field was frozen when deriving (3.2.96), where $k = 1$.

In fact, in the $k = 1$ case, projection to a particular symmetric representation arises due to the gauge field supported by the single D3-brane. Explicitly, the total action takes the form

$$S_{\text{defect}} = S_{g\text{F1}} + S_{\text{D3int}} , \tag{3.2.97}$$

$$S_{g\text{F1}} = \frac{1}{\hbar_{1\text{d}}} \int_{\mathbb{R}} \text{Tr}_{\mathbb{C}^N} (\varphi \text{d}_A \tilde{\varphi}) , \tag{3.2.98}$$

$$S_{\text{D3int}} = \frac{1}{\hbar_{1\text{d}}} \int_{\mathbb{R}} \mathcal{B} \left(\text{Tr}_{\mathbb{C}^N} (\varphi \tilde{\varphi}) - l \right) , \tag{3.2.99}$$

where the field $\mathcal{B}$ in (3.2.99) is the gauge field of the D3-brane worldvolume theory, and appears here because the quantum mechanics on $\mathbb{R}$ arises from a hypermultiplet that is coupled to both D5- and D3-brane worldvolume theories. The term involving the number l, which, as explained in Appendix C, determines the associated symmetric representation of $GL(N, \mathbb{C})$, arises from a Chern–Simons coupling on the D3-brane worldvolume theory.

Let us now consider D3$_f$-branes intersecting with D5-branes, and study how this configuration realizes Wilson lines in anti-symmetric representations of $GL(N, \mathbb{C})$. The D3$_f$-D5 strings are in fact described by the dimensional reduction of the D4-D6 I-brane system involving N chiral free fermions [115–117] (if we were to consider a stack of k D3$_f$ branes instead, we would obtain fermionic fields in the bifundamental representation of $U(N) \times U(k)$), with the action

$$\frac{1}{\hbar_{1\text{d}}} \int_{\mathbb{R}} \text{Tr}_{\mathbb{C}^N} \left(\psi \text{d}_A \tilde{\psi} \right) , \tag{3.2.100}$$

which is of the same form as (3.2.96), except that we now have fermionic fields ψ and $\tilde{\psi}$ instead of bosonic ones. As explained in Appendix C, the associated representation of the line operator is a direct sum of anti-symmetric representations, and the projection to a fixed anti-symmetric representation arises via a coupling of the form (3.2.99) (with bosonic fields replaced by their fermionic counterparts), where $\mathcal{B}$ is the gauge field of the D3$_f$-brane worldvolume theory.

When considering a *stack* of k D3-branes with $k > 0$, one can in fact realize *any* finite-dimensional irreducible representation of $GL(N, \mathbb{C})$.[6] Let us focus on D3$_b$-branes in what follows. As shown in the table below, the brane configuration of interest consists of the k D3-branes D3$_\alpha^{l_\alpha}$ ($1 \leq \alpha \leq k$),

---

[6]This construction is analogous to the realization of Wilson lines in 4d $\mathcal{N} = 4$ SYM by Gomis and Passerini [89, 90].

where l_α indicates the amount of fundamental string charge dissolved in the α-th D3-brane, giving rise to a Chern–Simons term on the D3-brane worldvolume theory. N D5-branes $\mathrm{D}5_i$ $(1 \leq i \leq N)$ share the 3d spacetime $\mathbb{R} \times \mathbb{R}^2_{+\epsilon}$ with the D3-branes:

	Σ		$\mathbb{R}^2_\epsilon$		C		$N\Sigma \subset T^*\Sigma$		$\mathbb{R}^2_{-\epsilon}$	
	0	**1**	**2**	**3**	**4**	**5**	**6**	**7**	**8**	**9**
$\mathbf{D5}_i$	×	×	×	×	×	×				
$\mathbf{D3}^{l_\alpha}_\alpha$	×		×	×			×			

$$(3.2.101)$$

Before topological twisting and turning on an Ω-deformation, light excitations of fundamental strings stretched between D3- and D5-branes produce a 3d $\mathcal{N} = 4$ hypermultiplet in the bifundamental representation of $U(k) \times U(N)$ [126].[7] In order to give rise to the bifundamental hypermultiplet, each of the D3-branes ought to intersect all the D5-branes, as described, for example, in Section 3.4 of [127]. In this configuration, a D3-brane segment between two D5-branes can break and move independently of other segments, such that the value of a component of the bifundamental hypermultiplet corresponds to the position of a D3-brane that connects two D5-branes. After twisting and turning on Ω-deformation, the hypermultiplet localizes to a quantum mechanical system on $\mathbb{R}$ with the kinetic term [65]

$$S_{\mathrm{F1}} = \frac{1}{\hbar_{\mathrm{1d}}} \sum_{\alpha=1}^{k} \int_{\mathbb{R}} \mathrm{Tr}_{\mathbb{C}^N} \left(\varphi_\alpha \mathrm{d}_A \tilde{\varphi}^\alpha \right) , \qquad (3.2.102)$$

where the scalar fields $\tilde{\varphi}^\alpha_i$ and φ^i_α transform as vectors for $\mathbb{C}^N \otimes (\mathbb{C}^k)^*$ and $(\mathbb{C}^N)^* \otimes \mathbb{C}^k$ under $\mathrm{GL}(N, \mathbb{C}) \times \mathrm{GL}(k, \mathbb{C})$, respectively.

If we include the coupling to the gauge fields of the D3-branes, the action of the resulting line defect is given by

$$\tilde{S}_{\mathrm{defect}} = \tilde{S}_{g\mathrm{F1}} + \tilde{S}_{\mathrm{D3int}} , \qquad (3.2.103)$$

$$\tilde{S}_{g\mathrm{F1}} = \frac{1}{\hbar_{\mathrm{1d}}} \sum_{\alpha=1}^{K} \int_{\mathbb{R}} \mathrm{Tr}_{\mathbb{C}^N} \left(\varphi_\alpha \mathrm{d}_A \tilde{\varphi}^\alpha \right) , \qquad (3.2.104)$$

---

[7]The K coincident D3-branes give rise to 4d $\mathcal{N} = 4$ $U(k)$ SYM, and these end on the stack of N D5-branes, with a 3d $\mathcal{N} = 4$ hypermultiplet supported along the intersection of the two stacks.

$$\tilde{S}_{\text{D3int}} = \frac{1}{\hbar_{\text{1d}}} \sum_{\alpha,\beta,\gamma,\delta=1}^{K} \int_{\mathbb{R}} \left(\mathcal{B}^{\alpha\beta} \text{Tr}_{\mathbb{C}^N} \left(\varphi_\gamma \rho(E_{\alpha\beta})^\gamma{}_\delta \tilde{\varphi}^\delta \right) - \mathcal{B}^{\alpha\alpha} l_\alpha \right) ,$$

$$(3.2.105)$$

where $\rho(E_{\alpha\beta})\,(\alpha,\beta = 1,\ldots,k)$ denotes the fundamental representation of the $\mathfrak{gl}(k)$ generators $E_{\alpha\beta}$ satisfying the commutation relations

$$[E_{\alpha\beta}, E_{\gamma\delta}] = \delta_{\beta\gamma} E_{\alpha\delta} - \delta_{\delta\alpha} E_{\gamma\beta} . \qquad (3.2.106)$$

As before, the field $\mathcal{B}$ in (3.2.105) is the gauge field of the D3-brane worldvolume theory. The relationship between the quantum mechanical system with the action $\tilde{S}_{\text{defect}}$ and Wilson lines in finite-dimensional irreducible representations corresponding to a Young tableau with rows determined by l_α is explained in Appendix C.

We can analogously consider stacks of D3$_f$-branes, whereby we can realize constrained quantum mechanical systems with an action of the form given in $\tilde{S}_{\text{defect}}$, but with *fermionic* degrees of freedom. These can also be used to realize Wilson lines in finite-dimensional irreducible representations, as explained in Appendix C. In this case, the analogues of the parameters l_α determine the sizes of *columns* of the corresponding Young tableaux.

The last type of Wilson line we would like to realize using D3-branes are those in *infinite*-dimensional representations. To realize these line operators, we use the D3-D5-NS5 system studied in [114, 128], where one considers the *same* number of D3, D5 and NS5-branes, in a configuration whereby each D3-brane ends on a single D5-brane and a single NS5-brane. This ensures that the *s*-rule of Hanany and Witten [129] is satisfied.

The D3-D5-NS5 configuration is depicted in Figure 3.2. The support of each brane can be understood using the following table:

	Σ		$\mathbb{R}^2_\epsilon$		C		$N\Sigma \subset T^*\Sigma$		$\mathbb{R}^2_{-\epsilon}$	
	0	1	2	3	4	5	6	7	8	9
D5	×	×	×	×	×	×	A			
D3	×		×	×			×			
NS5	×		×	×			B	×	×	×

$$(3.2.107)$$

where the support of the D3-brane is on a finite-sized interval along the x^6 direction. To elucidate this, we have indicated the positions of the D5- and NS5-branes along the x^6 direction as A and B, respectively.

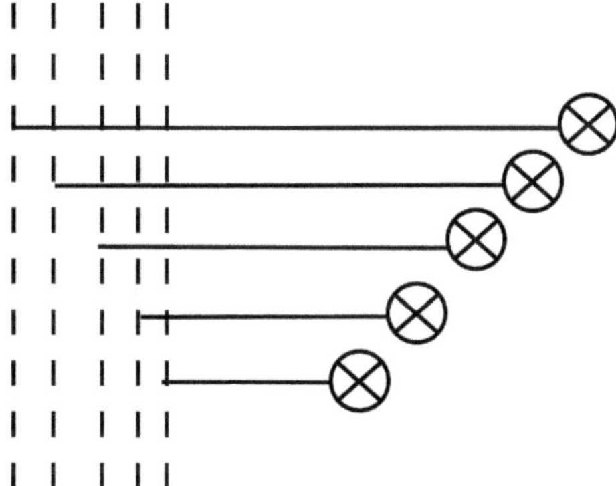

Figure 3.2. The configuration of D3-branes ending on D5-branes and NS5-branes necessary to realize Wilson lines in infinite-dimensional representations. Here, D5-branes are depicted by dashed lines, and NS5-branes are depicted by crossed circles.

One way to understand the line operator obtained via this construction is by noting that the D3-brane worldvolume theory can be described by a 3d $\mathcal{N} = 4$ supersymmetric gauge theory supported along the x^0, x^2 and x^3 directions. At low energy, the theory can be described in terms of a 3d sigma model whose target space is the moduli space of vacua of the theory, and in a B-type Ω-background, this localizes to topological quantum mechanics on this moduli space [65].

To be precise, the D3-D5-NS5 system is described by a quiver gauge theory known as $T[U(N)]$. The supersymmetric vacua of this theory form a space that can be described in terms of the nilpotent cone, $\mathcal{N}$, of $G_\mathbb{C} = GL(N, \mathbb{C})$, which admits the Springer resolution (achieved by turning on certain FI parameters, as described on page 103 of [130]) in terms of the cotangent space of a flag manifold, $T^*(G_\mathbb{C}/B)$, where B is the Borel subgroup of $GL(N, \mathbb{C})$.

Thus, the aforementioned topological quantum mechanics can be described in terms of an action of the form

$$\int_\mathbb{R} \beta_I \mathrm{d}_A \gamma^I \,, \tag{3.2.108}$$

where the summation convention is imposed on the indices I, where γ^I denote local coordinates on the flag manifold $G_\mathbb{C}/B$, where β_I parametrize the fiber directions of $T^*(G_\mathbb{C}/B)$, and where d_A is a covariant derivative coupled to the gauge field. The quantization of this 1d theory to furnish a Wilson line in an infinite-dimensional representation, i.e., a Verma module, is described in Appendix C. The aforementioned FI parameters are related to the Kähler parameters of the moduli space of vacua, and are thereby related to the highest weights of the Verma modules that arise from quantization of the line operators of interest.

Finally, one may consider more elaborate supersymmetric D5-D3-NS5-brane configurations, and these configurations are expected to correspond to 't Hooft and Wilson-'t Hooft lines in 4d Chern–Simons theory, as explained in [22, 128].

3.3 Embedding of 4d CS in 5d $\mathcal{N} = 2$ Gauge Theory with Boundary and Type IIA String Theory

In this section, we study another embedding of 4d CS in supersymmetric gauge theory and string theory, namely 5d $\mathcal{N} = 2$ Supersymmetric Yang–Mills with a boundary, which has an interpretaton as a D4-NS5 brane intersection in type IIA string theory. This approach was developed by Ashwinkumar and Tan in [53], based on earlier work by Ashwinkumar, Tan, and Zhao in [52].

3.3.1 *5d $\mathcal{N} = 2$ supersymmetric Yang–Mills*

We begin with the 5d maximally supersymmetric Yang–Mills theory (MSYM), obtained in the low-energy limit of a stack of D4-branes in type IIA string theory. The classical action may be written as

$$S = -\frac{1}{g_5^2} \int_{\mathcal{M}} d^5x \, \mathrm{Tr}\Big(\frac{1}{2}F_{MN}F^{MN} + D_M\phi_{\widehat{M}}D^M\phi^{\widehat{M}} + \frac{1}{2}[\phi_{\widehat{M}}, \phi_{\widehat{N}}][\phi^{\widehat{M}}, \phi^{\widehat{N}}]$$
$$- i\rho^{A\widehat{A}}(\Gamma^M)_A{}^B D_M\rho_{B\widehat{A}} - \rho^{A\widehat{A}}(\Gamma^{\widehat{M}})_{\widehat{A}}{}^{\widehat{B}}[\phi_{\widehat{M}}, \rho_{A\widehat{B}}]\Big),$$

$$(3.3.1)$$

which is invariant under the supersymmetry transformations

$$\delta A_M = i\zeta^{A\widehat{A}}(\Gamma_M)_A{}^B \rho_{B\widehat{A}}$$

$$\delta\phi^{\widehat{M}} = \zeta^{A\widehat{A}}(\Gamma^{\widehat{M}})_{\widehat{A}}{}^{\widehat{B}} \rho_{A\widehat{B}}$$

$$\delta\rho_{A\widehat{A}} = -i(\Gamma^M)_A{}^B D_M\phi^{\widehat{M}}(\Gamma_{\widehat{M}})_{\widehat{A}}{}^{\widehat{B}}\zeta_{B\widehat{B}} - \frac{1}{2}(\Gamma_{\widehat{M}})_{\widehat{A}}{}^{\widehat{B}}(\Gamma_{\widehat{N}})_{\widehat{B}\widehat{C}}[\phi^{\widehat{M}}, \phi^{\widehat{N}}]\zeta_A{}^{\widehat{C}}$$

$$+ \frac{1}{2}F^{MN}(\Gamma_{MN})_A{}^B\zeta_{B\widehat{A}},$$

$$(3.3.2)$$

where ζ denotes the supersymmetry transformation parameters, A denotes the gauge field, ρ denotes the gaugino field, and ϕ denotes the scalar fields of the theory. Here, $\mathcal{M}$ in the subscript of the integral in (3.3.1) denotes a flat 5d Euclidean manifold. Moreover, $(M, N, \ldots)$ and $(A, B, \ldots)$ are respectively vector and spinor indices for the $SO(5)$ rotation group, denoted $SO_E(5)$, with their hatted counterparts corresponding to the $SO_R(5)$ R-symmetry group, denoted $SO_R(5)$. Moreover, g_5 is the coupling constant of the theory, while $F_{MN} = \partial_M A_N - \partial_N A_M + [A_M, A_N]$ and $D_M = \partial_M + [A_M, \cdot]$ are the field strength and covariant derivative defined with respect to the gauge field, respectively.

The flat five-dimensional Euclidean metric

$$g = \begin{pmatrix} 1 & 0 & 0 & 0 & 0 \\ 0 & 1 & 0 & 0 & 0 \\ 0 & 0 & 1 & 0 & 0 \\ 0 & 0 & 0 & 1 & 0 \\ 0 & 0 & 0 & 0 & 1 \end{pmatrix}, \tag{3.3.3}$$

is used to lower and raise the vector indices for the $SO(5)$ rotation and R-symmetry groups. The gamma matrices obey the Clifford algebra

$$\{\Gamma_M, \Gamma_N\} = 2g_{MN}\mathbb{1}_{4\times4}, \tag{3.3.4}$$

where g_{MN} denote the components of the metric (3.3.3). The $SO(5)$ rotation group spinor indices are lowered and raised using a two index antisymmetric tensor Ω, i.e.,

$$\rho_A = \rho^B \Omega_{BA}, \qquad \rho^A = \Omega^{AB} \rho_B, \tag{3.3.5}$$

where ρ_A and ρ^A correspond to the representation $\mathbf{4}$ and its dual representation $\mathbf{4}^\vee$. Here, Ω^{AB} are the components of

$$\Omega = \begin{pmatrix} 0 & 1 \\ -1 & 0 \end{pmatrix} \otimes \begin{pmatrix} 0 & 1 \\ 1 & 0 \end{pmatrix}, \tag{3.3.6}$$

while Ω_{AB} are the components of its inverse. The $SO(5)$ R-symmetry group spinor indices are also lowered and raised using Ω and its inverse.

In addition, the Lie algebra of the $U(N)$ gauge group is taken to be generated by antihermitian matrices T_a, where $a = 1, \ldots, \dim \mathfrak{u}(N)$, meaning that the invariant quadratic form on this Lie algebra, denoted Tr, is negative-definite. In particular, we shall choose the matrices T_a such that $\mathrm{Tr}(T_a T_b) = -\delta_{ab}$.

The action and supersymmetry transformations of 5d $\mathcal{N} = 2$ supersymmetric Yang–Mills theory furnished above can be obtained from the action and supersymmetry transformations of 10d $\mathcal{N} = 1$ supersymmetric Yang–Mills theory, given in equations (3.2.5) and (3.2.6), through dimensional reduction. This is analogous to the derivation of the action and supersymmetry transformations of 6d $\mathcal{N} = (1,1)$ supersymmetric Yang–Mills theory from 10d, described in Section 3.2.1.

3.3.2   *Partial topological twist*

We shall now describe a partial topological twist of 5d $\mathcal{N} = 2$ supersymmetric Yang–Mills theory, in which we shall eventually embed 4d Chern–Simons theory.

We shall be interested in the case where $\mathcal{M}$ is the flat 5-manifold $\Sigma \times \mathbb{R}_+ \times C$, where Σ and C are 2-manifolds corresponding to the $\{x^1, x^2\}$ and $\{x^4, x^5\}$ directions, respectively, while $\mathbb{R}_+$ is half of the real line, $\mathbb{R}$, that corresponds to the x^3 direction.

The twist is performed along $V = \Sigma \times \mathbb{R}_+$, that is, by redefining its $SO_E(3)$ rotation group to be the diagonal subgroup $SO_E(3)'$ of $SO_E(3) \times SO_R(3)$, where $SO_R(3)$ is the subgroup of the R-symmetry group that rotates $\{\phi_{\hat{1}}, \phi_{\hat{2}}, \phi_{\hat{3}}\}$. In other words, we are considering the rotation subgroup $SO_E(3) \times SO_E(2) \subset SO_E(5)$ and the R-symmetry subgroup $SO_R(3) \times SO_R(2) \subset SO_R(5)$, and defining a new rotation group, $SO_E(3)'$, along $V = \Sigma \times \mathbb{R}_+$.

We shall utilize the following vector and spinor indices of the various symmetry groups:

	$SO_V(3)$	$SO_R(3)$	$SO_C(2)$	$SO_R(2)$
Vector	$\alpha, \beta, \gamma, \ldots$	$\hat{\alpha}, \hat{\beta}, \hat{\gamma}, \ldots$	$m, n, p, \ldots$	$\hat{m}, \hat{n}, \hat{p}, \ldots$
Spinor	$\bar{\alpha}, \bar{\beta}, \bar{\gamma}, \ldots$	$\hat{\bar{\alpha}}, \hat{\bar{\beta}}, \hat{\bar{\gamma}}, \ldots$	$\bar{m}, \bar{n}, \bar{p}, \ldots$	$\hat{\bar{m}}, \hat{\bar{n}}, \hat{\bar{p}}, \ldots$

The two factors in the tensor product (3.3.6) and their respective inverses can be used to raise and lower the $SO(3)$ and $SO(2)$ indices. In particular,

$$\epsilon = \begin{pmatrix} 0 & 1 \\ -1 & 0 \end{pmatrix} \tag{3.3.7}$$

can be used to raise $SO(3)$ spinor indices, e.g.,

$$\lambda^{\bar{\alpha}} = \epsilon^{\bar{\alpha}\bar{\beta}} \lambda_{\bar{\beta}}, \tag{3.3.8}$$

while the inverse of ϵ can be used to lower $SO(3)$ spinor indices. Furthermore,

$$B = \begin{pmatrix} 0 & 1 \\ 1 & 0 \end{pmatrix} \qquad (3.3.9)$$

can be used to lower and raise $SO(2)$ spinor indices. We shall use the fact that ϵ and its inverse act on the Pauli matrices

$$\sigma^1 = \begin{pmatrix} 0 & 1 \\ 1 & 0 \end{pmatrix}, \qquad \sigma^2 = \begin{pmatrix} 0 & -i \\ i & 0 \end{pmatrix}, \qquad \sigma^3 = \begin{pmatrix} 1 & 0 \\ 0 & -1 \end{pmatrix}, \qquad (3.3.10)$$

to give symmetric matrices, i.e., $(\sigma^\alpha)_{\bar\alpha}{}^{\bar\beta}\epsilon_{\bar\beta\bar\gamma} = (\sigma^\alpha)_{\bar\alpha\bar\gamma}$ and $\epsilon^{\bar\alpha\bar\beta}(\sigma^\alpha)_{\bar\beta}{}^{\bar\gamma} = (\sigma^\alpha)^{\bar\alpha\bar\gamma}$, where $(\sigma^\alpha)_{\bar\alpha\bar\gamma} = (\sigma^\alpha)_{\bar\gamma\bar\alpha}$ and $(\sigma^\alpha)^{\bar\alpha\bar\gamma} = (\sigma^\alpha)^{\bar\gamma\bar\alpha}$.

The partial twist amounts to the replacement of the hatted $SO_R(3)$ indices with unhatted $SO_E(3)$ indices. As a result of the twist, the scalar fields $\{\phi_{\hat 1}, \phi_{\hat 2}, \phi_{\hat 3}\}$ transform as the components $\{\phi_1, \phi_2, \phi_3\}$ of a one-form on $\Sigma \times \mathbb{R}_+$. Moreover, the twisting of the fermions which transform as $(\mathbf{2},\mathbf{2})$ under $SO_E(3) \times SO_R(3)$ results in fermions which transform as $\mathbf{1}$ and $\mathbf{3}$ under $SO_E(3)'$, i.e.,

$$\mathbf{2} \otimes \mathbf{2} = \mathbf{1} \oplus \mathbf{3}. \qquad (3.3.11)$$

To observe this explicitly, we expand the spinor fields $\rho_{A\hat A} = \rho_{\bar\alpha m \hat\alpha \hat m}$, after twisting, as

$$\rho_{\bar\alpha m \bar\beta \hat m} = \epsilon_{\bar\alpha\bar\beta}\eta_{m\hat m} + (\sigma^\alpha)_{\bar\alpha\bar\beta}\psi_{\alpha m \hat m}, \qquad (3.3.12)$$

where we have used the antisymmetric tensor $\epsilon_{\bar\alpha\bar\beta}$ and the symmetric tensor $(\sigma^\alpha)_{\bar\alpha\bar\beta}$. The supersymmetry transformation parameters $\zeta_{A\hat A} = \zeta_{\bar\alpha m \hat\alpha \hat m}$ can also be expanded, following the twist, in this manner:

$$\zeta_{\bar\alpha m \bar\beta \hat m} = \epsilon_{\bar\alpha\bar\beta}\zeta_{m\hat m} + (\sigma^\alpha)_{\bar\alpha\bar\beta}\zeta_{\alpha m \hat m}. \qquad (3.3.13)$$

One approach to deriving the partially-twisted supersymmetry transformations is to employ an explicit representation of the gamma matrices. We can use the following representation which involves tensor products of Pauli matrices:

$$\Gamma^1 = \sigma^1 \otimes \sigma^3, \quad \Gamma^2 = \sigma^2 \otimes \sigma^3, \quad \Gamma^3 = \sigma^3 \otimes \sigma^3,$$
$$\Gamma^4 = \mathbb{1} \otimes \sigma^1, \quad \Gamma^5 = \mathbb{1} \otimes \sigma^2. \qquad (3.3.14)$$

Using this representation for the gamma matrices, one can substitute (3.3.12) and (3.3.13) into (3.3.2) to derive the partially-twisted supersymmetry transformations. An alternative approach is to use the constraints on the supersymmetry parameters (involving gamma matrices) imposed by the partial twist and deduce the partially-twisted supersymmetry transformations, in analogy to the derivation of the 6d topological-holomorphic twist reviewed earlier in this chapter.

We would like to study a particular set of supersymmetry transformations generated by a supercharge, $\mathcal{Q}$, that is scalar along V. This supercharge must be associated with a subset of the parameters $\zeta_{\bar{m}\hat{m}}$ in (3.3.13), since they transform as scalars under $SO\,(3)'$. For our purposes, we shall pick only two of these parameters to be non-zero, namely ζ_{11} and ζ_{21}, and take a linear combination of their corresponding supercharges to be $\mathcal{Q}$.

Setting $\zeta_{11} \,=\, \kappa$ and $\zeta_{21} \,=\, \lambda$, where $\kappa, \lambda \,\in\, \mathbb{C}$, the associated supersymmetry transformations are

$$\delta A_\alpha = -\,2i\kappa\psi_{\alpha 22} + 2i\lambda\psi_{\alpha 12}$$

$$\delta\phi_\alpha = 2\kappa\psi_{\alpha 22} + 2\lambda\psi_{\alpha 12}$$

$$\delta A_4 = 2i\kappa\eta_{12} + 2i\lambda\eta_{22}$$

$$\delta A_5 = -\,2\kappa\eta_{12} + 2\lambda\eta_{22}$$

$$\delta\phi_{\hat{4}} = 2\kappa\eta_{21} + 2\lambda\eta_{11}$$

$$\delta\phi_{\hat{5}} = 2i\kappa\eta_{21} + 2i\lambda\eta_{11}$$

$$\delta\psi_{\alpha 11} = \kappa\varepsilon_{\alpha\beta\gamma}\left(\frac{i}{2}F^{\beta\gamma} - \frac{i}{2}\left[\phi^\beta, \phi^\gamma\right] - D^\beta\phi^\gamma\right)$$
$$+ \lambda\left(F_{\alpha 4} - iF_{\alpha 5} + i\left(D_4 - iD_5\right)\phi_\alpha\right)$$

$$\delta\psi_{\alpha 12} = \kappa\left(\left[\phi_\alpha, \phi_{\hat{4}} + i\phi_{\hat{5}}\right] - iD_\alpha\left(\phi_{\hat{4}} + i\phi_{\hat{5}}\right)\right)$$

$$\delta\psi_{\alpha 21} = \kappa\left(-F_{\alpha 4} - iF_{\alpha 5} + i\left(D_4 + iD_5\right)\phi_\alpha\right)$$
$$+ \lambda\varepsilon_{\alpha\beta\hat{m}}\left(\frac{i}{2}F^{\beta\gamma} - \frac{i}{2}\left[\phi^\beta, \phi^\gamma\right] + D^\beta\phi^\gamma\right)$$

$$\delta\psi_{\alpha 22} = \lambda\left(\left[\phi_\alpha, \phi_{\hat{4}} + i\phi_{\hat{5}}\right] + iD_\alpha\left(\phi_{\hat{4}} + i\phi_{\hat{5}}\right)\right)$$

$$\delta\eta_{11} = i\kappa\left(F_{45} + \left[\phi_{\hat{4}}, \phi_{\hat{5}}\right] + D_\beta\phi^\beta\right)$$

$$\delta\eta_{12} = -\,i\lambda\left(D_4 - iD_5\right)\left(\phi_{\hat{4}} + i\phi_{\hat{5}}\right)$$

$$\delta\eta_{21} = -\,i\lambda\left(F_{45} - \left[\phi_{\hat{4}}, \phi_{\hat{5}}\right] + D_\beta\phi^\beta\right)$$

$$\delta\eta_{22} = -\,i\kappa\left(D_4 + iD_5\right)\left(\phi_{\hat{4}} + i\phi_{\hat{5}}\right),$$

$$(3.3.15)$$

where $\epsilon_{\alpha\beta\gamma}$ is the three index Levi–Civita symbol.

We shall now make a convenient redefinition of the scalars $\phi_{\hat 4}, \phi_{\hat 5}$, Fermi fields ψ_α, η and the supersymmetry transformation parameters κ, λ. For the scalars, we define

$$\sigma = \frac{1}{\sqrt{2}}\left(\phi_{\hat 5} - i\phi_{\hat 4}\right), \quad \bar\sigma = \frac{1}{\sqrt{2}}\left(\phi_{\hat 5} + i\phi_{\hat 4}\right), \tag{3.3.16}$$

where σ is a complex scalar field. For the fermions, we define

$$\chi_\alpha = \frac{(1-i)}{2^{5/4}}\psi_{\alpha 11} + \frac{(-1-i)}{2^{5/4}}\psi_{\alpha 21}, \quad \widetilde{\chi}_\alpha = \frac{(-1-i)}{2^{5/4}}\psi_{\alpha 11} + \frac{(1-i)}{2^{5/4}}\psi_{\alpha 21}$$

$$\eta = \frac{(1+i)}{2^{1/4}}\eta_{11} + \frac{(1-i)}{2^{1/4}}\eta_{21}, \quad \widetilde{\eta} = \frac{(-1+i)}{2^{1/4}}\eta_{11} + \frac{(-1-i)}{2^{1/4}}\eta_{21}$$

$$\psi_\alpha = \frac{(1+i)}{2^{3/4}}\psi_{\alpha 12} + \frac{(-1+i)}{2^{3/4}}\psi_{\alpha 22}, \quad \widetilde{\psi}_\alpha = \frac{(-1+i)}{2^{3/4}}\psi_{\alpha 12} + \frac{(1+i)}{2^{3/4}}\psi_{\alpha 22}$$

$$\Upsilon = \frac{(1-i)}{2^{3/4}}\eta_{12} + \frac{(1+i)}{2^{3/4}}\eta_{22}, \quad \widetilde{\Upsilon} = \frac{(-1-i)}{2^{3/4}}\eta_{12} + \frac{(-1+i)}{2^{3/4}}\eta_{22}. \tag{3.3.17}$$

Finally, we redefine the supersymmetry transformation parameters as

$$u = \frac{1}{2^{1/4}}[(1+i)\kappa + (1-i)\lambda], \quad v = \frac{1}{2^{1/4}}[(-1+i)\kappa + (-1-i)\lambda]. \tag{3.3.18}$$

As a result, the supersymmetry transformations in (3.3.15) are equivalently expressed as

$$\delta A_\alpha = iu\psi_\alpha + iv\widetilde{\psi}_\alpha$$

$$\delta\phi_\alpha = iv\psi_\alpha - iu\widetilde{\psi}_\alpha$$

$$\delta A_4 = iu\Upsilon + iv\widetilde{\Upsilon}$$

$$\delta A_5 = iv\Upsilon - iu\widetilde{\Upsilon}$$

$$\delta\sigma = 0$$

$$\delta\bar\sigma = iu\eta + iv\widetilde{\eta}$$

$$\delta\chi_\alpha = \frac{1}{2}u\left[F_{\alpha 4} + D_5\phi_\alpha + \frac{1}{2}\varepsilon_{\alpha\beta\gamma}\left(F^{\beta\gamma} - [\phi^\beta, \phi^\gamma]\right)\right]$$

$$+ \frac{1}{2}v\left[F_{\alpha 5} - D_4\phi_\alpha + \varepsilon_{\alpha\beta\gamma}D^\beta\phi^\gamma\right]$$

$$\delta\widetilde{\chi}_\alpha = \frac{1}{2}v\left[F_{\alpha 4} + D_5\phi_\alpha - \frac{1}{2}\varepsilon_{\alpha\beta\gamma}\left(F^{\beta\gamma} - [\phi^\beta, \phi^\gamma]\right)\right]$$

$$- \frac{1}{2}u\left[F_{\alpha 5} - D_4\phi_\alpha - \varepsilon_{\alpha\beta\gamma}D^\beta\phi^\gamma\right]$$

$$\delta\eta = v\left(F_{45} + D_\alpha\phi^\alpha\right) + u[\bar\sigma, \sigma]$$

$$\delta\widetilde{\eta} = -u\left(F_{45} + D_\alpha\phi^\alpha\right) + v[\bar\sigma, \sigma]$$

$$\delta\psi_\alpha = uD_\alpha\sigma + v[\phi_\alpha, \sigma]$$

$$\delta\widetilde{\psi}_\alpha = vD_\alpha\sigma - u[\phi_\alpha, \sigma]$$

$$\delta\Upsilon = uD_4\sigma + vD_5\sigma$$

$$\delta\widetilde{\Upsilon} = vD_4\sigma - uD_5\sigma. \tag{3.3.19}$$

The above redefinitions facilitate a comparison of the partially-twisted supersymmetry transformations with the supersymmetry transformations of a twist of 4d $\mathcal{N} = 4$ supersymmetric Yang–Mills theory utilized by Kapusin and Witten in [54], in the context of deriving the geometric Langlands correspondence. This twist of 4d $\mathcal{N} = 4$ supersymmetric Yang–Mills theory is sometimes known as the geometric Langlands or GL-twist. We can show that the supersymmetry transformations in (3.3.19) are related to those of the GL-twisted theory. To be precise, the supersymmetry transformations in (3.3.19) reduce to those of the GL-twisted theory in 4d via a dimensional reduction.

Let us elucidate this dimensional reduction. Pick $C = \mathbb{R} \times S^1$, where S^1 is along the x^5 direction. We then dimensionally reduce along S^1, by demanding that the fields of the 5d theory are independent of the x^5 direction. Upon making the replacement $A_5 \to \phi_4$, we obtain the following reductions:

$$F_{\mu 5} \to D_\mu\phi_4$$

$$D_5 \to [\phi_4, \cdot]. \tag{3.3.20}$$

For the fermionic fields, we make the identification $\chi_\alpha = (\chi^+)_{\alpha 4}$ and $\widetilde{\chi}_\alpha = (\chi^-)_{\alpha 4}$, where $\chi^\pm$ are 4d self-dual/anti-self-dual tensors that satisfy

$$(\chi^\pm)_{\mu\nu} = \pm\frac{1}{2}\varepsilon_{\mu\nu}{}^{\rho\sigma}(\chi^\pm)_{\rho\sigma}. \tag{3.3.21}$$

To obtain the $\alpha\beta$ components of $\chi^{\pm}$ we use

$$\left(\chi^{+}\right)_{\alpha\beta} = \frac{1}{2}\varepsilon_{\alpha\beta}{}^{\gamma}\chi_{\gamma}$$
$$\left(\chi^{-}\right)_{\alpha\beta} = -\frac{1}{2}\varepsilon_{\alpha\beta}{}^{\gamma}\widetilde{\chi}_{\gamma}, \tag{3.3.22}$$

since this identification implies

$$\left(\chi^{\pm}\right)_{\alpha\beta} = \pm\frac{1}{2}\varepsilon_{\alpha\beta}{}^{\gamma 4}\left(\chi^{\pm}\right)_{\gamma 4}, \tag{3.3.23}$$

in agreement with (3.3.21). Finally, we identify Υ and $\widetilde{\Upsilon}$ as ψ_4 and $\widetilde{\psi}_4$, respectively.

The supersymmetry transformations (3.3.19) then reduce to

$$\delta A_{\mu} = iu\psi_{\mu} + iv\widetilde{\psi}_{\mu}$$
$$\delta\phi_{\mu} = iv\psi_{\mu} - iu\widetilde{\psi}_{\mu}$$
$$\delta\sigma = 0$$
$$\delta\bar{\sigma} = iu\eta + iv\widetilde{\eta}$$
$$\delta\left(\chi^{+}\right)_{\mu\nu} = u\left(F_{\mu\nu} - [\phi_{\mu},\phi_{\nu}]\right)^{+} + v\left(D_{\mu}\phi_{\nu}\right)^{+}$$
$$\delta\left(\chi^{-}\right)_{\mu\nu} = v\left(F_{\mu\nu} - [\phi_{\mu},\phi_{\nu}]\right)^{-} - u\left(D_{\mu}\phi_{\nu}\right)^{-} \tag{3.3.24}$$
$$\delta\eta = v\left(D_{\mu}\phi^{\mu}\right) + u\left[\bar{\sigma},\sigma\right]$$
$$\delta\widetilde{\eta} = -u\left(D_{\mu}\phi^{\mu}\right) + v\left[\bar{\sigma},\sigma\right]$$
$$\delta\psi_{\mu} = uD_{\mu}\sigma + v\left[\phi_{\mu},\sigma\right]$$
$$\delta\widetilde{\psi}_{\mu} = vD_{\mu}\sigma - u\left[\phi_{\mu},\sigma\right].$$

These are the supersymmetry transformations of GL-twisted 4d $\mathcal{N} = 4$ supersymmetric Yang–Mills theory that was studied by Kapustin and Witten in [54]. Therefore, we observe a relationship between the partially-twisted 5d $\mathcal{N} = 2$ theory and the GL-twisted 4d $\mathcal{N} = 4$ SYM theory. In other words, the partial twist that has been constructed is a five-dimensional analogue of the four-dimensional GL-twist.[8]

---

[8]This partial twist has also been discussed conceptually in [58].

3.3.2.1 *Partially-twisted Q-invariant Action*

We would like to implement the partial twist at the level of the action of the 5d $\mathcal{N} = 2$ supersymmetric Yang–Mills theory on $\Sigma \times \mathbb{R}_+ \times C$. As we shall observe, the action is $\mathcal{Q}$-exact (where $\mathcal{Q}$ is some complex linear combination of the scalar supercharges we have considered) up to a metric-independent $\mathcal{Q}$-invariant term which turns out to take the form of the 4d Chern–Simons action. In order to describe the partially-twisted action explicitly, some observations are in order.

From (3.3.19), note that the supersymmetry variation can be expressed as

$$\delta = u\delta_L + v\delta_R. \tag{3.3.25}$$

Equivalently, we may write the corresponding supercharge, $\mathcal{Q}$, in terms of supercharges $\mathcal{Q}_L$ and $\mathcal{Q}_R$, i.e.,

$$\mathcal{Q} = u\mathcal{Q}_L + v\mathcal{Q}_R. \tag{3.3.26}$$

The supercharge, $\mathcal{Q}$ acts on any field Φ as

$$[\mathcal{Q}, \Phi\} = \delta\Phi, \tag{3.3.27}$$

where $[.,.\}$ indicates a commutator or anti-commutator depending on whether the field Φ is bosonic or fermionic. Here, we have included the labels L and R to remind ourselves of the corresponding left- and right-handed supersymmetries in the 4d case which were denoted by ℓ and r in [54]. However, the reader should keep in mind that there is no notion of chirality in 5d. In addition, it is convenient to rescale the supersymmetry transformations such that they depend only on the ratio $t = v/u$, i.e., we obtain

$$\delta_t = \delta_L + t\delta_R \tag{3.3.28}$$

by dividing by u on both sides of (3.3.25), and defining $\frac{\delta}{u} = \delta_t$.

We would like to express the partially-twisted action in terms of the $\mathcal{Q}$-variation of another functional, that is, we want to express the action in $\mathcal{Q}$-exact form up to an additional $\mathcal{Q}$-invariant term. As we shall see, expressing the action of partially-twisted 5d $\mathcal{N} = 2$ SYM in this form facilitates the localization of the path integral of this theory to the path integral of 4d Chern–Simons theory. To this end, we require that the transformation δ_t is nilpotent off-shell (up to gauge transformations). In order to achieve this, we need to introduce some auxiliary fields. We shall first define two auxiliary

Lie algebra-valued 1-forms H and $\widetilde{H}$, that modify the supersymmetry transformations in (3.3.19) of χ and $\widetilde{\chi}$. The transformations of H and $\widetilde{H}$ may also be defined such that the nilpotency of $\mathcal{Q}$ is satisfied up to gauge transformations. Collectively, these supersymmetry transformations are

$$\delta_t \chi_\alpha = H_\alpha$$

$$\delta_t \widetilde{\chi}_\alpha = \widetilde{H}_\alpha$$

$$\delta_t H_\alpha = -i\left(1 + t^2\right)[\sigma, \chi_\alpha]$$

$$\delta_t \widetilde{H}_\alpha = -i\left(1 + t^2\right)[\sigma, \widetilde{\chi}_\alpha], \tag{3.3.29}$$

where $\alpha = 1, 2, 3$ as before. The factor of $(1 + t^2)$ in the supersymmetry variations of H_α and $\widetilde{H}_\alpha$ ensures that all fields satisfy the same supersymmetry algebra, as we shall see below. We can also define a Lie algebra-valued 0-form P, and use it to modify the supersymmetry transformations of η and $\widetilde{\eta}$ in (3.3.19). The supersymmetry transformation for P can also be defined in such a way that the nilpotency of $\mathcal{Q}$ is satisfied (up to gauge transformations). These supersymmetry transformations for η, $\widetilde{\eta}$ and P are given explicitly by

$$\delta\eta = tP + [\bar{\sigma}, \sigma]$$

$$\delta\widetilde{\eta} = -P + t[\bar{\sigma}, \sigma] \tag{3.3.30}$$

$$\delta P = -it[\sigma, \eta] + i[\sigma, \widetilde{\eta}].$$

The original supersymmetry transformations for the fields χ_α, $\widetilde{\chi}_\alpha$, η and $\widetilde{\eta}$ can be retrieved on-shell since the action we construct shall have equations of motion that ensure this. In particular, we will construct an action whose equations of motion include $H = \mathcal{V}$ and $\widetilde{H} = t\widetilde{\mathcal{V}}$, where $\mathcal{V}$ and $\widetilde{\mathcal{V}}$ are functions of t defined as

$$\mathcal{V}_\alpha(t) = \frac{1}{2}\left(\left[F_{\alpha 4} + D_5\phi_\alpha + \frac{1}{2}\varepsilon_{\alpha\beta\gamma}\left(F^{\beta\gamma} - [\phi^\beta, \phi^\gamma]\right)\right] \right.$$

$$\left. + t\left[F_{\alpha 5} - D_4\phi_\alpha + \varepsilon_{\alpha\beta\gamma}D^\beta\phi^\gamma\right] \right)$$

$$\widetilde{\mathcal{V}}_\alpha(t) = \frac{1}{2}\left(\left[F_{\alpha 4} + D_5\phi_\alpha - \frac{1}{2}\varepsilon_{\alpha\beta\gamma}\left(F^{\beta\gamma} - [\phi^\beta, \phi^\gamma]\right)\right] \right. \tag{3.3.31}$$

$$\left. - t^{-1}\left[F_{\alpha 5} - D_4\phi_\alpha - \varepsilon_{\alpha\beta\gamma}D^\beta\phi^\gamma\right] \right),$$

and by substituting these equations into (3.3.29) we retrieve the original supersymmetry transformations for χ_α and $\widetilde{\chi}_\alpha$ given in (3.3.19). Similarly, the action we shall construct furnishes the equation of motion $P = F_{45} + D_\alpha \phi^\alpha$, ensuring that we have on-shell agreement between the transformations of η and $\tilde{\eta}$ in (3.3.30) and (3.3.19).

Having introduced the auxiliary fields H_α, $\tilde{H}_\alpha$, and P, the following off-shell supersymmetry algebra holds for any field, Φ, of the partially-twisted theory:

$$\delta_t^2 \Phi = -i(1 + t^2)\mathcal{L}_\sigma(\Phi), \tag{3.3.32}$$

where $\mathcal{L}_\sigma(\Phi)$ is the variation in Φ due to a gauge transformation generated by σ, to first order. Moreover, we can use the relationship between δ_t, δ_L and δ_R given in (3.3.28) to show that this supersymmetry algebra is equivalent to

$$\delta_L^2 \Phi = \delta_R^2 \Phi = -i\mathcal{L}_\sigma(\Phi),$$
$$\{\delta_L, \delta_R\}\Phi = 0. \tag{3.3.33}$$

We shall now proceed to express the partially-twisted 5d $\mathcal{N} = 2$ SYM action in terms of the sum of a $\mathcal{Q}$-exact term and a $\mathcal{Q}$-invariant term. The $\mathcal{Q}$-exact term can be understood as the $\mathcal{Q}$-variation of a gauge invariant fermionic expression, $\widetilde{V}$. We shall express $\widetilde{V}$ as the sum of two terms, i.e., $\widetilde{V} = \widetilde{V}_1 + \widetilde{V}_2$. The first term is

$$\widetilde{V}_1 = \frac{2}{g_5{}^2} \int_\mathcal{M} d^5x \, \frac{4}{1 + t^2} \mathrm{Tr}\left(\chi_\alpha \left(\frac{1}{2}H^\alpha - \mathcal{V}^\alpha \right) + \tilde{\chi}_\alpha \left(\frac{1}{2}\tilde{H}^\alpha - t\tilde{\mathcal{V}}^\alpha \right) \right). \tag{3.3.34}$$

The corresponding contribution to the action takes the form

$$\begin{aligned} S_1 &= \delta_t \widetilde{V}_1 \\ &= \frac{1}{g_5{}^2} \int_\mathcal{M} d^5x \, \mathrm{Tr}\left(\frac{4}{1 + t^2} \left(\frac{1}{2}H^\alpha H_\alpha - H_\alpha \mathcal{V}^\alpha \right) \right. \\ &\qquad\qquad\qquad \left. + \frac{4}{1 + t^2} \left(\frac{1}{2}\tilde{H}^\alpha \tilde{H}_\alpha - t\tilde{H}_\alpha \tilde{\mathcal{V}}^\alpha \right) \right) + \cdots, \end{aligned} \tag{3.3.35}$$

where the ellipsis indicates terms involving fermionic fields. Integrating out the auxiliary fields H^α and $\tilde{H}^\alpha$, the resulting action, including terms involving fermionic fields, is

$$S_1 = \frac{1}{g_5{}^2} \int_{\mathcal{M}} d^5x \ \mathrm{Tr} \left(\frac{-4}{1+t^2} \left(\mathcal{V}^\alpha \mathcal{V}_\alpha + t^2 \widetilde{\mathcal{V}}^\alpha \widetilde{\mathcal{V}}_\alpha \right) + 4i\chi_\alpha[\sigma, \chi^\alpha] + 4i\widetilde{\chi}_\alpha[\sigma, \widetilde{\chi}^\alpha] \right.$$

$$+ 4\chi^\alpha \left[\ iD_\alpha \Upsilon - iD_4\psi_\alpha - iD_5\widetilde{\psi}_\alpha - i[\widetilde{\Upsilon}, \phi_\alpha] \right.$$

$$\left. + \frac{i}{2}\varepsilon_{\alpha\beta\gamma} \left(D^\beta \psi^\gamma + [\widetilde{\psi}^\beta, \phi^\gamma] - D^\gamma \psi^\beta - [\widetilde{\psi}^\gamma, \phi^\beta] \right) \right]$$

$$+ 4\widetilde{\chi}^\alpha \left[\ iD_\alpha \widetilde{\Upsilon} - iD_4\widetilde{\psi}_\alpha + iD_5\psi_\alpha + i[\Upsilon, \phi_\alpha] \right.$$

$$\left. \left. - \frac{i}{2}\varepsilon_{\alpha\beta\gamma} \left(D^\beta \widetilde{\psi}^\gamma - [\psi^\beta, \phi^\gamma] - D^\gamma \widetilde{\psi}^\beta + [\psi^\gamma, \phi^\beta] \right) \right] \right). \tag{3.3.36}$$

The second term in $\widetilde{V}$ can be written succinctly as $\widetilde{V}_2 = -\frac{1}{2t}(\delta_L - t\delta_R)\widetilde{V}_2'$ where $\widetilde{V}_2'$ is

$$\widetilde{V}_2' = \frac{2}{g_5{}^2} \int_{\mathcal{M}} d^5x \ \mathrm{Tr} \left(-\frac{1}{2}\eta\widetilde{\eta} - i\bar{\sigma} \left(F_{45} + D_\alpha \phi^\alpha \right) \right). \tag{3.3.37}$$

The corresponding contribution to the action is

$$S_2 = -\frac{1}{2t}(\delta_L + t\delta_R)(\delta_L - t\delta_R)\widetilde{V}_2' = \delta_L \delta_R \widetilde{V}_2'$$

$$= \frac{2}{g_5{}^2} \int_{\mathcal{M}} d^5x \ \mathrm{Tr} \left(\frac{1}{2}P^2 - P\left(F_{45} + D_\alpha \phi^\alpha \right) - D_M \bar{\sigma} D^M \sigma + \frac{1}{2}[\bar{\sigma}, \sigma]^2 \right.$$

$$- [\phi_\alpha, \sigma][\phi^\alpha, \bar{\sigma}] + \partial_\alpha(\bar{\sigma}D^\alpha\sigma) + i\widetilde{\eta}D_\alpha\widetilde{\psi}^\alpha + i\eta D_\alpha\psi^\alpha$$

$$+ i\widetilde{\eta} \left(D_4\widetilde{\Upsilon} + D_5\Upsilon \right) + i\eta \left(D_4\Upsilon - D_5\widetilde{\Upsilon} \right)$$

$$- \frac{i}{2}[\sigma, \widetilde{\eta}]\widetilde{\eta} - \frac{i}{2}[\sigma, \eta]\eta - i\widetilde{\eta}[\psi_\alpha, \phi^\alpha] + i\eta[\widetilde{\psi}_\alpha, \phi^\alpha]$$

$$\left. + i[\bar{\sigma}, \psi_\alpha]\psi^\alpha + i[\bar{\sigma}, \widetilde{\psi}_\alpha]\widetilde{\psi}^\alpha + i[\bar{\sigma}, \Upsilon]\Upsilon + i[\bar{\sigma}, \widetilde{\Upsilon}]\widetilde{\Upsilon} \right). \tag{3.3.38}$$

We can integrate out the auxiliary field P to obtain

$$S_2 = \frac{2}{g_5{}^2} \int_{\mathcal{M}} d^5x \ \mathrm{Tr} \left(-\frac{1}{2} \left(F_{45} + D_\alpha \phi^\alpha \right)^2 - D_M \bar{\sigma} D^M \sigma \right.$$

$$+ \frac{1}{2}[\bar{\sigma}, \sigma]^2 - [\phi_\alpha, \sigma][\phi^\alpha, \bar{\sigma}] + \partial_\alpha(\bar{\sigma}D^\alpha\sigma)$$

$$+ i\widetilde{\eta} D_\alpha \widetilde{\psi}^\alpha + i\eta D_\alpha \psi^\alpha + i\widetilde{\eta}\left(D_4\widetilde{\Upsilon} + D_5\Upsilon\right) + i\eta\left(D_4\Upsilon - D_5\widetilde{\Upsilon}\right)$$

$$- \frac{i}{2}[\sigma,\widetilde{\eta}]\widetilde{\eta} - \frac{i}{2}[\sigma,\eta]\eta - i\widetilde{\eta}[\psi_\alpha,\phi^\alpha] + i\eta[\widetilde{\psi}_\alpha,\phi^\alpha] + i[\bar{\sigma},\psi_\alpha]\psi^\alpha$$

$$+ i[\bar{\sigma},\widetilde{\psi}_\alpha]\widetilde{\psi}^\alpha + i[\bar{\sigma},\Upsilon]\Upsilon + i[\bar{\sigma},\widetilde{\Upsilon}]\widetilde{\Upsilon}\Big). \tag{3.3.39}$$

The complete $\mathcal{Q}$-exact action, denoted S, is the sum of the expressions given in (3.3.36) and (3.3.39):

$$S = S_1 + S_2$$

$$= \frac{1}{g_5{}^2}\int_\mathcal{M} d^5x \ \mathrm{Tr}\left(\frac{-4}{1+t^2}\left(\mathcal{V}^\alpha\mathcal{V}_\alpha + t^2\widetilde{\mathcal{V}}^\alpha\widetilde{\mathcal{V}}_\alpha\right) + 4i\chi_\alpha[\sigma,\chi^\alpha] + 4i\widetilde{\chi}_\alpha[\sigma,\widetilde{\chi}]\right.$$

$$+ 4\chi^\alpha\left[iD_\alpha\Upsilon - iD_4\psi_\alpha - iD_5\widetilde{\psi}_\alpha - i[\widetilde{\Upsilon},\phi_\alpha]\right.$$

$$\left.+ \frac{i}{2}\varepsilon_{\alpha\beta\gamma}\left(D^\beta\psi^\gamma + [\widetilde{\psi}^\beta,\phi^\gamma] - D^\gamma\psi^\beta - [\widetilde{\psi}^\gamma,\phi^\beta]\right)\right]$$

$$+ 4\widetilde{\chi}^\alpha\left[iD_\alpha\widetilde{\Upsilon} - iD_4\widetilde{\psi}_\alpha + iD_5\psi_\alpha + i[\Upsilon,\phi_\alpha]\right.$$

$$\left.- \frac{i}{2}\varepsilon_{\alpha\beta\gamma}\left(D^\beta\widetilde{\psi}^\gamma - [\psi^\beta,\phi^\gamma] - D^\gamma\widetilde{\psi}^\beta + [\psi^\gamma,\phi^\beta]\right)\right]$$

$$- (F_{45} + D_\alpha\phi^\alpha)^2 - 2D_M\bar{\sigma}D^M\sigma + [\bar{\sigma},\sigma][\bar{\sigma},\sigma]$$

$$- 2[\phi_\alpha,\sigma][\phi^\alpha,\bar{\sigma}] + 2\partial_\alpha(\bar{\sigma}D^\alpha\sigma)$$

$$+ 2i\widetilde{\eta}D_\alpha\widetilde{\psi}^\alpha + 2i\eta D_\alpha\psi^\alpha + 2i\widetilde{\eta}\left(D_4\widetilde{\Upsilon} + D_5\Upsilon\right) + 2i\eta\left(D_4\Upsilon - D_5\widetilde{\Upsilon}\right)$$

$$- i[\sigma,\widetilde{\eta}]\widetilde{\eta} - i[\sigma,\eta]\eta - 2i\widetilde{\eta}[\psi_\alpha,\phi^\alpha] + 2i\eta[\widetilde{\psi}_\alpha,\phi^\alpha] + 2i[\bar{\sigma},\psi_\alpha]\psi^\alpha$$

$$+ 2i[\bar{\sigma},\widetilde{\psi}_\alpha]\widetilde{\psi}^\alpha + 2i[\bar{\sigma},\Upsilon]\Upsilon + 2i[\bar{\sigma},\widetilde{\Upsilon}]\widetilde{\Upsilon}\Big). \tag{3.3.40}$$

We shall now utilize the identity

$$\frac{1}{g_5{}^2}\int_\mathcal{M} d^5x \ \mathrm{Tr}\left(\frac{-4t^{-1}}{t+t^{-1}}\left(\mathcal{V}^\alpha\mathcal{V}_\alpha + t^2\widetilde{\mathcal{V}}^\alpha\widetilde{\mathcal{V}}_\alpha\right) - (F_{45} + D_\alpha\phi^\alpha)^2\right)$$

$$= -\frac{1}{g_5{}^2}\int_\mathcal{M} d^5x \ \mathrm{Tr}\left(F_{\alpha m}F^{\alpha m} + F_{45}F^{45}\right.$$

$$+ \frac{1}{2}F_{\alpha\beta}F^{\alpha\beta} + D_m\phi_\alpha D^m\phi^\alpha + D_\alpha\phi_\beta D^\alpha\phi^\beta$$

$$+\frac{1}{2}[\phi_\alpha,\phi_\beta][\phi^\alpha,\phi^\beta] + \partial_\alpha\left(\phi^\alpha D_\beta\phi^\beta\right) - \partial_\gamma\left(\phi_\delta D^\delta\phi^\gamma\right) + 2\partial_\alpha(F_{45}\phi^\alpha)$$

$$-4\left(\frac{t-t^{-1}}{t+t^{-1}}\right)\left(\frac{1}{2}\varepsilon^{\alpha\beta\gamma}\right)\left(\frac{1}{2}F_{\alpha4}F_{\beta\gamma}+\frac{1}{2}\partial_\alpha\left(\phi_\beta D_4\phi_\gamma\right)+\partial_\alpha\left(F_{\beta5}\phi_\gamma\right)\right)$$

$$+\left(\frac{8}{t+t^{-1}}\right)\left(\frac{1}{2}\varepsilon^{\alpha\beta\gamma}\right)\left(\frac{1}{2}F_{\alpha5}F_{\beta\gamma}+\frac{1}{2}\partial_\alpha\left(\phi_\beta D_5\phi_\gamma\right)+\partial_\alpha\left(F_{\beta4}\phi_\gamma\right)\right)\bigg),$$

$$(3.3.41)$$

to rewrite the $\mathcal{Q}$-exact expression (3.3.40) in terms of a t-independent part and a t-dependent part. Crucially, apart from total derivative terms and t-dependent terms, the terms involving bosonic fields in (3.3.40) correspond to the standard kinetic and potential terms of 5d $\mathcal{N}=2$ supersymmetric Yang–Mills, partially-twisted along $\Sigma\times\mathbb{R}_+$.

The t-dependent term of the $\mathcal{Q}$-exact action takes the form

$$S_t = \frac{1}{g_5^2}\int_{\mathcal{M}} d^5x\,\varepsilon^{\alpha\beta\gamma}\mathrm{Tr}\left(2\left(\frac{t-t^{-1}}{t+t^{-1}}\right)\right.$$

$$\times\left(\frac{1}{2}F_{\alpha4}F_{\beta\gamma}+\frac{1}{2}\partial_\alpha\left(\phi_\beta D_4\phi_\gamma\right)+\partial_\alpha\left(F_{\beta5}\phi_\gamma\right)\right)$$

$$\left.-\left(\frac{4}{t+t^{-1}}\right)\left(\frac{1}{2}F_{\alpha5}F_{\beta\gamma}+\frac{1}{2}\partial_\alpha\left(\phi_\beta D_5\phi_\gamma\right)+\partial_\alpha\left(F_{\beta4}\phi_\gamma\right)\right)\right). \quad (3.3.42)$$

Such a t-dependent term should not arise from the physical 5d action, so the partial twist of the latter should involve an additional term that cancels (3.3.42). This term will be denoted S_3, where

$$S_3 = -S_t. \qquad (3.3.43)$$

Hence, the partially-twisted physical action should involve the sum $S_1 + S_2 + S_3$. However, this contribution to the physical action is not yet $\mathcal{Q}$-invariant, which leads us to discuss boundary conditions that will ensure $\mathcal{Q}$-invariance.

3.3.3 *Boundary conditions from NS5-branes*

We now specify boundary conditions at the origin of $\mathbb{R}_+$, i.e., $x^3 = 0$, such that we have a system that can be understood as the worldvolume theory of a stack of D4-branes ending on a (deformed) NS5-brane in type IIA string theory.[9] Specifically, we consider the following configuration in flat Euclidean space:

	1	2	3	4	5	6	7	8	9	10
D4	×	×	×	×	×					
$\widetilde{\textbf{NS5}}$	×	×		×	×	×	×			

where the column groups are Σ (1–2), $\mathbb{R}$ (3), C (4–5), and $NV' \subset T^*V'$ (6–7), with V' spanning Σ, $\mathbb{R}$, C.

This configuration is also depicted in Figure 3.3. The scalar fields $\{\phi_{\hat{1}}, \phi_{\hat{2}}, \phi_{\hat{3}}, \phi_{\hat{4}}, \phi_{\hat{5}}\}$ of the 5d theory parametrize the $\{6, 7, 8, 9, 10\}$ directions, respectively. The partial twist arises in this configuration because $V = (\Sigma \times \mathbb{R}_+) \subset V' = (\Sigma \times \mathbb{R})$, where V' is the zero section of the cotangent bundle T^*V', and 'coordinates' normal to V' in T^*V' ought to be components of one-forms, as obtained via twisting (this string-theoretical interpretation of twisting originated in [50]).

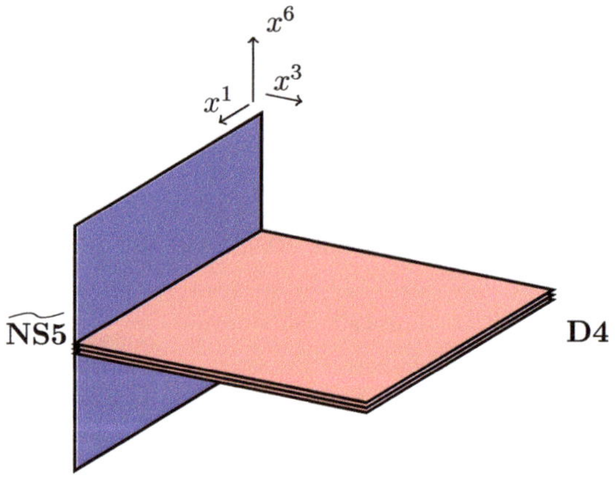

Figure 3.3. A stack of D4-branes ending on a deformed NS5-brane, denoted as $\widetilde{\text{NS5}}$. We shall eventually explain how the D4-brane worldvolume theory localizes to 4d Chern–Simons theory at the D4 $-\ \widetilde{\text{NS5}}$ intersection.

---

[9]By "deformed", we mean that the boundary condition will not be a pure Neumann condition as expected from the NS5-brane, but a generalization thereof. The reader will benefit from understanding analogous boundary conditions of 4d $N = 4$ supersymmetric Yang-Mills theory that arise from deformed NS5-branes. These boundary conditions are studied in detail in the work of Witten [55].

This brane configuration implies that the fields ϕ_3, $\sigma = \frac{1}{\sqrt{2}}\left(\phi_{\widehat{5}} - i\phi_{\widehat{4}}\right)$, and $\bar{\sigma} = \frac{1}{\sqrt{2}}\left(\phi_{\widehat{5}} + i\phi_{\widehat{4}}\right)$ (that parametrize the $\{8,9,10\}$ directions) obey Dirichlet boundary conditions that set them to zero at the boundary. In addition, the sets of fields $\{\phi_1,\phi_2\}$ and $\{A_1,A_2,A_4,A_5\}$ obey generalized Neumann boundary conditions that can be deduced from boundary couplings given by

$$
\begin{aligned}
S_{\partial\mathcal{M}} = \frac{1}{g_5{}^2} \int_{\partial\mathcal{M}} d^4x \, \mathrm{Tr} &\left((t + t^{-1}) \left(\frac{1}{2}\varepsilon^{\widetilde{\alpha}\widetilde{\beta}} D_5 \phi_{\widetilde{\alpha}}\phi_{\widetilde{\beta}} \right) \right. \\
&\left. + \left(\frac{t + t^{-1}}{t - t^{-1}} \right) \varepsilon^{ijk} \left(A_i \partial_j A_k + \frac{2}{3} A_i A_j A_k \right) \right),
\end{aligned}
\tag{3.3.44}
$$

by requiring that the equations of motion do not have boundary corrections when derived by performing a general variation of the fields. Here, we have defined the indices $i,j,k = 1,2,4$ and $\widetilde{\alpha}, \widetilde{\beta} = 1,2$, and the Levi–Civita symbols are defined such that $\varepsilon^{124} = 1$ and $\varepsilon^{12} = 1$.

One can also derive boundary conditions on the fermionic fields by demanding that the complete set of boundary conditions is supersymmetry invariant. It is also important for our application that the boundary conditions defined above imply that

$$
\delta_t(A_i + w\phi_i)|_{x_3=0} = 0
$$
$$
\delta_t(A_4 + wA_5)|_{x_3=0} = 0,
\tag{3.3.45}
$$

for $i = 1,2$ and $w = \frac{t-t^{-1}}{2}$. Finally, the boundary conditions restrict the complex parameter t such that $|t| = 1$.

When performing the dimensional reduction of the 5d partially-twisted theory to the 4d GL-twisted theory as described in the previous subsection, the boundary conditions and interactions defined above will reduce to the GL-twist of the deformed NS5 boundary conditions and interactions for 4d $\mathcal{N} = 4$ supersymmetric Yang–Mills theory that were studied in [55, 127]. The aforementioned constraint $|t| = 1$ is satisfied in the 4d theory since it obeys $t^2 = \frac{\bar{\tau}}{\tau}$, where τ is the complex coupling of this theory.

With the boundary conditions and boundary interactions at $x_3 = 0$ specified for the partially-twisted theory, we find that the total derivative terms

$$
\frac{2}{g_5^2} \int_{\mathcal{M}} d^5x \, \mathrm{Tr} \, \partial_\alpha(\bar{\sigma} D^\alpha \sigma)
\tag{3.3.46}
$$

in (3.3.40) and

$$
-\frac{1}{g_5^2} \int_{\mathcal{M}} d^5x \, \mathrm{Tr} \left(\partial_\alpha \left(\phi^\alpha D_\beta \phi^\beta \right) - \partial_\gamma \left(\phi_\delta D^\delta \phi^\gamma \right) + 2\partial_\alpha(F_{45}\phi^\alpha) \right)
\tag{3.3.47}
$$

in (3.3.41) vanish via Stokes' theorem and the boundary conditions $\bar{\sigma} = 0$ and $\phi_3 = 0$.

We also ought to specify the boundary conditions at $x^3 = \infty$ on the half-line. These boundary conditions correspond to a critical point of the 4d Chern–Simons action that satisfies

$$D_{\widetilde{\alpha}}\phi^{\widetilde{\alpha}} + F_{45} = 0, \tag{3.3.48}$$

together with Neumann boundary conditions on ϕ_3, σ and $\bar{\sigma}$. Note that (3.3.48) and the Neumann boundary condition for σ also set to zero the total derivative terms (3.3.46) and (3.3.47). This ensures that the contribution of the boundary theory at $x^3 = \infty$ is just an overall constant in the path integral, which can be absorbed into the measure.

3.3.4   *5d topological-holomorphic theory*

In this subsection, we shall begin to see how the partially-twisted 5d $\mathcal{N} = 2$ SYM action is related to 4d Chern–Simons theory. Recall that along the boundary at $x_3 = 0$, we have t-dependent boundary interactions given by (3.3.44). We shall show that when these boundary interactions are combined with the t-dependent action S_3 in (3.3.43), a term proportional to a 4d Chern–Simons action is obtained.

To arrive at this observation, we first need to rewrite the coordinates x^4, x^5 on C in terms of complex ones. We shall define the complex coordinates

$$z_w = 2\frac{(wx^4 - x^5)}{w - \overline{w}}$$

$$\bar{z}_w = 2\frac{(\overline{w}x^4 - x^5)}{\overline{w} - w} \tag{3.3.49}$$

(recall that w was defined previously to be $w = \frac{t - t^{-1}}{2}$). The partial derivatives corresponding to these complex coordinates are

$$\partial_{z_w} = \frac{1}{2}(\partial_4 + \overline{w}\partial_5)$$

$$\partial_{\bar{z}_w} = \frac{1}{2}(\partial_4 + w\partial_5). \tag{3.3.50}$$

We shall also introduce the complexified gauge fields

$$\mathcal{A}_{w\widetilde{\alpha}} = A_{\widetilde{\alpha}} + w\phi_{\widetilde{\alpha}} \tag{3.3.51}$$

(for $\widetilde{\alpha} = 1, 2$) and

$$\mathcal{A}_{w\bar{z}_w} = \frac{1}{2}\left(A_4 + wA_5\right) \tag{3.3.52}$$

that are $\mathcal{Q}$-invariant along the boundary at $x_3 = 0$.

The sum of the boundary interactions and the t-dependent action S_3 is

$$S_3 + S_{\partial\mathcal{M}}$$

$$= -\frac{1}{g_5{}^2}\int_{\mathcal{M}} d^5x\, \varepsilon^{\alpha\beta\gamma}$$

$$\times \mathrm{Tr}\left(2\left(\frac{t - t^{-1}}{t + t^{-1}}\right)\left(\frac{1}{2}F_{\alpha 4}F_{\beta\gamma} + \frac{1}{2}\partial_\alpha\left(\phi_\beta D_4\phi_\gamma\right) + \partial_\alpha\left(F_{\beta 5}\phi_\gamma\right)\right)\right.$$

$$\left. -\left(\frac{4}{t + t^{-1}}\right)\left(\frac{1}{2}F_{\alpha 5}F_{\beta\gamma} + \frac{1}{2}\partial_\alpha\left(\phi_\beta D_5\phi_\gamma\right) + \partial_\alpha\left(F_{\beta 4}\phi_\gamma\right)\right)\right)$$

$$+ \frac{1}{g_5{}^2}\int_{\partial\mathcal{M}} d^4x\, \mathrm{Tr}\left((t + t^{-1})\left(\frac{1}{2}\varepsilon^{\widetilde{\alpha}\widetilde{\beta}}D_5\phi_{\widetilde{\alpha}}\phi_{\widetilde{\beta}}\right)\right.$$

$$\left. +\left(\frac{t + t^{-1}}{t - t^{-1}}\right)\varepsilon^{ijk}\left(A_i\partial_j A_k + \frac{2}{3}A_i A_j A_k\right)\right). \tag{3.3.53}$$

It can be shown that

$$\mathrm{Tr}(\varepsilon^{\alpha\beta\gamma 4}F_{\alpha\beta}F_{\gamma 4}) = \mathrm{Tr}\left(\frac{1}{4}\varepsilon^{\mu\nu\rho\sigma}F_{\mu\nu}F_{\rho\sigma}\right) \tag{3.3.54}$$

for $\mu, \nu, \rho, \sigma = 1, 2, 3, 4$ where the Levi–Civita symbol is defined such that $\varepsilon^{1234} = 1$, and analogously

$$\mathrm{Tr}(\varepsilon^{\alpha\beta\gamma 5}F_{\alpha\beta}F_{\gamma 5}) = \mathrm{Tr}\left(\frac{1}{4}\varepsilon^{\widetilde{\mu}\widetilde{\nu}\widetilde{\rho}\widetilde{\sigma}}F_{\widetilde{\mu}\widetilde{\nu}}F_{\widetilde{\rho}\widetilde{\sigma}}\right) \tag{3.3.55}$$

for $\widetilde{\mu}, \widetilde{\nu}, \widetilde{\rho}, \widetilde{\sigma} = 1, 2, 3, 5$ and $\varepsilon^{1235} = 1$. In addition, these expressions can be written as total derivatives, i.e.,

$$\mathrm{Tr}\left(\frac{1}{4}\varepsilon^{\mu\nu\rho\sigma}F_{\mu\nu}F_{\rho\sigma}\right) = \varepsilon^{\mu\nu\rho\sigma}\partial_\mu\mathrm{Tr}\left(A_\nu\partial_\rho A_\sigma + \frac{2}{3}A_\nu A_\rho A_\sigma\right) \tag{3.3.56}$$

and

$$\mathrm{Tr}\left(\frac{1}{4}\varepsilon^{\widetilde{\mu}\widetilde{\nu}\widetilde{\rho}\widetilde{\sigma}}F_{\widetilde{\mu}\widetilde{\nu}}F_{\widetilde{\rho}\widetilde{\sigma}}\right) = \varepsilon^{\widetilde{\mu}\widetilde{\nu}\widetilde{\rho}\widetilde{\sigma}}\partial_{\widetilde{\mu}}\mathrm{Tr}\left(A_{\widetilde{\nu}}\partial_{\widetilde{\rho}}A_{\widetilde{\sigma}} + \frac{2}{3}A_{\widetilde{\nu}}A_{\widetilde{\rho}}A_{\widetilde{\sigma}}\right). \tag{3.3.57}$$

Therefore, all the terms in S_3 can be expressed as total derivatives, and via Stokes' theorem, one finds that the entire t-dependent action (3.3.53) can be written as the boundary action

$$\frac{1}{g_5^2}\int_{\partial\mathcal{M}} d^4x\,\mathrm{Tr}\left(T\varepsilon^{ijk}\left(A_i\partial_j A_k + \frac{2}{3}A_i A_j A_k\right) + Tw\left(2\varepsilon^{\widetilde{\alpha}\widetilde{\beta}}F_{\widetilde{\beta}4}\phi_{\widetilde{\alpha}}\right)\right.$$

$$- Tw^2\varepsilon^{\widetilde{\alpha}\widetilde{\beta}}\phi_{\widetilde{\alpha}}D_4\phi_{\widetilde{\beta}} + Tw\varepsilon^{\widetilde{i}\widetilde{j}\widetilde{k}}\left(A_{\widetilde{i}}\partial_{\widetilde{j}}A_{\widetilde{k}} + \frac{2}{3}A_{\widetilde{i}}A_{\widetilde{j}}A_{\widetilde{k}}\right)$$

$$\left. + Tw^2\left(2\varepsilon^{\widetilde{\alpha}\widetilde{\beta}}F_{\widetilde{\beta}5}\phi_{\widetilde{\alpha}}\right) - Tw^3\varepsilon^{\widetilde{\alpha}\widetilde{\beta}}\phi_{\widetilde{\alpha}}D_5\phi_{\widetilde{\beta}}, \right)$$

$$\tag{3.3.58}$$

where $\widetilde{i},\widetilde{j},\widetilde{k}=1,2,5$, and

$$T = -\frac{t-t^{-1}}{t+t^{-1}} + \frac{t+t^{-1}}{t-t^{-1}}.\tag{3.3.59}$$

Here, we have used

$$Tw = \frac{2}{t+t^{-1}},\tag{3.3.60}$$

$$Tw^2 = \frac{t-t^{-1}}{t+t^{-1}},\tag{3.3.61}$$

and

$$Tw^3 = \frac{t+t^{-1}}{2} - \frac{2}{t+t^{-1}}.\tag{3.3.62}$$

The boundary action (3.3.58) is equal to

$$S_3 + S_{\partial\mathcal{M}} = \frac{i\widetilde{\Psi}}{4\pi}\int_{\partial\mathcal{M}} d^4x\,\mathrm{Tr}\,\left(\mathcal{A}_{w1}(\partial_2 A_4 - \partial_4\mathcal{A}_{w2}) + \frac{2}{3}\mathcal{A}_{w1}[\mathcal{A}_{w2},A_4]\right.$$

$$+ \mathcal{A}_{w2}(\partial_4\mathcal{A}_{w1} - \partial_1 A_4) + \frac{2}{3}\mathcal{A}_{w2}[A_4,\mathcal{A}_{w1}]$$

$$\left. + A_4(\partial_1\mathcal{A}_{w2} - \partial_2\mathcal{A}_{wi}) + \frac{2}{3}A_4[\mathcal{A}_{w1},\mathcal{A}_{w2}]\right)$$

$$+ w\frac{i\widetilde{\Psi}}{4\pi}\int_{\partial\mathcal{M}} d^4x\,\mathrm{Tr}\,\left(\mathcal{A}_{w1}(\partial_2 A_5 - \partial_5\mathcal{A}_{w2}) + \frac{2}{3}\mathcal{A}_{w1}[\mathcal{A}_{w2},A_5]\right.$$

$$+ \mathcal{A}_{w2}(\partial_5\mathcal{A}_{w1} - \partial_1 A_5) + \frac{2}{3}\mathcal{A}_{w2}[A_5,\mathcal{A}_{w1}]$$

$$\left. + A_5(\partial_1\mathcal{A}_{w2} - \partial_2\mathcal{A}_{w1}) + \frac{2}{3}A_5[\mathcal{A}_{w1},\mathcal{A}_{w2}]\right),$$

$$\tag{3.3.63}$$

where $\widetilde{\Psi}$ is the t-dependent parameter

$$\widetilde{\Psi} = \frac{4\pi i}{g_5{}^2}\left(\frac{t - t^{-1}}{t + t^{-1}} - \frac{t + t^{-1}}{t - t^{-1}}\right), \tag{3.3.64}$$

or equivalently, $\widetilde{\Psi} = -\frac{4\pi i}{g_5^2}T$. This can be checked by expanding (3.3.63) in terms of A_1, A_2, ϕ_1 and ϕ_2. Crucially, the expression (3.3.63) can be written more concisely as

$$S_3 + S_{\partial M} = \frac{w - \bar{w}}{4}\frac{i\widetilde{\Psi}}{2\pi}\int_{\partial M} dz_w \wedge \mathrm{Tr}\left(\mathcal{A}_w \wedge d\mathcal{A}_w + \frac{2}{3}\mathcal{A}_w \wedge \mathcal{A}_w \wedge \mathcal{A}_w\right), \tag{3.3.65}$$

which is a 4d Chern–Simons action.

Hence, the partially-twisted 5d $\mathcal{N} = 2$ supersymmetric Yang–Mills theory on $\Sigma \times \mathbb{R}_+ \times C$ has an action that is $\mathcal{Q}$-exact up to the $\mathcal{Q}$-invariant 4d Chern–Simons action (3.3.65), which is given by

$$S = \delta_t \widetilde{V}_1 - \frac{1}{2t}\delta_t(\delta_L - t\delta_R)\widetilde{V}_2'$$

$$+ \frac{w - \bar{w}}{4}\frac{i\widetilde{\Psi}}{2\pi}\int_{\partial \mathcal{M}} dz_w \wedge \mathrm{Tr}\left(\mathcal{A}_w \wedge d\mathcal{A}_w + \frac{2}{3}\mathcal{A}_w \wedge \mathcal{A}_w \wedge \mathcal{A}_w\right). \tag{3.3.66}$$

This partially-twisted 5d Yang–Mills theory is topological-holomorphic, since the boundary action depends on the complex structure defined on C via (3.3.49), and because of this, the partial twist is often referred to as a topological-holomorphic twist.

3.3.4.1 *Localization To 4d Chern–Simons theory*

Let us now proceed to describe how the path integral of the 5d topological-holomorphic twisted Yang–Mills theory is equivalent to the path integral of 4d Chern–Simons theory. In the process, we shall find a description of 4d Chern–Simons theory that is valid beyond perturbation theory.

Firstly, in what follows, we shall exclude $t = \pm 1$ in the classical theory, as the parameter $w = \frac{t - t^{-1}}{2}$ ought to be nonzero in order to ensure that (3.3.65) is a well-defined 4d Chern–Simons action. In fact, since $|t| = 1$, $w = \frac{t - t^{-1}}{2}$ is purely imaginary, i.e., $w = i\mathrm{Im}(w)$. Also, since $|t| = 1$, $\widetilde{\Psi}$ is real.

Moreover, we shall also only consider $t \neq \pm i$. As a result, one can show that the t-dependence of δ_t can be eliminated via rescaling of δ_t as well as fermion redefinitions (analogous manipulations can be found in page 36 of

[54]). Consequently, t only appears in the Q-exact sector of the action, as well as the definitions of $\widetilde{\Psi}$ and w. We can thus consider $\widetilde{\Psi}$ and w to be parameters independent of the t that appears in the Q-exact sector, which can be tuned to any convenient value in the quantum theory.

Now, the path integral localizes to field configurations that obey $\delta_t \lambda = 0$, for all fermionic fields, denoted collectively as λ. However, not all of these configurations play an equal role, as we shall observe below. Utilizing some field redefinitions, we can find fermionic fields whose (on-shell) variations are $\mathcal{V}_\alpha(t)$, $\widetilde{\mathcal{V}}_\alpha(t)$ and $\mathcal{V}_0 = F_{45} + D_\alpha \phi^\alpha$, which we refer to as χ'_α, $\widetilde{\chi}'_\alpha$ and η', respectively. Localization of the path integral to $\mathcal{V}_\alpha = \widetilde{\mathcal{V}}_\alpha = \mathcal{V}_0 = 0$ can be achieved by scaling up the Q-exact terms in (3.3.66), since these equations are among the conditions for the Q-exact terms to vanish. Thus, we find that the path integral is supported on the solution space of the equations

$$\mathcal{V}_\alpha(t) = 0$$

$$\widetilde{\mathcal{V}}_\alpha(t) = 0 \qquad (3.3.67)$$

$$\mathcal{V}_0 = 0.$$

Since $t \neq \pm i$, we find that the remaining (bosonic) field configurations necessary for the Q-exact terms to vanish are

$$D_M \sigma = 0$$

$$[\phi_\alpha, \sigma] = 0 \qquad (3.3.68)$$

$$[\sigma, \overline{\sigma}] = 0,$$

and these just imply that $\sigma = 0$ everywhere on $\mathcal{M}$, since we have imposed $\sigma = 0$ at the boundary $x_3 = 0$.

Now, localization of the path integral of the 5d partially-twisted theory reduces it to a path integral over the bosonic fields $\mathcal{A}_{w\widetilde{\alpha}}$ and $\mathcal{A}_{w\overline{z}}$ along the boundary, with the integral being restricted to solutions of the equations (3.3.67). This follows since the bulk modes contained in the Q-exact action can be integrated out to give bosonic and fermionic one-loop determinants that cancel due to the Q-symmetry, leaving only the path integral over the boundary 4d Chern–Simons action (assuming a certain anomaly vanishes, as we shall explain below).

However, to ensure that the resulting path integral actually converges, we require that the parameter t takes a suitable value in the Q-exact sector of the action prior to localization, via the addition of Q-exact terms. As we shall see, $t = \pm 1$ are such suitable values. Note that although we exclude these values classically, the freedom to add Q-exact terms to the action in the quantum theory allows us to set $t = \pm 1$ in the Q-exact sector.

Let us further elucidate the localization procedure by expressing the partially-twisted theory as a 1d gauged sigma model. This sigma model can be understood as a dimensional reduction of a 2d gauged A-model, and is therefore referred to as a 1d gauged A-model. This 1d gauged A-model was studied previously in the work of Witten [56].

The partially-twisted 5d gauge theory is equivalent to the 1d gauged A-model with target space being the space $\mathfrak{A}$ of all possible $\mathcal{A}_{w\tilde{\alpha}}$ and $\mathcal{A}_{w\bar{z}}$ fields, and gauge group the space H of maps from $\Sigma \times C$ to the worldvolume gauge group, G (assuming that we formulate our 5d theory using a trivial G-bundle). The essential observation is that a 4d Chern–Simons action serves as the superpotential of the 1d theory, via which 5d bulk terms involving fields in $\mathfrak{A}$ can be obtained as standard terms of the 1d theory.

To see this explicitly, let us first pick, on $\mathfrak{A}$, the metric

$$g = -\frac{1}{2g_5^2} \int_{\Sigma \times C} d^2z d^2x \ \text{Tr}(\delta \mathcal{A}_{\tilde{\alpha}} \otimes \delta \overline{\mathcal{A}}^{\tilde{\alpha}} + \delta \overline{\mathcal{A}}_{\tilde{\alpha}} \otimes \delta \mathcal{A}^{\tilde{\alpha}}$$

$$+ 4\delta \mathcal{A}_{\bar{z}} \otimes \delta \mathcal{A}_z + 4\delta \mathcal{A}_z \otimes \delta \mathcal{A}_{\bar{z}}), \tag{3.3.69}$$

and the complex structure where $\mathcal{A}_{\tilde{\alpha}}$ and $\mathcal{A}_{\bar{z}}$ are holomorphic, which implies the following moment map for the H-action:

$$\mu = -\frac{1}{g_5^2}(D_{\tilde{\alpha}} \phi^{\tilde{\alpha}} + F_{45}). \tag{3.3.70}$$

Here, we have defined the complex coordinates $z = x^4 + ix^5$ and $\bar{z} = x^4 - ix^5$, and the complex gauge fields

$$\mathcal{A}_\alpha = A_\alpha + i\phi_\alpha, \quad \overline{\mathcal{A}}_\alpha = A_\alpha - i\phi_\alpha, \tag{3.3.71}$$

and

$$\mathcal{A}_z = \frac{1}{2}(A_4 - iA_5), \quad \mathcal{A}_{\bar{z}} = \frac{1}{2}(A_4 + iA_5). \tag{3.3.72}$$

Let us also define the superpotential

$$W = -\frac{e^{i\alpha}}{g_5^2} \int_{\Sigma \times C} dz \wedge \text{Tr}\left(\mathcal{A} \wedge d\mathcal{A} + \frac{2}{3}\mathcal{A} \wedge \mathcal{A} \wedge \mathcal{A}\right), \tag{3.3.73}$$

which is proportional to a 4d Chern–Simons action. Here, W is only defined up to an arbitrary phase factor (that can be changed by an R-symmetry rotation), which we have denoted by $e^{i\alpha}$.

The bosonic sector of the bulk action of the 1d gauged A-model model has the form

$$S_{1d}^{Bose} = \int d\tau \left(g_{i\bar{j}} \partial_\tau^A x^i \partial_\tau^A x^{\bar{j}} + g_{i\bar{j}} V_a^i \widetilde{\sigma}^a \overline{V}_b^{\bar{j}} \overline{\widetilde{\sigma}}^b + g_{i\bar{j}} V_a^i \overline{\widetilde{\sigma}}^a \overline{V}_b^{\bar{j}} \widetilde{\sigma}^b + g_{i\bar{j}} V_a^i \widetilde{\phi}^a \overline{V}_b^{\bar{j}} \widetilde{\phi}^b \right.$$

$$- g_{i\bar{j}} F^i \overline{F}^{\bar{j}} + \frac{1}{2} F^i \partial_i W + \frac{1}{2} \overline{F}^{\bar{j}} \partial_{\bar{j}} \overline{W} \Big)$$

$$- \frac{1}{e^2} \int d\tau \, \mathrm{Tr}' \left(D_\tau \widetilde{\phi} D_\tau \widetilde{\phi} + 2 D_\tau \widetilde{\sigma} D_\tau \overline{\widetilde{\sigma}} + [\widetilde{\sigma}, \overline{\widetilde{\sigma}}][\overline{\widetilde{\sigma}}, \widetilde{\sigma}] + 2[\widetilde{\phi}, \widetilde{\sigma}][\widetilde{\phi}, \overline{\widetilde{\sigma}}] \right.$$

$$- D^2 + 2e^2 \mu D \Big), \tag{3.3.74}$$

where $\partial_\tau^A x^i = \partial_\tau x^i + A_\tau^a V_a^i$. Here, x is a map from $\mathbb{R}_+$ to $\mathfrak{A}$, V_a for $a = 1, \ldots, \dim H$ are the Killing vector fields generating the action of H on $\mathfrak{A}$, $\widetilde{\phi}^a$ is a real scalar field, $\widetilde{\sigma}^a$ and $\overline{\widetilde{\sigma}}^a$ are complex scalar fields, F^i and D are auxiliary fields, e^2 is a coupling constant, and Tr' is a negative-definite quadratic form on the Lie algebra of H.

By integrating out the auxiliary fields F^i and D, we find that the potential energy terms of the form $\int d\mathcal{T} \left(\frac{1}{4} g^{i\bar{j}} \partial_i W \partial_{\bar{j}} \overline{W} - e^4 \mathrm{Tr}' \mu^2 \right)$ of the 1d gauged sigma model are (for $\mathcal{T} = x^3$)

$$-\frac{1}{g_5^2} \int_{\mathcal{M}} d^5 x \, \mathrm{Tr} \left(\frac{1}{2} \mathcal{F}^{\widetilde{\alpha}\widetilde{\beta}} \overline{\mathcal{F}}_{\widetilde{\alpha}\widetilde{\beta}} + 4 \mathcal{F}^{\widetilde{\alpha}}_{\bar{z}} \overline{\mathcal{F}}_{\widetilde{\alpha} z} + (-2i\mathcal{F}_{z\bar{z}} + D_{\widetilde{\alpha}} \phi^{\widetilde{\alpha}})^2 \right), \tag{3.3.75}$$

where the covariant derivatives

$$\mathcal{D}_\alpha = \partial_\alpha + [\mathcal{A}_\alpha, \cdot\,], \quad \overline{\mathcal{D}}_\alpha = \partial_\alpha + [\overline{\mathcal{A}}_\alpha, \cdot\,], \tag{3.3.76}$$

and

$$\mathcal{D}_z = \partial_z + [\mathcal{A}_z, \cdot\,], \quad \mathcal{D}_{\bar{z}} = \partial_{\bar{z}} + [\mathcal{A}_{\bar{z}}, \cdot\,], \tag{3.3.77}$$

have been used to define the field strengths $\mathcal{F}_{\beta\gamma} = [\mathcal{D}_\beta, \mathcal{D}_\gamma]$, $\mathcal{F}_{\alpha\bar{z}} = [\mathcal{D}_\alpha, \mathcal{D}_{\bar{z}}]$ and $\mathcal{F}_{z\bar{z}} = [\mathcal{D}_z, \mathcal{D}_{\bar{z}}]$. Upon integration by parts, (3.3.75) is equal to

$$-\frac{1}{g_5^2} \int_{\mathcal{M}} d^5 x \, \mathrm{Tr} \left(\frac{1}{2} F^{\widetilde{\alpha}\widetilde{\beta}} F_{\widetilde{\alpha}\widetilde{\beta}} + D^{\widetilde{\alpha}} \phi^{\widetilde{\beta}} D_{\widetilde{\alpha}} \phi_{\widetilde{\beta}} \right.$$

$$+ \frac{1}{2} [\phi^{\widetilde{\alpha}}, \phi^{\widetilde{\beta}}][\phi_{\widetilde{\alpha}}, \phi_{\widetilde{\beta}}] + 4 F^{\widetilde{\alpha}}_{\bar{z}} F_{\widetilde{\alpha} z} + 4 D_z \phi_{\widetilde{\alpha}} D_{\bar{z}} \phi^{\widetilde{\alpha}} - 4 \mathcal{F}_{z\bar{z}} \mathcal{F}_{z\bar{z}} \Big). \tag{3.3.78}$$

These are just the terms involving bosonic fields in the bulk sector of the partially-twisted 5d action that do not involve the x^3 direction nor the fields ϕ_3, σ and $\bar{\sigma}$. The remaining bulk terms of the 1d theory involving bosonic fields correspond to the bulk terms of the 5d theory (involving bosonic fields) that do not appear in (3.3.78), i.e., by identifying e^2, $\widetilde{\phi}$, $\widetilde{\sigma}$ and $\widetilde{\overline{\sigma}}$ with g_5^2, ϕ_3, σ and $\overline{\sigma}$ respectively, the remaining terms of the 1d gauged A-model can be expressed as

$$-\frac{1}{g_5^2}\int_{\mathcal{M}} d^5x\,\mathrm{Tr}\,\Big(F_{3\widetilde{\alpha}}F^{3\widetilde{\alpha}} + D_3\phi_{\widetilde{\alpha}}D^3\phi^{\widetilde{\alpha}} + F_{3x}F^{3x} + 2D_{\widetilde{\alpha}}\sigma D^{\widetilde{\alpha}}\overline{\sigma}$$

$$+ 2[\phi_{\widetilde{\alpha}},\sigma][\phi^{\widetilde{\alpha}},\overline{\sigma}] + 2D_{\widetilde{z}}\sigma D^{\widetilde{z}}\overline{\sigma}$$

$$+ D_{\widetilde{\alpha}}\phi_3 D^{\widetilde{\alpha}}\phi^3 + [\phi_{\widetilde{\alpha}},\phi_3][\phi^{\widetilde{\alpha}},\phi^3] + D_{\widetilde{z}}\phi_3 D^{\widetilde{z}}\phi^3$$

$$+ D_3\phi_3 D^3\phi^3 + 2D_3\sigma D^3\overline{\sigma} + [\sigma,\overline{\sigma}][\overline{\sigma},\sigma] + 2[\phi_3,\sigma][\phi^3,\overline{\sigma}]\Big), \quad (3.3.79)$$

where the index $\widetilde{z} = z,\bar{z}$. In this manner, the entire 5d topological-holomorphic partially-twisted theory can be shown to be equivalent to the 1d gauged A-model with target $\mathfrak{A}$, and boundary action (3.3.65).

Hence, from Ref. [56], we know that this model should localize to the boundary action. However, note that there is an anomaly arising from the residual $SO(2)$ R-symmetry that does not enter the twisting. We shall assume that this anomaly vanishes; even when it does not, we may insert suitable operators to guarantee a non-vanishing path integral.

We thus arrive at

$$\int_{\widetilde{\Gamma}} D\mathcal{A}_w\,\exp\left(-\frac{w-\bar{w}}{4}\frac{i\widetilde{\Psi}}{2\pi}\int_{\partial\mathcal{M}} dz_w\wedge\mathrm{Tr}\left(\mathcal{A}_w\wedge d\mathcal{A}_w + \frac{2}{3}\mathcal{A}_w\wedge\mathcal{A}_w\wedge\mathcal{A}_w\right)\right),$$
$$(3.3.80)$$

where $\widetilde{\Gamma}$ is a subspace of $\mathfrak{A}$ defined by solutions of (3.3.67). Recalling that the argument of the exponent in (3.3.80) is the negative of (3.3.63), we shall perform the coordinate redefinition

$$x^5 \to \mathrm{Im}(w)x^5, \quad (3.3.81)$$

that implies $d^4x \to \mathrm{Im}(w)d^4x$, $\mathrm{Im}(w)\partial_5 \to \partial_5$ and $\mathrm{Im}(w)A_5 \to A_5$, whereby this argument becomes

$$-\frac{i\widetilde{\Psi}}{2\pi}\,\mathrm{Im}(w)\int_{\partial\mathcal{M}} d^4x\,\mathrm{Tr}\left(\mathcal{A}_{w1}(\partial_2\mathcal{A}_{\bar{z}} - \partial_{\bar{z}}\mathcal{A}_{w2}) + \frac{2}{3}\mathcal{A}_{w1}[\mathcal{A}_{w2},\mathcal{A}_{\bar{z}}]\right.$$

$$+ \mathcal{A}_{w2}(\partial_{\bar{z}}\mathcal{A}_{w1} - \partial_1 A_{\bar{z}}) + \frac{2}{3}\mathcal{A}_{w2}[A_{\bar{z}},\mathcal{A}_{w1}]$$

$$\left.+ A_{\bar{z}}(\partial_1\mathcal{A}_{w2} - \partial_2\mathcal{A}_{w1}) + \frac{2}{3}A_{\bar{z}}[\mathcal{A}_{w1},\mathcal{A}_{w2}]\right). \quad (3.3.82)$$

Now, apart from a factor multiplying the action, and a factor multiplying the path integral measure (that we can renormalize away), the path integral only depends on $\mathrm{Im}(w)$ in the definitions of $\mathcal{A}_{w1}$ and $\mathcal{A}_{w2}$. Hence, we can conveniently fix it in these fields, whereby we obtain the path integral

$$\int_{\widetilde{\Gamma}} D\mathcal{A} \, \exp\left(\frac{\widetilde{\Psi}\mathrm{Im}(w)}{4\pi} \int_{\partial\mathcal{M}} dz \wedge \mathrm{Tr}\left(\mathcal{A} \wedge d\mathcal{A} + \frac{2}{3}\mathcal{A} \wedge \mathcal{A} \wedge \mathcal{A}\right)\right). \quad (3.3.83)$$

This is the path integral for 4d Chern–Simons theory with the coupling constant

$$\hbar = -\frac{2}{i\widetilde{\Psi}\mathrm{Im}(w)}, \quad (3.3.84)$$

defined with an integration cycle determined by $\widetilde{\Gamma}$.

This path integral for 4d Chern–Simons theory is valid for all values of $\hbar$, and therefore allows us to define 4d Chern–Simons theory beyond infinitesimal values of $\hbar$, i.e., beyond perturbation theory. This is because the integration cycle ensures the convergence of the path integral, as long as we tune the value of t in the $\mathcal{Q}$-exact terms of the action (prior to localization) to 1 or -1. To see this, note that the equations $\delta\chi'_\alpha = \mathcal{V}_\alpha = 0$ and $\delta\widetilde{\chi}'_\alpha = \widetilde{\mathcal{V}}_\alpha = 0$ can be rewritten (for any $t \in \mathbb{R}$) via

$$t = \frac{\cos\alpha - 1}{\sin\alpha} \quad (3.3.85)$$

as the single equation

$$\mathcal{F}_{\alpha\bar{z}} = -\frac{1}{4}e^{-i\alpha}\varepsilon_{\alpha\beta\gamma}\overline{\mathcal{F}}^{\beta\gamma}. \quad (3.3.86)$$

The equation (3.3.86) is equivalent to

$$\mathcal{F}_{3\widetilde{\gamma}} = -e^{-i\alpha}2\varepsilon_{\widetilde{\gamma}}^{\ \widetilde{\alpha}}\overline{\mathcal{F}}_{\widetilde{\alpha}z}$$

$$\mathcal{F}_{3\bar{z}} = -\frac{1}{4}e^{-i\alpha}\varepsilon^{\widetilde{\beta}\widetilde{\gamma}}\overline{\mathcal{F}}_{\widetilde{\beta}\widetilde{\gamma}}, \quad (3.3.87)$$

where $\widetilde{\alpha}, \widetilde{\beta}, \widetilde{\gamma} = 1, 2$. These equations correspond to gradient flow equations, since they can be written in the gauge $A_3 = 0$ (with $x^3 = \mathcal{T}$) as

$$\frac{dx^i}{d\mathcal{T}} = -g^{i\bar{j}}\frac{\partial\overline{W}}{\partial x^{\bar{j}}} \quad (3.3.88)$$

(using the field-space metric (3.3.69), and the 4d Chern–Simons functional given in (3.3.73)),[10] which are gradient flow equations for a Morse function

---

[10]Note that in relating (3.3.87) and (3.3.88), as well as $\mathcal{V}_0 = 0$ and $\mu = 0$ below, one requires the condition $\phi_3 = 0$. This condition can be shown to be a consequence of the localization equations together with the boundary conditions on ϕ_3, using an argument analogous to that given in Section 4.1 of [57].

that is $2\text{Re}(cW)$, where $c \in \mathbb{R}$. This factor of c is inconsequential as we are free to rescale $\mathcal{T}$ in (3.3.88).

Now, for $t = \pm 1$, we have $e^{i\alpha} = \mp i$, and

$$W = \pm \frac{i}{g_5^2} \int_{\Sigma \times C} \mathrm{d}z \wedge \text{Tr}\left(\mathcal{A} \wedge \mathrm{d}\mathcal{A} + \frac{2}{3} \mathcal{A} \wedge \mathcal{A} \wedge \mathcal{A} \right). \tag{3.3.89}$$

Since $\text{Im}(W)$ is conserved along a gradient flow [57], $\text{Re}(iW)$ is conserved along the gradient flow; in fact, $\text{Re}(\widetilde{c}iW)$ is conserved for any $\widetilde{c} \in \mathbb{R}$. Hence, the real part of the argument of the exponent in (3.3.83) is conserved along $\widetilde{\Gamma}$. The gradient flow starts from a critical point, or more precisely, a critical H-orbit, given by $\delta W = 0$ at $x^3 = \infty$, that ensures the argument of the exponent is a constant at this boundary. Thus, we find that this argument is appropriately bounded to ensure the convergence of the path integral. In addition, the boundary condition $\mu = 0$ at $x^3 = \infty$ means that the critical H-orbit is semistable,[11] and the fact that $\mu = 0$ (that is equivalent to $\mathcal{V}_0 = 0$) is also a localization condition is consistent with μ being conserved along gradient flows.

Therefore, the localization equations in the form of the gradient flow equations (3.3.88) together with the condition $\mu = 0$ define an integration cycle for 4d Chern–Simons theory that ensures its convergence. This integration cycle is the Lefschetz thimble associated with the critical point $\delta W = 0$ that is a boundary condition at $x^3 = \infty$.

In order to obtain lattice models from the D4-NS5 brane construction, we may repeat the above derivation with fundamental strings ending on the D4-brane boundary at $x^3 = 0$. The worldlines of the endpoints of these strings realize the desired Q-invariant Wilson lines, given by

$$W = \text{Tr}(P \, e^{\int_L \mathcal{A}_w}), \tag{3.3.90}$$

where L is a line along $\Sigma \subset \partial \mathcal{M}$. In this manner, we may reproduce R-matrices, the Yang–Baxter equation with spectral parameter, and partition functions of integrable lattice models, all from a 5d partially-twisted supersymmetric gauge theory obtained from a type IIA string theory configuration involving branes and fundamental strings.

In this chapter, we have reviewed two approaches to deriving 4d Chern–Simons theory from supersymmetric gauge theories associated with 6d and 5d D-brane worldvolume theories, and it is natural to wonder if

---

[11]The importance of semistable critical orbits in the analytic continuation of path integrals using gradient flow equations is explained in the work of Witten [57].

there is a relationship between these approaches. In fact, there is indeed a relationship, described, for example, in [146]. This follows essentially because the Ω-deformed plane of the 6d gauge theory can be replaced by a cigar geometry [48], and this equivalence was a feature of the string theory embedding described in Section 3.2.4. T-dualizing along the circle fibers of the cigar results in a 5d gauge theory on a half-line. The boundary condition at the origin of the half-line must be associated with an NS5-brane, simply because T-dualizing along the circle fiber of a Taub-NUT results in an NS5-brane at the point where the fiber becomes degenerate.

3.3.5   *Relation to 3d Chern–Simons theory and the geometric Langlands correspondence*

At the beginning of this chapter, we studied the quantum field theoretic T-duality between 4d Chern–Simons theory and 3d analytically-continued Chern–Simons theory, as well as the embedding of this T-duality in topological string theory. In this section, we shall explain how the T-duality between 3d and 4d Chern–Simons theory can be deduced from the D4-NS5 brane realization of the latter. In the process, we shall explain how integrable lattice models realized by 4d Chern–Simons theory can be related to invariants of 3d analytically-continued Chern–Simons theory, as well as the geometric Langlands program.

3.3.5.1   *T-duality and 3d Chern–Simons theory*

To relate 4d Chern–Simons theory to 3d analytically-continued Chern–Simons theory, we recall the D4-NS5 configuration described in Section 3.3.3. Here, we further specify C to be $\mathbb{R} \times S^1$, with S^1 (parametrized by x^5) having infinitesimal radius. Taking T-duality along this infinitesimal S^1 decompactifies it to $\mathbb{R}$.

As a result, we arrive at the following D3-NS5 configuration:

| | $\tilde{V}'$ | | | | $N\tilde{V}'{\subseteq}T^*\tilde{V}'$ | | | | | |
| | Σ | | $\mathbb{R}$ | $\mathbb{R}$ | $\mathbb{R}$ | $N\tilde{V}{\subseteq}T^*\tilde{V}$ | | | | |
	1	2	3	4	5	6	7	8	9	10
D3	$\times$	$\times$	$\times$	$\times$						
$\widetilde{\text{NS5}}$	$\times$	$\times$		$\times$	$\times$	$\times$	$\times$			

This is a special case of the system studied by Witten [55], that realizes 3d analytically-continued Chern–Simons theory on $\Sigma \times \mathbb{R}$ at $x^3 = 0$, with an appropriate integration cycle defined by the 4d supersymmetric field configurations satisfying the dimensional reduction of (3.3.67), known as the Kapustin–Witten equations. Thus, we find that T-duality of the D4-NS5 and D3-NS5 configurations describes a relationship between 4d and 3d analytically-continued Chern–Simons theories.

Now, in the string theory picture, if we include fundamental strings along the D4-brane boundary at $x^3 = 0$ to realize a lattice, they remain invariant under the operation of T-duality. As a result, we would have a lattice of Wilson lines along Σ in 3d analytically-continued Chern–Simons theory on $\Sigma \times \mathbb{R}$. Now, if, for example, we pick Σ to be T^2, these lattices form links in $\Sigma \times \mathbb{R}$, since, in general, the Wilson lines are located at arbitrary points along $\mathbb{R}$.

In this manner, we find a relationship between lattice models realized by 4d Chern–Simons theory and link invariants of analytically-continued 3d Chern–Simons theory.

3.3.5.2 *S-duality and the geometric Langlands program*

The geometric Langlands correspondence is realized via 4d GL-twisted $\mathcal{N} = 4$ SYM on a product of Riemann surfaces, $\tilde{C} \times (I \times \mathbb{R})$, as shown by Kapustin and Witten [54]. Concisely, shrinking the Riemann surface $\tilde{C}$ leads to a sigma model governing maps from $I \times \mathbb{R}$ to Hitchin's moduli space, and S-duality of boundary conditions in the 4d gauge theory descends to (homological) mirror symmetry of branes in the sigma model, which essentially furnishes the geometric Langlands correspondence.

It is thus natural to expect that a generalization of the geometric Langlands correspondence arises from the topological-holomorphic twist of 5d $\mathcal{N} = 2$ supersymmetric Yang–Mills theory reviewed in this chapter. Indeed, such a generalization was conjectured by Elliot and Pestun in [58], where the relevant generalization of Hitchin's moduli space is the multiplicative Hitchin system. Although this has not been explored much in the physics literature, some progress in describing the sigma model on the multiplicative Hitchin system (which can be identified with the moduli space of solutions to the Bogomolny equations) has been made in the work of Ashwinkumar *et al.* [59], that shall be described briefly at the end of the next chapter.

Chapter 4

4d Chern–Simons Theory with Boundary and Holography

The topic of holography is inherently linked to Chern–Simons theory, particularly since the action for 3d Chern–Simons theories with non-compact gauge groups can, in certain cases, be reformulated as the Einstein–Hilbert action for 3d gravity. Moreover, the duality between 3d Chern–Simons theory and the 2d WZW model, where the conformal blocks of the latter form the Hilbert space of the former, is one of the earliest examples of holographic duality.

In this chapter, we will explore two distinct instances of holography within the context of 4d Chern–Simons theory. The first involves a duality with 2d BF theory coupled to 1d bosonic or fermionic quantum mechanics, a system that has been shown to realize the Yangian algebra, as demonstrated by Ishtiaque *et al.* [60]. The second concerns a duality with a 3d analogue of the chiral WZW model, which captures the quasi-classical expansion of rational R-matrices, as established by Ashwinkumar [61]. Furthermore, we will review how the modification of boundary conditions results in the reduction of the 3d WZW model to a 3d analogue of Toda field theory, based on the work of Ashwinkumar *et al.* [59].

4.1 2d BF Theory Coupled to 1d Quantum Mechanics

In this section, we shall review the holographic duality between 2d BF theory coupled to 1d quantum mechanics and 4d Chern-Simons theory.

The motivation for describing the holographic dual of 4d Chern-Simons theory in terms of this coupled 2d-1d system can be understood from the perspective of intersecting 4-branes and 2-branes in the topological string

165

theory realization of 4d Chern-Simons theory. Therefore, we shall first describe how the coupled 2d-1d system arises from this brane configuration, and why it is a natural candidate for a holographic dual of 4d Chern-Simons theory. The coupled 2d-1d system also arises from intersecting D5-branes and D3-branes in the type IIB string theory realization of 4d Chern-Simons theory, and this shall be elucidated as well. Subsequently, we shall review the realization of the Yangian in the coupled 2d-1d system, and how it leads to an example of *twisted* holography,[1] with 4d Chern-Simons theory (with a Wilson line insertion) as the holographic dual.

4.1.1 *Brane Realization in Topological String Theory*

Recall, from Section 3.1.1 the following configuration of branes in a mix of A-type topological string theory along (x, y, p_x, p_y) and B-type topological string theory along (t, p_t).

$$
\begin{array}{c|cc|c|cc|c}
 & x & y & t & p_x & p_y & p_t \\
\hline
L'_{\text{base}} & \times & \times & \times & & & \times \\
\hline
L'_x & \times & & & & \times &
\end{array}
\qquad (4.1.1)
$$

Here, the 4-branes supported along (x, y, p_x, p_y) realize 4d Chern-Simons theory, while the open topological strings stretched between the 2-brane and 4-branes give rise to a quantum mechanical theory realizing a Wilson line at the intersection of these branes.[2]

We shall generalize this configuration to include a *stack* of 2-branes, and we will also be interested in the theory on the stack of 2-branes itself. This theory can be identified as 2d BF theory, a gauge theory which is the dimensional reduction of 3d Chern-Simons theory:

$$
\int_{\mathbb{R}^2 \times S^1} CS(\mathcal{A}) \xrightarrow{\text{Reduction along } S^1} \int_{\mathbb{R}^2} \text{Tr} \mathcal{B} \mathcal{F}, \qquad (4.1.2)
$$

where $\mathcal{B}$ has been used to denote the component of $\mathcal{A}$ along S^1, and $\mathcal{F} = d\mathcal{A}_{\mathbb{R}^2} + \mathcal{A}_{\mathbb{R}^2} \wedge \mathcal{A}_{\mathbb{R}^2}$ is the field strength of $\mathcal{A}$ along $\mathbb{R}^2$. This identification can be made by applying T-duality to the 3-brane configuration that realizes 3d

---

[1] Other examples of twisted holography, also known as topological holography, can be found in [147, 148].

[2] Note that in this section, 2-branes and 4-branes shall be understood to be branes with two-dimensional and four-dimensional worldvolumes, respectively. We shall also encounter D3-branes and D5-branes, which have four-dimensional and six-dimensional worldvolumes, respectively.

Chern-Simons theory, along one of the directions of support of the 3-brane, resulting in a stack of 2-branes.

In what follows, we shall relabel p_x as the coordinate v and p_y as the coordinate w, while t and p_t shall be understood as real coordinates on the complex plane $C = \mathbb{C}$, which is parametrized by complex coordinates z and $\bar{z}$. In addition, the position of the 2-branes shall be specified to be at $y = 0$, $v = 0$ and $z = 0$, while the 4-branes are located at $w = 0$ and $v = 0$. We shall be interested in the configuration of (4.1.1) with K 4-branes and N 2-branes. This brane configuration is depicted in Figure 4.1.

Since we are interested in holographic duality, it is also important to consider the closed string sector of the mixed A-B topological string theory. This is described by a mix of Kodaira-Spencer/BCOV theory (closed B-model) along the complex plane $\mathbb{C}$ parametrized by $(z, \bar{z})$ [149, 150], and Kähler gravity (closed A-model) along the $\mathbb{R}^4$ parametrized by (x, y, v, w) [151]. The set of fields of this theory, $\mathcal{F}$, are differential forms on $\mathbb{R}^4$ and (p, q)-forms on $\mathbb{C}$. The equation of motion of this closed topological string theory, for any $\alpha \in \mathcal{F}$, is

$$(\mathrm{d}_{\mathbb{R}^4} + \bar{\partial}_{\mathbb{C}})\alpha = 0, \qquad (4.1.3)$$

where $\mathrm{d}_{\mathbb{R}^4}$ is the de Rham differential on $\mathbb{R}^4$ and $\bar{\partial}_{\mathbb{C}}$ is the Dolbeault differential on $\mathbb{C}$.

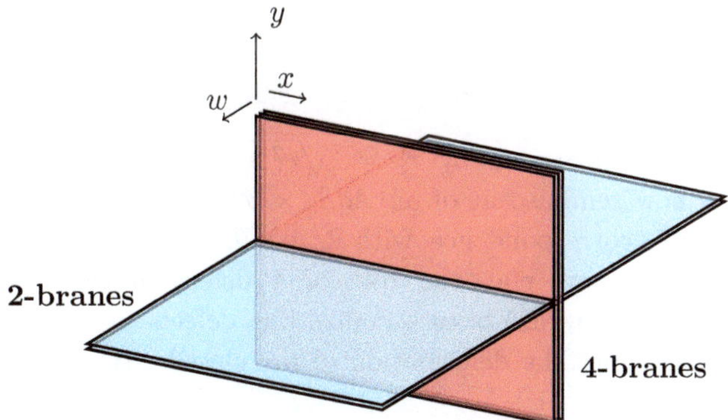

Figure 4.1.   A stack of N 2-branes intersecting a stack of K 4-branes in a mix of A-type topological string theory along $\mathbb{R}^4_{v,w,x,y}$ and B-type topological string theory along $\mathbb{C}_z$. Note that the $\mathbb{R}_v$ and $\mathbb{C}_z$ directions are suppressed. The theory on the 4-branes is 4d Chern-Simons theory, the theory on the 2-branes is 2d BF theory, while a one-dimensional quantum mechanical system lives at their intersection.

The N 2-branes in the configuration of interest source a 3-form field in this closed topological string theory.[3] Along a topological S^3 surrounding the stack of D2-branes, the 3-form field, denoted F_3, gives rise to a flux that can be normalized as

$$\int_{S^3} F_3 = N. \tag{4.1.4}$$

Here, the S^3 being topological means that its continuous deformations should not modify (4.1.4). The difference between an integral of F_3 over S^3 and an integral over its continuous deformation can be written as an integral of F_3 over a closed 3-manifold that does not enclose the support of the 2-branes. Using Stokes' theorem, this is equivalent to requiring that

$$\mathrm{d}_{\mathbb{R}^4 \times \mathbb{C}} F_3(\mathsf{a}) = 0, \tag{4.1.5}$$

where a is any point in the complement of the support of the 2-branes, and $\mathrm{d}_{\mathbb{R}^4 \times \mathbb{C}}$ is the de Rham differential on $\mathbb{R}^4 \times \mathbb{C}$. Since F_3 satisfies (4.1.5) as well as the equation of motion (4.1.3), while being translation invariant along the directions parallel to the stack of 2-branes, it must take the form

$$F_3 = \frac{iN}{2\pi \left(v^2 + y^2 + z\bar{z}\right)^2}\left(v\mathrm{d}y \wedge \mathrm{d}z \wedge \mathrm{d}\bar{z} - y\mathrm{d}v \wedge \mathrm{d}z \wedge \mathrm{d}\bar{z} - 2\bar{z}\mathrm{d}v \wedge \mathrm{d}y \wedge \mathrm{d}z\right).$$

$$\tag{4.1.6}$$

The requirement that an S^3 supporting the flux of F_3 exists, together with demanding translation invariance along the 2-branes supported on the $\mathbb{R}^2_{w,x}$-plane, implies that the closed string background $\mathbb{R}^4_{v,w,x,y} \times \mathbb{C}_z$ is deformed to

$$\mathbb{R}^2_{w,x} \times \mathbb{R}_+ \times S^3, \tag{4.1.7}$$

where $\mathbb{R}_+$ is parametrized by $r := \sqrt{v^2 + y^2 + z\bar{z}}$.[4] The background geometry is now reminiscent of an $AdS_3 \times S^3$ geometry that appears in the $AdS_3/\mathrm{CFT2}$ correspondence, with $\mathbb{R}^2_{w,x} \times \mathbb{R}_+$ playing the role of AdS_3.

The 4-branes now play the role of a defect in the closed string theory, similar to how D5-branes realize line defects in the $AdS_5/\mathrm{CFT4}$ correspondence [90]. The deformation of the closed string background to

---

[3]This is analogous to D3-branes (with four-dimensional worldvolume) sourcing a 5-form field in the AdS/CFT duality between $\mathcal{N} = 4$ super Yang-Mills and string theory on $AdS_5 \times S^5$.

[4]Recall that in the AdS/CFT correspondence between string theory on $AdS_5 \times S^5$ and $\mathcal{N} = 4$ super Yang-Mills theory, the S^5 in the bulk geometry arises due to the 5-form flux sourced by the D3-branes.

the geometry given in (4.1.7) implies that the 4-brane geometry is deformed to

$$\mathbb{R}_x \times \mathbb{R}_+ \times S^2, \tag{4.1.8}$$

where $\mathbb{R}_+$ is parametrized by $r' := \sqrt{y^2 + z\bar{z}}$.

It is thus natural to investigate whether there exists a holographic duality between the effective gravitational theory on $\mathbb{R}^2_{w,x} \times \mathbb{R}_+$ with a defect supported on $\mathbb{R}_x \times \mathbb{R}_+$, and a dual theory on $\mathbb{R}^2_{w,x}$. As we shall see, such a duality indeed exists in the limit where the number of 2-branes is large, also known as the large-N limit. Notably, in this limit, one is in the regime of *rigid* holography where there is no longer any coupling between the 4-branes and the closed string modes [64].

Thus, the gravitational theory is entirely governed by the theory on the defect on $\mathbb{R}_x \times \mathbb{R}_+$, or equivalently, the 4-brane worldvolume theory to which it is related via Kaluza-Klein compactification. This is just 4d Chern-Simons theory on $\mathbb{R}_x \times \mathbb{R}_+ \times S^2$ with N units of flux supported along S^2. In practice, it shall be convenient to work with an equivalent description as 4d Chern-Simons theory on $\mathbb{R}^2_{x,y} \times \mathbb{C}_z$, with a Wilson line at $y = 0$, $z = 0$ which is associated with a representation of $G = GL(K, \mathbb{C})$ that is specified by the number, N, of 2-branes. In this description, the asymptotic boundary $r' = \infty$ is identified with $y = \infty$.

The duality we shall review will thus be between

- $GL(K, \mathbb{C})$ 4d Chern-Simons theory on $\mathbb{R}^2_{x,y} \times \mathbb{C}_z$ with a distinguished boundary at $y = \infty$, and with a Wilson line insertion at $y = 0$, $z = 0$, and
- $GL(N, \mathbb{C})$ 2d BF theory on $\mathbb{R}^2_{w,x}$ with a 1d quantum mechanical system at $w = 0$,

in the large-N limit. In particular, at $y = \infty$, the gauge field will be restricted to have a fixed profile. It shall be useful to define a line at $y = \infty$ for some fixed coordinate z in the complex plane as

$$\ell_\infty(z) := \mathbb{R}_x \times \{y = \infty\} \times \{z\}. \tag{4.1.9}$$

It is important to note that although the 2-branes were initially located at $y = v = z = 0$, corresponding to $r = 0$ in the backreacted geometry, the holographic duality involves an identification between local variations of boundary values of fields in the bulk theory at the asymptotic boundary $r = \infty$ and local operators in the 2-brane worldvolume theory. In fact, the 2-branes do not actually exist in the backreacted geometry in the large-N

limit, but are replaced by black branes, while the D2-brane worldvolume theory (2d BF theory coupled to 1d quantum mechanics) becomes the description of the boundary dual of the holographic duality.

4.1.2 *Brane Realization in Type IIB String Theory*

We shall now review how the 2d-1d coupled system of 2d BF theory and 1d quantum mechanics arises in a configuration involving D3-branes intersecting D5-branes that realize 4d Chern-Simons theory, in type IIB string theory.

Recall, from Section 3.2.4, the following brane configuration in a Taub-NUT background of type IIB string theory that realizes 4d Chern-Simons theory with a Wilson line insertion:

	$\overbrace{\qquad\Sigma\qquad}$		$\overbrace{\quad\mathbb{R}^2_\epsilon\quad}$		$\overbrace{\quad C\quad}$		$\overbrace{N\Sigma\subset T^*\Sigma}$		$\overbrace{\mathbb{R}^2_{-\epsilon}}$	
	0	**1**	**2**	**3**	**4**	**5**	**6**	**7**	**8**	**9**
D5	×	×	×	×	×	×				
D3	×		×	×				×		

The configuration considered there was a stack of D3-branes ending on a stack of D5-branes. In this section, we shall modify this configuration slightly, such that the N D3-branes *intersect* with the K D5-branes. The resulting brane configuration is depicted in Figure 4.2.

In what follows, we shall study this D3-D5 brane system, and explain how the D3-brane worldvolume theory localizes to 2d BF theory, while the strings stretched between the D3- and D5-branes give rise to 1d quantum mechanics at the intersection of the D3- and D5-branes.

Firstly, the D3-brane worldvolume theory is naturally twisted in terms of the 4d geometric Langlands (GL) twist [54]. This is because the equations (3.2.17) can be used to obtain the following six equations, of which three are independent:

$$(\Gamma_{02} + \Gamma_{68})\epsilon = 0, \qquad (\Gamma_{03} + \Gamma_{69})\epsilon = 0, \qquad (\Gamma_{23} + \Gamma_{89})\epsilon = 0,$$
$$(\Gamma_{07} + \Gamma_{16})\epsilon = 0, \qquad (\Gamma_{27} + \Gamma_{18})\epsilon = 0, \qquad (\Gamma_{37} + \Gamma_{19})\epsilon = 0. \tag{4.1.10}$$

These are precisely the equations that define a scalar supercharge in the GL twist of 4d $\mathcal{N} = 4$ super Yang-Mills theory on $\mathbb{R}^4_{0237}$, as explained in [54].

In fact, it was shown in [54] that there is a family of GL twists parameterized by $t \in \mathbb{CP}^1$. The specific twist of the D3-brane worldvolume

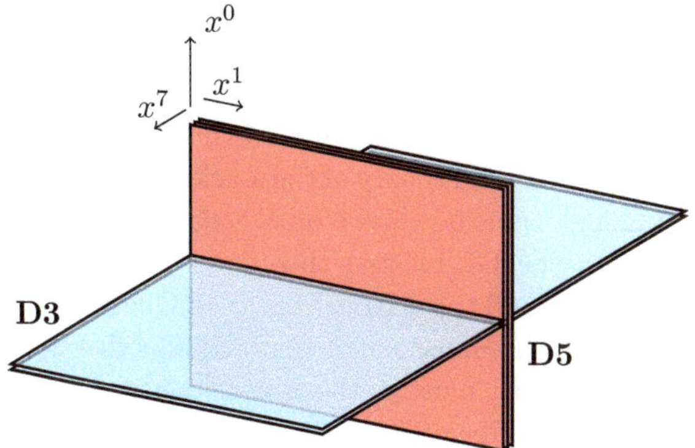

Figure 4.2. A stack of D3-branes intersecting a stack of D5-branes. When this system is subject to a topological-holomorphic twist and Ω-deformation, the D3- and D5-brane worldvolume theories will localize to 2d BF theory and 4d Chern-Simons theory, respectively, with both theories coupled to a 1d quantum mechanical system at the brane intersection.

theory we are interested in is a particular member of this family. With some relatively straightforward analysis [60], it can be shown that it corresponds to

$$t = i. \tag{4.1.11}$$

Next, note that at the three-dimensional D3-D5 intersection there exists a 3d $\mathcal{N} = 4$ theory consisting of bifundamental hypermultiplets coupled to background gauge fields which are restrictions of the gauge fields from the D3 and the D5 branes [62]. Working at the level of Q-cohomology, we find that the 3d theory is also a topological theory.

For the twisting parameter $t = i$, the 4d GL-twisted super Yang-Mills theory is an analogue of a 2d B-model [54].[5] This means that the 3d theory at the D3-D5 intersection must be a 3d analogue of the 2d B-model, which is a B-type topological twist of a 3d $\mathcal{N} = 4$ theory known as Rozansky-Witten theory [63]. To be precise, this is a gauged version of Rozansky-Witten theory with $U(N) \times U(K)$ gauge symmetry.

Now, recall that the theory on the stack of D5-branes is a topological-holomorphic twist of 6d $\mathcal{N} = (1,1)$ supersymmetric gauge theory, subject to an Ω-deformation due to the Taub-NUT background we are considering.

---

[5] Notably, the 4d GL-twisted Super Yang-Mills theory on $\mathbb{R}^2 \times T^2$ is related via a dimensional reduction on T^2 to a 2d B-model on $\mathbb{R}^2$.

The theories on the D3-brane worldvolume as well as the D3-D5 intersection are also Ω-deformed for the same reason. As a result, the 4d GL-twisted theory on the stack of D3-branes and the 3d Rozansky-Witten theory at the D3-D5 intersection localize to 2d and 1d theories, respectively.

To identify the effective 2d theory of the stack of D3-branes, we note that the topological-holomorphic 6d theory on $\Sigma \times \mathbb{R}_\epsilon^2 \times C$ can be dimensionally reduced along C to the GL-twisted theory on $\Sigma \times \mathbb{R}_\epsilon^2$. Given that the topological-holomorphic 6d theory localizes to 4d Chern-Simons theory due to the Ω-deformation, this must mean that the effective 2d theory arising from the GL-twisted theory must be the dimensional reduction of 4d Chern-Simons theory along C. Recall from the beginning of Chapter 3 that 4d Chern-Simons theory dimensionally reduces along one of the directions of C to 3d Chern-Simons theory, and from (4.1.2) that the latter dimensionally reduces to 2d BF theory. Thus, we conclude that the effective 2d theory arising from the GL-twisted theory must be 2d BF theory, with $GL(N, \mathbb{C})$ gauge symmetry.

The effective 1d theory at the D3-D5 intersection can also be identified. Yagi showed that an Ω-deformation of the Rozansky-Witten twist of a 3d $\mathcal{N} = 4$ theory with only hypermultiplets localizes to a 1d bosonic quantum mechanical system [65]. A straightforward generalization of this result involves the gauging of the flavor symmetry of the hypermultiplets, resulting in a 1d gauged bosonic quantum mechanics. This means that the 1d theory at the D3-D5 intersection is just $GL(N, \mathbb{C}) \times GL(K, \mathbb{C})$ gauged quantum mechanics, with the action

$$S_{\mathrm{QM}} := \int_{\mathbb{R}_x} \left(\bar{\chi}_{ia} \mathrm{d}\chi^{ia} + \bar{\chi}_{ib} \mathcal{A}_b^a \chi_a^i + \bar{\chi}_{ja} A_i^j \chi^{ia} \right). \tag{4.1.12}$$

Here, χ and $\bar{\chi}$ are bosonic fields that transform in the fundamental representations of both $GL(K, \mathbb{C})$ and $GL(N, \mathbb{C})$, where $(i,j) = 1, \ldots, K$ and $(a,b) = 1, \ldots, N$, $\mathcal{A}$ is the $\mathfrak{gl}_N$ gauge field of 2d BF theory, while A is the $\mathfrak{gl}_K$ gauge field of 4d Chern-Simons theory. Although bosonic quantum mechanics has been derived here, replacing the bosonic fields by fermionic fields will not change the main result that we shall review, and we shall in fact focus on fermionic fields in what follows, and only briefly comment on bosonic fields.

In summary, the effective description of the intersecting D3-D5 brane system depicted in Figure 4.2 is 4d Chern-Simons theory with gauge group $GL(K, \mathbb{C})$ intersecting with 2d BF theory with gauge group $GL(N, \mathbb{C})$, with $GL(K, \mathbb{C}) \times GL(N, \mathbb{C})$ gauged quantum mechanics at the intersection.

4.1.3 *The Yangian from 2d BF Theory coupled to 1d quantum mechanics*

The holographic duality that shall be reviewed involves matching the algebra of local operators defined along the intersection of the coupled 2d-1d system of $GL(N, \mathbb{C})$ 2d BF theory and 1d quantum mechanics with the algebra arising from local variations of boundary values of fields in 4d Chern-Simons theory with a Wilson line insertion, in the large-N limit. As we shall see, this algebra is isomorphic to the Yangian of $\mathfrak{gl}_K$.

Let us proceed to review the first half of this result, which states that the algebra of local operators at the intersection of the 2d-1d coupled system is isomorphic to the Yangian of $\mathfrak{gl}_K$, in the large-N limit.

The 2d BF theory coupled to the gauged fermionic quantum mechanics has the action

$$S_{2d-1d} = S_{\mathrm{BF}} + S_{\mathrm{QM}}, \tag{4.1.13}$$

where

$$S_{\mathrm{BF}} = \int_{\mathbb{R}^2_{w,x}} \mathrm{Tr}(\mathcal{B}\mathcal{F}), \tag{4.1.14}$$

$$S_{\mathrm{QM}} = \int_{\mathbb{R}_x} \left(\overline{\chi}_{ia}\mathrm{d}\chi^{ia} + \overline{\chi}_{ib}\mathcal{A}^\alpha \rho(T_\alpha)^b_a \chi^{ia}\right). \tag{4.1.15}$$

Here, we have utilized a basis $\{T_\alpha\}$ of $\mathfrak{gl}_N$ which is orthonormal with respect to the trace, Tr, i.e.,

$$\mathrm{Tr}(T_\alpha T_\beta) = \delta_{\alpha\beta}. \tag{4.1.16}$$

We have also used ρ to denote the fundamental representation of $\mathfrak{gl}_N$. Note that for the following analysis, the coupling to the $\mathfrak{gl}_K$ gauge field is not important, and has thus been suppressed.

The 2d BF propagator is the 2-point correlation function given by

$$P^{\alpha\beta}(\mathsf{a}, \mathsf{b}) := \left\langle \mathcal{B}^\alpha(\mathsf{a})\mathcal{A}^\beta(\mathsf{b})\right\rangle, \tag{4.1.17}$$

where a and b are points on $\mathbb{R}^2$. The two point correlation function takes the form

$$P^{\alpha\beta}(\mathsf{a}, \mathsf{b}) = \delta^{\alpha\beta} P(\mathsf{a}, \mathsf{b}), \tag{4.1.18}$$

where $P(\mathsf{a}, \mathsf{b})$ is determined by the equation

$$\frac{1}{\hbar}\mathrm{d}P(0, \mathsf{a}) = \delta^2(\mathsf{a})\mathrm{d}w \wedge \mathrm{d}x. \tag{4.1.19}$$

We shall employ the Lorentz gauge

$$\partial_w \mathcal{A}_w + \partial_y \mathcal{A}_y = 0, \tag{4.1.20}$$

which means that the propagator 1-form should satisfy

$$(\partial_w \iota_w + \partial_y \iota_y) P(0, \mathsf{a}) = 0. \tag{4.1.21}$$

The propagator 1-form of 2d BF theory in this gauge can be shown to be

$$P(0, \mathsf{a}) = \frac{\hbar}{2\pi} \left(\frac{w}{w^2 + y^2} \mathrm{d}y - \frac{y}{w^2 + y^2} \mathrm{d}w \right). \tag{4.1.22}$$

This can be verified just as in the case of the 4d Chern-Simons propagator two-form in Chapter 1. We can check that, away from the origin, $\mathrm{d}P(0, \mathsf{a}) = 0$, since

$$
\begin{aligned}
\mathrm{d}P(0, \mathsf{a}) &= \frac{\hbar}{2\pi} \left(\frac{\mathrm{d}w \wedge \mathrm{d}y}{w^2 + y^2} - \frac{\mathrm{d}y \wedge \mathrm{d}w}{w^2 + y^2} - \frac{2w^2}{(w^2 + y^2)^2} \mathrm{d}w \wedge \mathrm{d}y \right. \\
&\quad \left. + \frac{2y^2}{(w^2 + y^2)^2} \mathrm{d}y \wedge \mathrm{d}w \right) \\
&= \frac{\hbar}{2\pi} \left(\frac{2\mathrm{d}w \wedge \mathrm{d}y}{w^2 + y^2} - 2 \frac{(w^2 + y^2)}{(w^2 + y^2)^2} \mathrm{d}w \wedge \mathrm{d}y \right) \\
&= 0.
\end{aligned}
\tag{4.1.23}
$$

We also can check that

$$
\begin{aligned}
&(\partial_w \iota_w + \partial_y \iota_y) P(0, \mathsf{a}) \\
&= \frac{\hbar}{2\pi} \left(-\partial_w \left(\frac{y}{w^2 + y^2} \right) + \partial_y \left(\frac{w}{w^2 + y^2} \right) \right) \\
&= \frac{\hbar}{2\pi} \left(\frac{2yw}{(w^2 + y^2)^2} - \frac{2yw}{(w^2 + y^2)^2} \right) \\
&= 0.
\end{aligned}
\tag{4.1.24}
$$

Finally, we can verify the normalization of the propagator by checking (4.1.19) at the origin. Noting that

$$\frac{\partial}{\partial y} \left(\tan^{-1} \left(\frac{y}{w} \right) \right) = \frac{w}{w^2 + y^2} \tag{4.1.25}$$

$$\frac{\partial}{\partial w} \left(\tan^{-1} \left(\frac{y}{w} \right) \right) = -\frac{y}{w^2 + y^2}, \tag{4.1.26}$$

we find that

$$\frac{1}{\hbar}\int_{w^2+y^2\leq 1} \mathrm{d}P(0,\mathsf{a}) = \frac{1}{\hbar}\int_{w^2+y^2=1} P(0,\mathsf{a}) = \frac{1}{2\pi}\int_{w^2+y^2=1} \mathrm{d}\varphi(0,\mathsf{a}) = 1$$

(4.1.27)

where $\varphi(0,\mathsf{a})$ is the angle (measured counter-clockwise) between the line joining the origin and a and a reference line passing through the origin.

Using (4.1.25) and (4.1.26), we note that the propagator $P(0,\mathsf{a})$ can be reexpressed as

$$P(0,\mathsf{a}) = \frac{\hbar}{2\pi}\,\mathrm{d}\varphi(0,\mathsf{a})\,.$$

(4.1.28)

Using translation invariance to replace the origin with an arbitrary point, we thus find that for two arbitrary points a and b, the propagator is

$$P(\mathsf{a},\mathsf{b}) = \frac{\hbar}{2\pi}\,\mathrm{d}\varphi(\mathsf{a},\mathsf{b})\,,$$

(4.1.29)

where $\varphi(\mathsf{a},\mathsf{b})$ is the angle (measured counter-clockwise) between a reference line passing through a and the line joining the points a and b. The Feynman diagram used for the propagator shall be

$$P(\mathsf{a},\mathsf{b}) = \mathsf{a} \longrightarrow \mathsf{b}\,.$$

(4.1.30)

The propagator for the fermionic quantum mechanical system can be defined using

$$\frac{1}{\hbar}\partial_{x_2}\left\langle \overline{\chi}_i^a(x_1)\chi_b^j(x_2)\right\rangle = \delta_b^a\,\delta_i^j\,\delta^1(x_1 - x_2)\,,$$

(4.1.31)

which has the solution

$$\left\langle \overline{\chi}_i^a(x_1)\chi_b^j(x_2)\right\rangle = \hbar\delta_b^a\,\delta_i^j\,\theta(x_2 - x_1)\,,$$

(4.1.32)

where $\theta(x_2 - x_1)$ is a unit step function, defined as

$$\theta(x) = \frac{1}{2}\mathrm{sgn}(x) = \begin{cases} 1/2 & \text{for } x > 0 \\ 0 & \text{for } x = 0 \\ -1/2 & \text{for } x < 0 \end{cases}.$$

(4.1.33)

Using the anticommutativity of the fermionic fields χ and $\overline{\chi}$, one finds that χ and $\overline{\chi}$ can be interchanged in the propagator, and it only depends on the order of the fields.[6] The Feynman diagram for this propagator is denoted

---

[6] If we χ and $\overline{\chi}$ were bosonic instead of fermionic, then interchanging these fields in the propagator would incur a minus sign.

as

$$\hbar\theta(x_2 - x_1) = \underset{x_1 \qquad x_2}{\overbrace{}}, \tag{4.1.34}$$

where the curved line corresponds to the propagator, where the pink and cyan dots represent the fields $\overline{\chi}_j^a$ and χ_b^i, respectively, and the horizontal line denotes $\mathbb{R}_x$.

We shall now proceed to derive the algebra arising from the operator products of the local operator

$$O_j^i[m] = \frac{1}{\hbar}\overline{\chi}_j^a\,(\mathcal{B}^m)_a^b\,\chi_b^i, \tag{4.1.35}$$

which is gauge invariant (with respect to $GL(N,\mathbb{C})$), and involves the m-th power of the local operator B. The operator $O_j^i[m]$ is represented by the symbol $\textcircled{\circ\bullet\circ}$ where the orange dot represents the field $(\mathcal{B}^m)_a^b$, for any m. We shall now compute the operator product between $O_j^i[m]$ and $O_l^k[n]$ by first considering contributions from the free part of the theory, and subsequently taking interactions into account.

In what follows, the operator product shall be denoted as

$$O_j^i[m] \cdot O_l^k[n]. \tag{4.1.36}$$

We shall consider operators at points x_1 and x_2 on $\mathbb{R}_x$, and subsequently take the limit where these points approach each other. The first contribution to the operator product comes from the following Feynman diagram:

$$\underset{x_1 \qquad\qquad x_2}{\overbrace{}} \tag{4.1.37}$$

This diagram can be evaluated to be

$$\frac{1}{\hbar}\overline{\chi}_j^a(x_1)(\mathcal{B}(x_1)^m)_a^b\,\frac{1}{2}\hbar\delta_b^c\delta_l^i\,\frac{1}{\hbar}(\mathcal{B}(x_2)^n)_c^d\chi_d^k(x_2), \tag{4.1.38}$$

$$= \frac{1}{2\hbar}\delta_l^i\overline{\chi}_j^a(x_1)(\mathcal{B}(x_1)^m)_a^b(\mathcal{B}(x_2)^n)_b^c\chi_c^k(x_2). \tag{4.1.39}$$

In the limit where x_1 approaches x_2, the diagram has the value

$$\frac{1}{2\hbar}\delta_l^i\overline{\chi}_j\mathcal{B}^{m+n}\chi^k = \frac{1}{2}\delta_l^i O_j^k[m+n], \tag{4.1.40}$$

where the contracted $\mathfrak{gl}_N$ indices have been suppressed for convenience (we shall continue to suppress such indices in what follows).

The next contribution to the operator product comes from the Feynman diagram

$$(4.1.41)$$

which evaluates to

$$\frac{1}{2\hbar}\delta_j^k \chi^i \mathcal{B}^{m+n}\overline{\chi}_l = -\frac{1}{2\hbar}\delta_j^k \overline{\chi}_l \mathcal{B}^{m+n}\chi^i = -\frac{1}{2}\delta_j^k O_l^i[m+n]. \qquad (4.1.42)$$

The final contribution to the operator product that does not involve interactions arises from the Feynman diagram

$$(4.1.43)$$

which evaluates to

$$\frac{1}{4}\delta_l^i \delta_j^k \mathrm{Tr}\mathcal{B}^{m+n}. \qquad (4.1.44)$$

The sum of the contributions of these three diagrams together with the contribution of the (commuting) c-number product between the operators is

$$O_j^i[m]\cdot O_l^k[n] = O_j^i[m]O_l^k[n]+\frac{1}{2}\delta_l^i O_j^k[m+n]-\frac{1}{2}\delta_j^k O_l^i[m+n]+\frac{1}{4}\delta_l^i \delta_j^k \mathrm{Tr}\mathcal{B}^{m+n}. \qquad (4.1.45)$$

Using (4.1.45), we can compute the Lie bracket of the algebra of local operators in the limit where interactions are ignored, which is

$$\left[O_j^i[m],O_l^k[n]\right]_{OPE}^{\mathrm{free}} = \delta_l^i O_j^k[m+n] - \delta_j^k O_l^i[m+n], \qquad (4.1.46)$$

where we have used the notation $[A,B]_{OPE} = A\cdot B - B\cdot A$. This is the polynomial loop algebra, $\mathfrak{gl}_K[[z]]$.[7]

We shall now derive corrections to this algebra from Feynman diagrams with higher-order loops arising from the interaction term in 2d BF theory, which takes the form

$$f_{\alpha\beta\gamma}\int_{\mathbb{R}^2}\mathcal{B}^\alpha \mathcal{A}^\beta \wedge \mathcal{A}^\gamma. \qquad (4.1.47)$$

---

[7]The algebra (4.1.46) can also be derived using bosonic quantum mechanics.

This interaction is represented in terms of a Feynman diagram by a trivalent vertex with two incoming edges and one outgoing edge. The Feynman diagram for the corresponding amplitude with the inclusion of propagators for the edges and integration over the vertex point, denoted as

$$V^{\alpha\beta\gamma}(\mathsf{b}_1,\mathsf{b}_2,\mathsf{b}_3) =: \frac{\hbar^2}{(2\pi)^3} f^{\alpha\beta\gamma} \int_{\mathsf{a}\in\mathbb{R}^2} \mathrm{d}_{\mathsf{b}_1}\varphi(\mathsf{a},\mathsf{b}_1) \wedge \mathrm{d}_{\mathsf{b}_2}\varphi(\mathsf{a},\mathsf{b}_2) \wedge \mathrm{d}_{\mathsf{b}_3}\varphi(\mathsf{a},\mathsf{b}_3)\,,$$

$$(4.1.48)$$

is

$$(4.1.49)$$

Feynman diagrams with this interaction involve another operator, denoted as

$$= \frac{1}{\hbar} \int_{\mathbb{R}_x} \overline{\chi}_{ib}\mathcal{A}_a^b\chi_a^i\,, \qquad (4.1.50)$$

which arises from the interaction term in the quantum mechanical action (4.1.15) that couples to the $\mathfrak{gl}_N$ gauge field. Here, $\overline{\chi}$ and χ are taken to be constant when performing the integration over $\mathbb{R}_x$, i.e., derivatives of the fermions are set to zero.[8] The only possible interaction diagram turns out to be

$$(4.1.51)$$

Let us now proceed to consider diagrams of the form (4.1.51), with the inclusion of all possible fermionic propagators.

We shall first focus on the case of diagrams without fermionic propagators. We shall be mainly interested in computing products of $O_j^i[1]$, since, together with $O_j^i[0]$, they turn out to generate the entire algebra of local operators. In the absence of any fermionic propagators, the only diagram

---

[8]This can be justified by noting that the equations of motion for the fermionic fields, which are $\mathrm{d}\chi^i = -A\chi^i$ and $\mathrm{d}\overline{\chi}_i = A\overline{\chi}_i$, imply that derivatives of the fermions are not gauge-invariant quantities. However, the operator product of gauge-invariant operators should only involve other gauge-invariant operators.

is (4.1.51). The amplitude corresponding to this diagram is

$$\frac{1}{\hbar^3}\overline{\chi}_j T_\alpha \chi^i \overline{\chi} T_\beta \chi \overline{\chi}_l T_\gamma \chi^k \int_{\mathbb{R}_x} V^{\alpha\beta\gamma}(x_1, x, x_2)\,. \tag{4.1.52}$$

Here, we have employed the expansions of $\mathcal{B} = \mathcal{B}^\alpha T_\alpha$ and $\mathcal{A} = \mathcal{A}^\beta T_\beta$ in the orthonormal basis $\{T_\alpha\}$ of $\mathfrak{gl}_N$.

Let us divide the integral of the vertex function $V^{\alpha\beta\gamma}$ into three integrals, which each depend on the location of the point x relative to x_1 and x_2, i.e., we write

$$\int_{\mathbb{R}_x} V^{\alpha\beta\gamma}(x_1, x, x_2) = \tilde{V}_{\mathsf{A}}^{\alpha\beta\gamma}(x_1, x_2) + \tilde{V}_{\mathsf{B}}^{\alpha\beta\gamma}(x_1, x_2) + \tilde{V}_{\mathsf{C}}^{\alpha\beta\gamma}(x_1, x_2)\,,$$

$$\tag{4.1.53}$$

where

$$\tilde{V}_{\mathsf{A}}^{\alpha\beta\gamma}(x_1, x_2) := \int_{x<x_1} V^{\alpha\beta\gamma}(x_1, x, x_2) = \frac{\hbar^2}{24} f^{\alpha\beta\gamma}\,, \tag{4.1.54a}$$

$$\tilde{V}_{\mathsf{B}}^{\alpha\beta\gamma}(x_1, x_2) := \int_{x_1<x<x_2} V^{\alpha\beta\gamma}(x_1, x, x_2) = \frac{\hbar^2}{24} f^{\alpha\beta\gamma}\,, \tag{4.1.54b}$$

$$\tilde{V}_{\mathsf{C}}^{\alpha\beta\gamma}(x_1, x_2) := \int_{x_2<x} V^{\alpha\beta\gamma}(x_1, x, x_2) = \frac{\hbar^2}{24} f^{\alpha\beta\gamma}\,. \tag{4.1.54c}$$

We thus find that (4.1.52) is

$$\frac{1}{8\hbar}\overline{\chi}_j T_\alpha \chi^i \overline{\chi} T_\beta \chi \overline{\chi}_l T_\gamma \chi^k f^{\alpha\beta\gamma}\,. \tag{4.1.55}$$

Let us now use the basis of elementary matrices as the basis for $\mathfrak{gl}_N$, which leads us to rewrite (4.1.55) as

$$\frac{\pi^2}{2\hbar}\overline{\chi}_j e_b^a \chi^i \overline{\chi} e_d^c \chi \overline{\chi}_l e_f^e \chi^k f_{ace}^{bdf}\,, \tag{4.1.56}$$

where $f_{ace}^{bdf} = \delta_a^d \delta_c^f \delta_e^b - \delta_c^b \delta_e^d \delta_a^f$, or equivalently,

$$\frac{1}{8\hbar}\left(\overline{\chi}_j \chi^m \overline{\chi}_m \chi^k \overline{\chi}_l \chi^i - \overline{\chi}_l \chi^m \overline{\chi}_m \chi^i \overline{\chi}_j \chi^k\right)\,, $$

$$\tag{4.1.57}$$

$$= \frac{1}{8}\hbar^2 \left(O_j^m[0] O_m^k[0] O_l^i[0] - O_l^m[0] O_m^i[0] O_j^k[0]\right)\,.$$

Since this expression is anti-symmetric under the exchange $(i, j) \leftrightarrow (k, l)$, the contribution of the diagram (4.1.51) to the quantum correction of the Lie bracket (4.1.46) is twice the value of the diagram.

Let us now proceed to consider the possible Feynman diagrams with a single fermionic propagator:

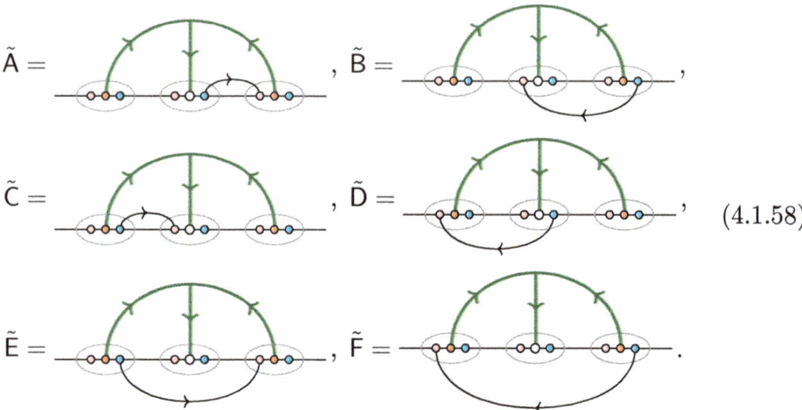

In these diagrams and all subsequent ones, the leftmost and rightmost operators are $O^i_j[1]$ located at x_1 and $O^k_l[1]$ located at x_2, respectively. Let us first evaluate the diagram $\tilde{\mathsf{A}}$. This diagram corresponds to

$$\tilde{\mathsf{A}} = \frac{1}{\hbar^3} \int_{\mathbb{R}_x} \overline{\chi}_j(x_1) T_\alpha \chi^i(x_1) \overline{\chi}^a_m(x) \left(T_\beta\right)^b_a \langle \chi^m_b(x) \overline{\chi}^c_l(x_2) \rangle \\ \times \left(T_\gamma\right)^d_c \chi^k_d(x_2) V^{\alpha\beta\gamma}(x_1, x, x_2), \tag{4.1.59}$$

where the two-point function is the quantum mechanical propagator given in (4.1.32), which depends on the sign of $x_2 - x$. Taking this into account, the amplitude is

$$\frac{1}{\hbar^2} \overline{\chi}_j T_\alpha \chi^i \overline{\chi}_l T_\beta T_\gamma \chi^k \left(\tilde{V}^{\alpha\beta\gamma}_{\mathsf{A}} + \tilde{V}^{\alpha\beta\gamma}_{\mathsf{B}} - \tilde{V}^{\alpha\beta\gamma}_{\mathsf{C}} \right), \\ = \frac{1}{24} \overline{\chi}_j T_\alpha \chi^i \overline{\chi}_l T_\beta T_\gamma \chi^k \, f^{\alpha\beta\gamma} = \frac{1}{24} \overline{\chi}_j T_\alpha \chi^i \overline{\chi}_l T_\delta \chi^k \, f^\delta_{\beta\gamma} f^{\alpha\beta\gamma}. \tag{4.1.60}$$

Since $f^\delta_{\beta\gamma} f^{\alpha\beta\gamma} = f^\alpha_{\beta\gamma} f^{\delta\beta\gamma}$, this amplitude is symmetric under the exchange $(i,j) \leftrightarrow (k,l)$, implying that it does not contribute quantum corrections to the Lie bracket (4.1.46). The diagrams $\tilde{\mathsf{B}}$, $\tilde{\mathsf{C}}$, and $\tilde{\mathsf{D}}$ do not contribute quantum corrections to the Lie bracket (4.1.46) for the same reason.

The remaining two diagrams have amplitudes that can be computed to be

$$\tilde{\mathsf{E}} = \frac{1}{8\hbar} f^{\alpha\beta\gamma} \delta^i_l \overline{\chi}_j T_\alpha T_\gamma \chi^k \overline{\chi} T_\beta \chi, \tag{4.1.61a}$$

$$\tilde{\mathsf{F}} = -\frac{1}{8\hbar} f^{\alpha\beta\gamma} \delta^k_j \overline{\chi}_l T_\gamma T_\alpha \chi^i \overline{\chi} T_\beta \chi. \tag{4.1.61b}$$

Since the sum of these amplitudes is symmetric under the exchange $(i,j) \leftrightarrow (k,l)$, they do not contribute quantum corrections to the Lie bracket (4.1.46).

In this way, one finds that none of the diagrams with one fermionic propagator contribute quantum corrections to the Lie bracket (4.1.46).

Let us now proceed to consider diagrams with two and three fermionic propagators. There are nine diagrams involving two fermionic propagators:

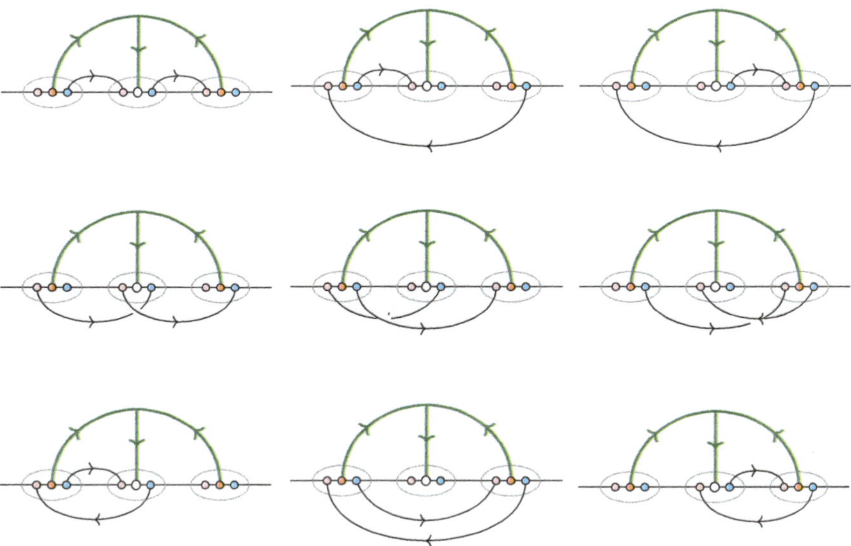

In addition, there are two diagrams with three fermionic propagators:

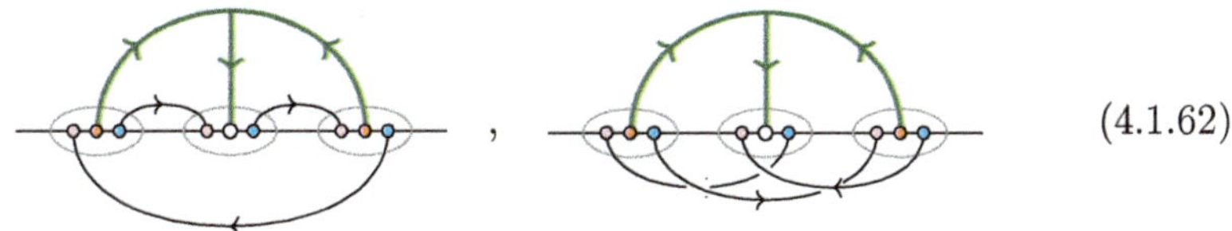

$$(4.1.62)$$

None of these diagrams with two and three fermionic propagators contribute quantum corrections to (4.1.46), and this can be shown using the symmetry properties of the associated amplitudes. The reader is invited to check this as an exercise, or consult the original reference [60] for details.

Thus, the quantum corrected version of the Lie bracket (4.1.46) can be written as

$$[O_j^i[1], O_l^k[1]]_{OPE} = \delta_l^i O_j^k[2] - \delta_j^k O_l^i[2]$$

$$+ \frac{\hbar^2}{4} \left(O_j^m[0]O_m^k[0]O_l^i[0] - O_l^m[0]O_m^i[0]O_j^k[0] \right).$$

$$(4.1.63)$$

This is an exact result, since there are no other non-vanishing Feynman diagrams.

Now, let us rewrite (4.1.63) using

$$O_j^m[0]O_m^k[0]O_l^i[0] = \{O_j^m[0], O_m^k[0], O_l^i[0]\}, \qquad (4.1.64)$$

where, for any three operators O_1, O_2 and O_3, one has

$$\{O_1, O_2, O_3\} = \frac{1}{3!} \sum_{s \in S_3} O_{s(1)} O_{s(2)} O_{s(3)}, \qquad (4.1.65)$$

where S_3 is the symmetric group of order 3!. Let us define

$$Q_{jl}^{ik} := f_{jvm}^{iun} f_{uor}^{vpq} f_{qsl}^{rtk} \{O_n^m[0], O_p^o[0], O_t^s[0]\}, \qquad (4.1.66)$$

where $f_{lmn}^{ijk} = \delta_l^j \delta_m^k \delta_n^i - \delta_m^i \delta_n^j \delta_l^k$ are the $\mathfrak{gl}_K$ structure constants in the basis of elementary matrices. The expression (4.1.66) can be rewritten as

$$Q_{jl}^{ik} = 3 \{O_l^i, O_j^m, O_m^k\} - 3 \{O_j^k, O_l^m, O_m^i\}$$

$$+ \delta_j^k \{O_l^m, O_m^n, O_n^i\} - \delta_l^i \{O_j^m, O_m^n, O_n^k\}, \qquad (4.1.67)$$

where the notation $[0]$ in each of the operators has been suppressed.

Now, performing the redefinition

$$\tilde{O}_j^k[2] := O_j^k[2] - \frac{\hbar^2}{12} \{O_j^m, O_m^n, O_n^k\}, \qquad (4.1.68)$$

one finds that (4.1.63) can be rewritten as the algebra

$$[O_j^i[1], O_l^k[1]]_{OPE} = \delta_l^i \tilde{O}_j^k[2] - \delta_j^k \tilde{O}_l^i[2] + \frac{\hbar^2}{12} Q_{jl}^{ik}. \qquad (4.1.69)$$

This Lie bracket appears to correspond to a presentation of the Yangian, which, in particular, appears in the work of Costello, Witten and Yamazaki [7], reviewed in Chapter 1.[9]

---

[9]This result can also be derived if one replaces the fermionic quantum mechanics with bosonic quantum mechanics.

However, note that as long as N is finite, $\mathcal{B}$ is a finite-dimensional $N \times N$ matrix, which implies that $\mathcal{B}^N$ can be written as a linear combination of matrices of the form $\mathcal{B}^p$, where $p < N$. Thus, in general, for finite N, there can be further relations between the operators $O_j^i[m]$. Such relations reduce the algebra to a *quotient* of the Yangian, and to avoid this and obtain the standard Yangian, one ought to take the large-N limit.

4.1.4　*Bulk algebra via anomaly of Wilson line*

In holographic dualities like AdS/CFT, correlation functions of local boundary operators are identified with functional derivatives (with respect to boundary values of the bulk fields) of the partition function of the bulk theory. In some cases involving supersymmetry/twisting, one can identify non-singular, position-independent *algebras* derived from the two dual theories. Indeed, if there is such an algebra of operators of the form

$$\lim_{x_1 \to x_2} O_i(x_1)O_j(x_2) = C_{ij}^k O_k(x_2) \tag{4.1.70}$$

in the boundary theory, for points x_1 and x_2 on the boundary, this algebra can be computed in the bulk theory as

$$\lim_{x_1 \to x_2} \frac{\delta}{\delta\phi^i(x_1)} \frac{\delta}{\delta\phi^j(x_2)} Z_{\text{bulk}}(\phi) = C_{ij}^k \frac{\delta}{\delta\phi^k(x_2)} Z_{\text{bulk}}(\phi), \tag{4.1.71}$$

where $Z_{\text{bulk}}(\phi)$ is the bulk partition function computed with particular boundary values of the bulk fields, ϕ^i.

In the present context of 4d Chern-Simons theory, one considers functional derivatives of the correlation function of a single Wilson line at $y = 0$, $z = 0$, associated with a representation $\rho : \mathfrak{gl}_K \to \text{End}(V)$, of the form

$$\frac{\delta}{\delta\partial_z^{n_1} A_{i_1}^{j_1}(\mathsf{a}_1)} \cdots \frac{\delta}{\delta\partial_z^{n_m} A_{i_m}^{j_m}(\mathsf{a}_m)} \langle W_\rho \rangle_A , \quad \mathsf{a}_1, \cdots, \mathsf{a}_m \in \ell_\infty(z), \tag{4.1.72}$$

where $\ell_\infty(z)$ is defined in (4.1.9), $n_1, n_2, \ldots, n_m \in \mathbb{N}$, and the subscript A in $\langle W_\rho \rangle_A$ denotes the dependence of expectation values on the fixed boundary profile of the $\mathfrak{gl}_k$ gauge field. In particular, $\frac{\delta}{\delta\partial_z^n A_i^j(\mathsf{a})} \langle W_\rho \rangle_A$ is identified with an operator $T_j^i[n](\mathsf{a})$ that acts on the representation space V, while (4.1.72) corresponds to the composition

$$T_{j_1}^{i_1}(\mathsf{a}_1) \cdots T_{j_m}^{i_m}(\mathsf{a}_m) \tag{4.1.73}$$

of such operators.

In the original work of Ishtiaque, Moosavian and Zhou [60], holographic duality was proved by deriving the algebra associated with the bulk 4d Chern-Simons theory (with Wilson line insertion) using Witten diagrams, and showing that it takes the same form as the algebra (4.1.69). This involved the evaluation of diagrams that involved bulk-boundary propagators. The main Witten diagrams that contribute nontrivially consist of the diagram

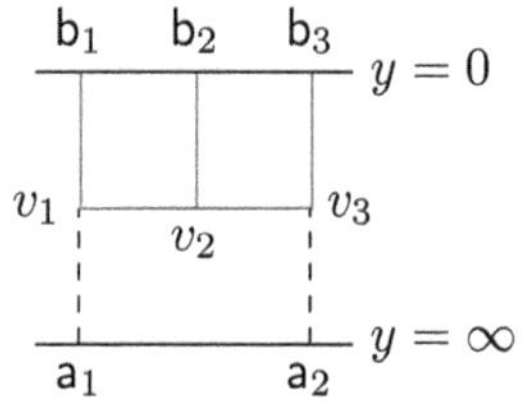

$$\text{(4.1.74)}$$

and five other diagrams that can be obtained from (4.1.74) by permuting the points b_1, b_2 and b_3. In (4.1.74), the solid line at $y = 0$ denotes the Wilson line, while the solid line at $y = \infty$ denotes the boundary of interest. The interaction vertices v_1, v_2, and v_3 are connected to the familiar 4d CS bulk propagators, as well as bulk-boundary propagators that are denoted by dashed lines. These bulk-boundary propagators are explicitly derived in [60]. We note the similarity between the diagram (4.1.74) and the leftmost diagram in Figure 1.12 that was used to derive the order $\hbar^2$ terms of the Yangian in Chapter 1.

The evaluation of the six Witten diagrams that can be obtained from (4.1.74) led to the result that the algebra associated with the bulk theory is the Yangian of $\mathfrak{gl}_K$, as long as one takes the large-N limit. The large-N limit is important here because the representation, ρ, of the Wilson line depends on N since it was engineered using the N 2-branes, and therefore operators that enter the bulk algebra will satisfy additional relations unless one takes the large-N limit.

We shall not evaluate the Witten diagrams explicitly here, but instead we shall review an indirect argument given in [60] that proves that the bulk algebra must be the Yangian.

Observe that as we vary the background value of the gauge field A along the boundary line $\ell_\infty(z)$, the variation of the expectation value of the Wilson line, $\langle W_\rho \rangle_A$, is

$$\delta \langle W_\rho \rangle_A = \sum_{n=0}^{\infty} \int_{\mathsf{a} \in \ell_\infty(z)} \frac{\delta}{\delta \partial_z^n A^\mu(\mathsf{a})} \langle W_\rho \rangle_A \, \delta \partial_z^n A^\mu(\mathsf{a}) \,. \tag{4.1.75}$$

Here, and in what follows, $\mu, \nu, \xi, \ldots$ are Lie algebra indices for an orthonormal basis $\{t^\mu\}$ of $\mathfrak{gl}_K$. If one makes the following choice of variation:

$$\delta \partial_z A^\mu(x) = \delta^1(x - \mathsf{a})k^\mu = \mathrm{d}_x \tilde\theta(x - \mathsf{a})k^\mu \,, \tag{4.1.76}$$

for a specific Lie algebra element $k^\mu t_\mu \in \mathfrak{gl}_K$, where

$$\tilde\theta(x) = \begin{cases} 1 & \text{for } x > 0 \\ 1/2 & \text{for } x = 0 \\ 0 & \text{for } x < 0, \end{cases} \tag{4.1.77}$$

one finds

$$\delta \langle W_\rho \rangle_A = \frac{\delta}{\delta \partial_z A^\mu(\mathsf{a})} \langle W_\rho \rangle_A \, k^\mu \,, \tag{4.1.78}$$

by integrating over the delta function.

A variation of the boundary value of the gauge field of this type can be thought of as a gauge transformation that does not vanish at the boundary. Let us denote the support of the Wilson line at $y = 0$ and $z = 0$ as L. From results of [7], reviewed in Chapter 1, such a variation leads to a variation of the Wilson line of the form

$$\delta \langle W_\rho \rangle_A = \left([\rho_{\mu,1}, \rho_{\nu,1}] + \Theta_{\mu,1,\nu,1}\right) \int_L \partial_z A^\mu \partial_z \mathsf{c}^\nu \,, \tag{4.1.79}$$

which arises from diagrams (A) and (B) in Figure 1.13, where c is the ghost field satisfying

$$\partial_z \mathrm{d} \mathsf{c}^\mu = \delta \partial_z A^\mu, \tag{4.1.80}$$

$\rho_{\mu,1} \in \mathrm{End}(V)$ is the part of the representation of $\mathfrak{gl}_K[[z]]$ that couples the derivative of the gauge field, $\partial_z A^\mu$, to the Wilson line, and $\Theta_{\mu,1,\nu,1}$, which is antisymmetric in μ and ν, is a matrix that acts on V, whose explicit expression can be deduced from equation (1.5.27).[10]

---

[10] In Chapter 1, field variations of this sort were understood to measure the gauge anomaly associated with the Wilson line. However, in the present context, one is not measuring an anomaly, since gauge transformations arising from varying the gauge field at the boundary are not actually part of the gauge symmetry of the theory.

From (4.1.76) and (4.1.80), we observe that choosing the ghost field to satisfy $\partial_z c^\mu(x) = \tilde{\theta}(x - \mathsf{a})k^\mu$ leads to:

$$\delta \langle W_\rho \rangle_A = \left(f^\xi_{\mu\nu}\rho_{\xi,2} + \Theta_{\mu,1,\nu,1}\right) \int_{x>\mathsf{a}} \partial_z A^\mu k^\nu$$

$$= \frac{\delta}{\delta\partial_z A^\nu(\mathsf{a})} \langle W_\rho \rangle_A\, k^\nu,$$

(4.1.81)

where we have used (4.1.78) and the fact that the matrices $\rho_{\mu,1}$ satisfy the polynomial loop algebra.

Since A is a background gauge field that satisfies the equations of motion, it takes the form of a flat connection. We shall assume that there are no non-contractible loops on the the manifold on which 4d CS is defined. The symmetry along Σ implies that the gauge field must be translation invariant along the direction of the Wilson line, which is the line L. Now, consider the integral of A along the following contour:

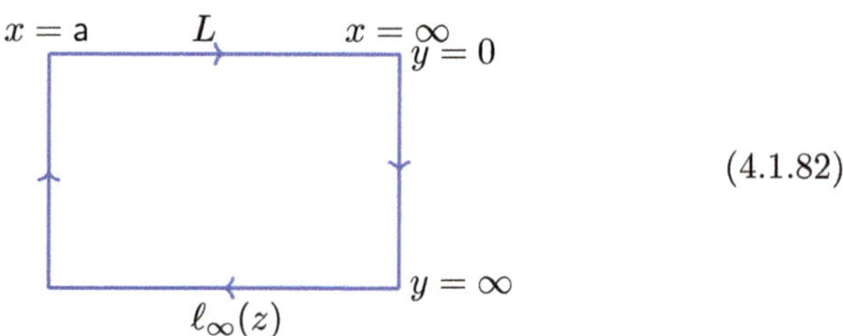

(4.1.82)

Since A is flat, this integral must be equal to zero. Moreover, translation invariance in the x-direction means that the contributions from the vertical segments of the contour cancel. This means that we can modify the support of the integral in (4.1.81) from L to $\ell_\infty(z)$, to obtain:

$$\delta \langle W_\rho \rangle_A = \left(f^\xi_{\mu\nu}\rho_{\xi,2} + \Theta_{\mu,1,\nu,1}\right) \int_{x>\mathsf{a},x\in\ell_\infty(z)} \partial_z A^\mu k^\nu\,.$$

(4.1.83)

Then, from (4.1.81), we find

$$\frac{\delta}{\delta\partial_z A^\nu(\mathsf{a})} \langle W_\rho \rangle_A = \left(f^\xi_{\mu\nu}\rho_{\xi,2} + \Theta_{\mu,1,\nu,1}\right) \int_{x>\mathsf{a},x\in\ell_\infty(z)} \partial_z A^\mu\,.$$

(4.1.84)

We would like to compute the algebra

$$[T_\mu[1], T_\nu[1]]$$

$$= \lim_{\epsilon\to 0}\left[\frac{\delta}{\delta\partial_z A^\mu(\mathsf{a}+\epsilon)}\frac{\delta}{\delta\partial_z A^\nu(\mathsf{a})} - \frac{\delta}{\delta\partial_z A^\nu(\mathsf{a})}\frac{\delta}{\delta\partial_z A^\mu(\mathsf{a}+\epsilon)}\right] \langle W_\rho \rangle_A\,.$$

(4.1.85)

From (4.1.84), we find that the RHS of (4.1.85) is $f^{\xi}_{\mu\nu}\rho_{\xi,2} + \Theta_{\mu,1,\nu,1}$. A simple Witten diagram computation identifies $\rho_{\xi,2}$ with $T_{\xi}[2]$. Then, using the explicit form of $\Theta_{\mu,1,\nu,1}$ given in (1.5.27), one finds the algebra

$$[T_\mu[1], T_\nu[1]] = f^{\xi}_{\mu\nu} T_{\xi}[2] + \tfrac{\hbar^2}{12} f^{\sigma\tau}_{\nu} f^{\upsilon\psi}_{\sigma} f^{\omega}_{\mu\nu} \{T_\tau[0], T_\psi[0], T_\omega[0])\}. \qquad (4.1.86)$$

Working in the large-N limit, this means that the bulk algebra of operators $T^\mu[n]$ is the Yangian of $\mathfrak{gl}_K$. This confirms the holographic duality between 4d Chern-Simons theory with a Wilson line insertion and 2d BF theory coupled to 1d quantum mechanics, via the identification of bulk and boundary algebras.

4.2 3d WZW Model

We shall now pursue the study of an alternate holographic dual to 4d Chern–Simons theory, namely, the 3d "chiral" WZW model. The results reviewed in this section are based on the work of Ashwinkumar [61].

The setup in which we are interested is 4d Chern–Simons theory with a complex gauge group, G, defined on $\Sigma \times C$, where Σ is a disk, denoted D, and C is the complex plane, $\mathbb{C}$. Its action is

$$S = \frac{1}{2\pi\hbar} \int_{D\times\mathbb{C}} \mathrm{d}z \wedge \mathrm{Tr}\left(\mathcal{A} \wedge \mathrm{d}\mathcal{A} + \frac{2}{3}\mathcal{A} \wedge \mathcal{A} \wedge \mathcal{A} \right), \qquad (4.2.1)$$

where $\mathcal{A}$ is the partial connection

$$\mathcal{A} = \mathcal{A}_r \mathrm{d}r + \mathcal{A}_\varphi \mathrm{d}\varphi + \mathcal{A}_{\bar{z}}\mathrm{d}\bar{z}, \qquad (4.2.2)$$

and where (r, φ) are polar coordinates on D and $(z, \bar{z})$ are complex coordinates on $\mathbb{C}$.

Varying the action, one finds

$$\delta S = \frac{1}{2\pi\hbar} \int_{D\times\mathbb{C}} \mathrm{d}z \wedge \mathrm{Tr}\left(\delta\mathcal{A} \wedge \mathcal{F} + \mathrm{d}(\delta\mathcal{A} \wedge \mathcal{A}) \right). \qquad (4.2.3)$$

Using Stokes' theorem, the total derivative term can be converted to a boundary term. To ensure that the equations of motion are local and free from boundary corrections, the boundary condition $\mathcal{A}_{\bar{z}} = 0|_{\partial D}$ is imposed, whereby the boundary term vanishes.

To prove that the action is gauge invariant, we extend $\mathcal{A}$ to a *full* connection over $D \times \mathbb{C}$, i.e., $\mathcal{A} = \mathcal{A}_r \mathrm{d}r + \mathcal{A}_\varphi \mathrm{d}\varphi + \mathcal{A}_z \mathrm{d}z + \mathcal{A}_{\bar{z}}\mathrm{d}\bar{z}$, and rewrite

(4.2.1) as

$$S = -\frac{1}{2\pi\hbar} \int_{D\times\mathbb{C}} z\mathrm{Tr}\left(\mathcal{F} \wedge \mathcal{F}\right) + \frac{1}{2\pi\hbar} \int_{\partial D\times\mathbb{C}} z\mathrm{Tr}\left(\mathcal{A} \wedge d\mathcal{A} + \frac{2}{3}\mathcal{A} \wedge \mathcal{A} \wedge \mathcal{A}\right).$$

$$(4.2.4)$$

The boundary term in this expression depends only on the components $\mathcal{A}_\varphi$, $\mathcal{A}_z$ and $\mathcal{A}_{\bar{z}}$, and vanishes using the boundary condition $\mathcal{A}_{\bar{z}}|_{\partial D} = 0$ that was imposed previously, as well as the boundary condition $\mathcal{A}_z|_{\partial D} = 0$. The remaining term is gauge invariant under large gauge transformations of the form

$$\mathcal{A} \to U\mathcal{A}U^{-1} - dUU^{-1}. \tag{4.2.5}$$

Here, we ought to restrict U such that the boundary conditions $\mathcal{A}_{\bar{z}}|_{\partial D} = 0$ and $\mathcal{A}_z|_{\partial D} = 0$ are preserved. This can be achieved by insisting that U tends to the identity element of G at the boundary.

Now, using the boundary condition $\mathcal{A}_{\bar{z}}|_{\partial D} = 0$, the action (4.2.1) can be shown to be equivalent to

$$\frac{1}{2\pi\hbar} \int dz \wedge dr \wedge d\varphi \wedge d\bar{z} \,\, \mathrm{Tr}\left(2\mathcal{A}_{\bar{z}}\mathcal{F}_{r\varphi} - \mathcal{A}_r\partial_{\bar{z}}\mathcal{A}_\varphi + \mathcal{A}_\varphi\partial_{\bar{z}}\mathcal{A}_r\right), \quad (4.2.6)$$

upon integration by parts. Varying $\mathcal{A}_{\bar{z}}$ results in the constraint $\mathcal{F}_{r\varphi} = 0$, which is solved by

$$\mathcal{A}_r = -\partial_r gg^{-1}, \quad \mathcal{A}_\varphi = -\partial_\varphi gg^{-1}, \tag{4.2.7}$$

where g is a G-valued field on $D \times \mathbb{C}$.

At the level of the functional integral, when changing variables from $\mathcal{A}_r$ and $\mathcal{A}_\varphi$ to g, no Jacobian appears when transforming the measure, i.e.,

$$\frac{1}{vol\,G} \int D\mathcal{A}_r D\mathcal{A}_\varphi \,\, \delta(\mathcal{F}_{r\varphi}) = \frac{1}{vol\,G} \int Dg, \tag{4.2.8}$$

where the expression on the RHS is the functional integral for g defined using the Haar measure for G, divided by the volume of G. This is analogous to 3d Chern–Simons theory defined on $D \times \mathbb{R}$ [66].

Now, substituting the solutions (4.2.7) into (4.2.6), we obtain the action

$$S(g) = \frac{1}{2\pi\hbar} \int_{S^1\times\mathbb{C}} d\varphi \wedge dz \wedge d\bar{z} \,\, \mathrm{Tr}(\partial_\varphi gg^{-1}\partial_{\bar{z}}gg^{-1})$$

$$+ \frac{1}{6\pi\hbar} \int_{D\times\mathbb{C}} dz \wedge \mathrm{Tr}(dgg^{-1} \wedge dgg^{-1} \wedge dgg^{-1}),$$

$$(4.2.9)$$

which is a three-dimensional analogue of the 2d chiral WZW model. The focus of the rest of this section will be the study of this model, that shall be referred to as the 3d "chiral" WZW model.

The large gauge transformation (4.2.5) amounts to $g \to Ug$ in (4.2.7). As a result, we may change the value of g in the interior without changing its value at the boundary, so (4.2.9) only depends on g at the boundary.[11] This implies that we can divide out the volume of the gauge group to obtain the path integral

$$\int Dg \ e^{iS(g)}, \qquad (4.2.10)$$

where g is now a G-valued field on $\partial D \times \mathbb{C}$.

The 3d "chiral" WZW model action (4.2.9) can be varied to give us

$$\delta S = -\frac{1}{\pi \hbar} \int d\varphi \wedge dz \wedge d\bar{z} \ \mathrm{Tr}(g^{-1}\delta g \partial_\varphi(g^{-1}\partial_{\bar{z}}g)), \qquad (4.2.11)$$

whereby we obtain the classical equation of motion

$$\partial_\varphi(g^{-1}\partial_{\bar{z}}g) = 0, \qquad (4.2.12)$$

which is equivalent to $\partial_{\bar{z}}(\partial_\varphi g g^{-1}) = 0$, and has the solution

$$g(z, \bar{z}, \varphi) = A(z, \varphi)B(z, \bar{z}). \qquad (4.2.13)$$

The equations $\partial_\varphi(g^{-1}\partial_{\bar{z}}g) = 0$ and $\partial_{\bar{z}}(\partial_\varphi g g^{-1}) = 0$ can be understood to be current conservation equations for the symmetry of the action under the transformation

$$g(\varphi, z, \bar{z}) \to \tilde{\Omega}(\varphi, z)g\Omega(z, \bar{z}), \qquad (4.2.14)$$

where $\tilde{\Omega}$ and Ω give rise to the conserved currents $J_\varphi = -\frac{1}{\pi\hbar}\partial_\varphi g g^{-1}$ and $J_{\bar{z}} = -\frac{1}{\pi\hbar}g^{-1}\partial_{\bar{z}}g$, respectively.

---

[11]For the 2d chiral WZW model dual to 3d Chern–Simons theory on $D \times \mathbb{R}$ with a compact gauge group studied in [66], there is quantization of the WZW level/Chern–Simons coupling. However, the coupling $\hbar$ that appears in (4.2.9) is not quantized. This difference stems from the fact that invariance of the 3d Chern–Simons action under large gauge transformations of the form (4.2.5) (which leads to the Wess–Zumino term being independent of the choice of extension) requires such a quantization, while large gauge invariance of the 4d Chern–Simons action does not require such a quantization, as shown below (4.2.4).

4.2.1 *Toroidal Lie Algebra as a Current Algebra via Canonical Quantization*

It is well-known that the 2d WZW model has conserved currents that realize an affine Kac–Moody algebra. It is thus natural to ask whether a current algebra can be generated from the conserved currents of the 3d "chiral" WZW model.

It can be shown that this is indeed the case, by computing a Poisson bracket involving the conserved current $J_\varphi = -\frac{1}{\pi\hbar}\partial_\varphi g g^{-1}$. This Poisson bracket can be used to obtain a quantum mechanical commutation relation in the form of a current algebra. In what follows, we shall take $\bar{z}$ to be the (complex) time direction.

Firstly, recall that if we have an arbitrary action that is first order in the time derivative with dynamical variables ϕ^i, i.e.,

$$I = \int dt\, \mathscr{A}_i(\phi)\frac{d\phi^i}{dt}, \qquad (4.2.15)$$

its variation takes the form

$$\delta I = \int dt\, \omega_{ij}\delta\phi^i\frac{d\phi^j}{dt}, \qquad (4.2.16)$$

where $\omega_{ij} = \frac{\partial}{\partial\phi^i}\mathscr{A}_j - \frac{\partial}{\partial\phi^j}\mathscr{A}_i$ defines the symplectic structure on the classical phase space. The Poisson bracket of any two functions X and Y on the phase space can be defined by

$$[X,Y]_{PB} = \omega^{ij}\frac{\partial X}{\partial\phi^i}\frac{\partial Y}{\partial\phi^j}, \qquad (4.2.17)$$

where $\omega^{jk}\omega_{kl} = \delta^j_l$.

The variation (4.2.11) implies that the phase space symplectic structure of the 3d "chiral" WZW model is given by $\omega = 1_{\mathfrak{g}} \otimes \frac{(-1)}{\pi\hbar}\frac{\partial}{\partial\varphi} \otimes 1_z$, where $1_{\mathfrak{g}}$ acts on the Lie algebra index, $\frac{(-1)}{\pi\hbar}\frac{\partial}{\partial\varphi}$ acts on the φ coordinate, and 1_z acts on the z coordinate. Its inverse is

$$\omega^{-1} = 1_{\mathfrak{g}} \otimes (-\pi\hbar)\left(\frac{\partial}{\partial\varphi}\right)^{-1} \otimes 1_z, \qquad (4.2.18)$$

which can be used to compute Poisson brackets in this theory.

We would like to compute the Poisson bracket of $X = \mathrm{Tr}A\frac{\partial g}{\partial\varphi}g^{-1}(\varphi, z)$ and $Y = \mathrm{Tr}B\frac{\partial g}{\partial\varphi'}g^{-1}(\varphi', z')$, where A and B are arbitrary elements of the Lie algebra of the group G. In general, using the notation of (4.2.15), one can compute the Poisson bracket of two functions X and Y on phase space by evaluating $\delta X \delta Y = \frac{\partial X}{\partial \phi^i} \frac{\partial Y}{\partial \phi^j} \delta\phi^i \delta\phi^j$, and subsequently replacing $\delta\phi^i \delta\phi^j$ by ω^{ij}.

In the case at hand, we find

$$\delta X \delta Y = \mathrm{Tr}\ g^{-1}(\varphi, z) A g(\varphi, z) \frac{\partial}{\partial \varphi}(g^{-1}\delta g(\varphi, z))$$

$$\cdot \mathrm{Tr}\ g^{-1}(\varphi', z') B g(\varphi', z') \frac{\partial}{\partial \varphi'}(g^{-1}\delta g(\varphi', z')). \qquad (4.2.19)$$

The Poisson bracket can then be obtained by replacing $(g^{-1}\delta g(\varphi, z))^a$ $(g^{-1}\delta g(\varphi', z'))^b$ (where a and b are Lie algebra indices) by

$$\delta^{ab}(-\pi\hbar)\theta(\varphi - \varphi')\delta(z - z'), \qquad (4.2.20)$$

where $\theta(\varphi - \varphi')$ is the inverse of $\frac{\partial}{\partial \varphi}$. This is equivalent to saying that $\frac{\partial}{\partial \varphi}(g^{-1}\delta g(\varphi, z))^a \cdot \frac{\partial}{\partial \varphi'}(g^{-1}\delta g(\varphi', z'))^b$ in (4.2.19) ought to be replaced by $\delta^{ab}\pi\hbar\delta'(\varphi - \varphi')\delta(z - z')$. By doing so, we arrive at the Poisson bracket

$$[X, Y]_{PB} = \pi\hbar\delta'(\varphi - \varphi')\delta(z - z')\mathrm{Tr}\ g^{-1}(\varphi, z) A g(\varphi, z) g^{-1}(\varphi', z') B g(\varphi', z')$$

$$= \pi\hbar\delta(\varphi - \varphi')\delta(z - z')\mathrm{Tr}\left([A, B]\frac{\partial g}{\partial \varphi}g^{-1}\right)$$

$$+ \pi\hbar\delta'(\varphi - \varphi')\delta(z - z')\mathrm{Tr}\ AB. \qquad (4.2.21)$$

Rescaling both sides of this equation by $\left(\frac{1}{-\pi\hbar}\right)^2$, we arrive at a classical current algebra for J_φ :

$$[\mathrm{Tr}A J_\varphi(\varphi, z), \mathrm{Tr}B J_\varphi(\varphi', z')]_{PB} = -\delta(\varphi - \varphi')\delta(z - z')\mathrm{Tr}[A, B]J_\varphi(\varphi, z)$$

$$+ \frac{1}{\pi\hbar}\delta'(\varphi - \varphi')\delta(z - z')\mathrm{Tr}AB. \qquad (4.2.22)$$

The second term of (4.2.22) is reminiscent of the central extension of an affine Kac–Moody algebra, which itself arises in the Poisson bracket for currents in the 2d chiral WZW model, as shown in [67].

In the quantum 3d "chiral" WZW model, the Poisson bracket (4.2.22) is replaced by the canonical commutation relation

$$[\mathrm{Tr}A J_\varphi(\varphi, z), \mathrm{Tr}B J_\varphi(\varphi', z')] = i\tilde{\hbar}\delta(\varphi - \varphi')\delta(z - z')\mathrm{Tr}[A, B]J_\varphi(\varphi, z)$$

$$- \frac{i\tilde{\hbar}}{\pi\hbar}\delta'(\varphi - \varphi')\delta(z - z')\mathrm{Tr}AB + O(\tilde{\hbar}^2), \qquad (4.2.23)$$

where $\tilde{\hbar}$ is Planck's constant (recall that $\hbar$ here is a coupling constant, based on the definition of the action in (4.2.1)), and where we have indicated the possibility of terms that are higher order in Planck's constant.

To express (4.2.23) in a more standard form, one can expand the currents in terms of their Fourier modes along S^1,

$$J_\varphi(\varphi, z) = \frac{1}{2\pi} \sum_{n=-\infty}^{\infty} J_\varphi^n(z) e^{in\varphi}, \tag{4.2.24}$$

and utilize the orthogonality of these modes, which furnishes

$$\left[\mathrm{Tr}AJ_\varphi^n(z), \mathrm{Tr}BJ_\varphi^m(z')\right] = i\mathrm{Tr}[A, B]J_\varphi^{n+m}(z)\delta(z - z')$$

$$+ \frac{2}{\hbar}(n\delta_{m+n,0})\delta(z - z')\mathrm{Tr}AB + \cdots. \tag{4.2.25}$$

where we have indicated the terms that are higher order in $\tilde{\hbar}$ by an ellipsis, and set $\tilde{\hbar} = 1$ for convenience. This algebra has the form of an affine Kac–Moody algebra whose generators have holomorphic dependence on the complex plane, $\mathbb{C}$. Unlike the current algebra derived from the 2d WZW model, there is no quantization condition on $\hbar$ here.

To further understand this algebra, we shall write $z = \epsilon t + i\theta$, compactify the θ direction to be valued in $[0, 2\pi]$, and subsequently take $\epsilon \to 0$. Upon doing so, we may perform another expansion in Fourier modes, i.e.,

$$J_\varphi^n(\theta) = \frac{1}{2\pi} \sum_{\tilde{n}=-\infty}^{\infty} J_\varphi^{n,\tilde{n}} e^{i\tilde{n}\theta}. \tag{4.2.26}$$

Then, employing the orthogonality of these modes, the resulting algebra takes the form

$$\left[\mathrm{Tr}AJ_\varphi^{n,\tilde{n}}, \mathrm{Tr}BJ_\varphi^{m,\tilde{m}}\right] = i\mathrm{Tr}[A, B]J_\varphi^{n+m,\tilde{n}+\tilde{m}}$$

$$+ \frac{4\pi}{\hbar}n\delta_{m+n,0}\delta_{\tilde{m}+\tilde{n},0}\mathrm{Tr}AB + \cdots. \tag{4.2.27}$$

This algebra has the form of a two-toroidal Lie algebra (or a centrally-extended double loop algebra).[12] Hence, the algebra (4.2.25) can be understood to be an 'analytical continuation' of the two-toroidal Lie algebra (4.2.27), with one of the two 'loop directions' decompactified.[13] This is not

---

[12] A two-toroidal Lie algebra also arises as the current algebra of the four-dimensional WZW model studied in [68–71].

[13] In the preceding analysis, we could have alternatively taken the radial direction on the complex plane to be the time direction, and we would have arrived at the current algebra (4.2.27).

surprising, considering the fact that 4d Chern–Simons theory for $C = \mathbb{R} \times S^1$ can be understood to be 3d Chern–Simons theory for the loop group, but with the 'loop direction' complexified [72].

4.2.2 *Wilson lines, boundary local operators and R-matrices*

In this subsection, we shall use the 3d "chiral" WZW model to derive the quasi-classical expansion of the rational R-matrix from correlation functions of local operators, up to linear order in $\hbar$. These local operators correspond to the endpoints of Wilson lines in the bulk, and therefore, what we shall describe is evidence of a holographic duality between 4d Chern–Simons theory and the 3d "chiral" WZW model.

Let us first relate Wilson lines in 4d Chern–Simons theory to local operators of its boundary dual, the 3d "chiral" WZW model. This relationship arises due to the flatness condition that restricts the gauge field components on D to be pure gauge configurations, as shown in (4.2.7). To observe this explicitly, recall that a Wilson line along a curve $\mathcal{C}$, starting at t_i and ending at t_f, and in a representation R, satisfies

$$\mathcal{P}e^{\int_{\mathcal{C}} \mathcal{A}} = g_R^{-1}(t_f)\mathcal{P}e^{\int_{\mathcal{C}} \mathcal{A}'} g_R(t_i) \tag{4.2.28}$$

where $\mathcal{A} = g\mathcal{A}'g^{-1} - dgg^{-1}$. Setting $\mathcal{A}' = 0$ so that $\mathcal{A}$ is pure gauge, one finds that

$$\mathcal{P}e^{\int_{\mathcal{C}}(-dgg^{-1})} = g_R^{-1}(t_f)g_R(t_i). \tag{4.2.29}$$

If t_f and t_i are points on $\partial D = S^1$, this implies that a bulk Wilson line operator corresponds to a pair of local boundary operators. In fact, such boundary-anchored Wilson lines are automatically gauge invariant, since all gauge transformations are trivial at the boundary.

This suggests that correlation functions of Wilson line operators in 4d Chern–Simons theory can be captured by correlation functions of local operators in the 3d "chiral" WZW model. This includes correlators of crossed Wilson lines which compute the R-matrices of integrable lattice models. In the following, evidence for this proposal shall be elucidated.

In the bulk computation that was reviewed in Chapter 1, the order $\hbar$ contribution to the R-matrix was found by performing perturbation theory around the trivial field configuration, $\mathcal{A} = 0$, and evaluating interactions between crossed Wilson lines using the gluon propagator. In the same vein, one can consider perturbation theory around the field configuration $g = \mathbb{1}$, and use the propagator of the 3d "chiral" WZW model to compute this correlation function using local operators that appear on the right hand side of (4.2.29). In other words, we shall use the equivalence

$$\left\langle \mathcal{P}e^{\int_{\varphi=\pi}^{\varphi=0} \mathcal{A}_{R_1}(z_1,\bar{z}_1)} \otimes \mathcal{P}e^{\int_{\varphi=3\pi/2}^{\varphi=\pi/2} \mathcal{A}_{R_2}(z_2,\bar{z}_2)} \right\rangle$$

$$= \left\langle g_{R_1}^{-1}(0, z_1, \bar{z}_1) g_{R_1}(\pi, z_1, \bar{z}_1) \otimes g_{R_2}^{-1}(\pi/2, z_2, \bar{z}_2) g_{R_2}(3\pi/2, z_2, \bar{z}_2) \right\rangle,$$
$$\tag{4.2.30}$$

to compute the R-matrix holographically, using the 3d "chiral" WZW model. The configuration of (4.2.30) is depicted in Figure 4.3.

In perturbation theory, the field g may be expanded as

$$g = e^{\phi_a T^a} = \mathbb{1} + \phi_a T^a + \dots \tag{4.2.31}$$

The free part of the 3d WZW action is then

$$\frac{1}{2\pi\hbar} \int_{S^1 \times \mathbb{C}} \mathrm{d}\varphi \wedge \mathrm{d}z \wedge \mathrm{d}\bar{z}\ \mathrm{Tr}\left(\partial_\varphi g g^{-1} \partial_{\bar{z}} g g^{-1}\right)$$

$$= -\frac{1}{2\pi\hbar} \int_{S^1 \times \mathbb{C}} \mathrm{d}\varphi \wedge \mathrm{d}z \wedge \mathrm{d}\bar{z}\ \phi^a \partial_\varphi \partial_{\bar{z}} \phi_a + \cdots, \tag{4.2.32}$$

where we have performed integration by parts after expanding the field g.

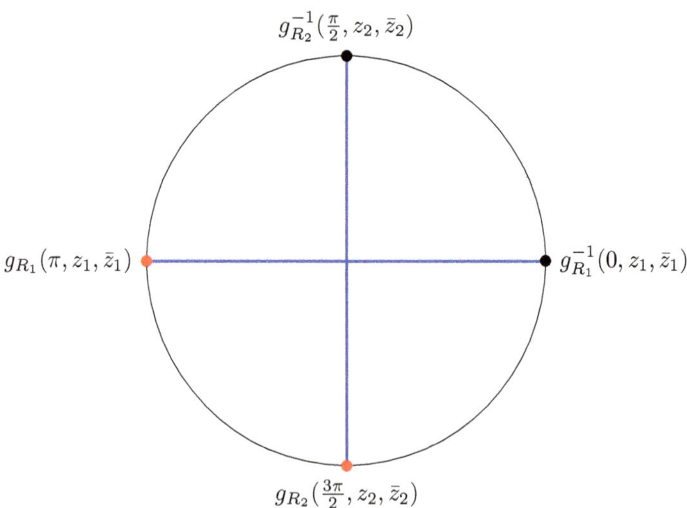

Figure 4.3. Perpendicular Wilson lines on D, that are located at distinct points, denoted z_1 and z_2, on $C = \mathbb{C}$.

From this free action, one may construct the generating functional

$$Z_0[J] = \frac{\int D\phi\, e^{\frac{i}{2\pi\hbar}\int_{S^1\times\mathbb{C}} d\varphi\wedge dz\wedge d\bar{z}(-\phi^a\partial_\varphi\partial_{\bar{z}}\phi_a + 2\pi i\hbar J_a\phi^a)}}{\int D\phi\, e^{\frac{i}{2\pi\hbar}\int_{S^1\times\mathbb{C}} d\varphi\wedge dz\wedge d\bar{z}(-\phi^a\partial_\varphi\partial_{\bar{z}}\phi_a)}}$$

$$= \exp\left(-\frac{2\pi i\hbar}{4}\int d^3x\int d^3y\, J_a(x)\Delta^{ab}(x-y)J_b(y)\right), \tag{4.2.33}$$

where $x = (\varphi, z, \bar{z})$, $y = (\varphi', z', \bar{z}')$, and Δ^{ab} is the propagator which obeys

$$\partial_\varphi\partial_{\bar{z}}\Delta^{ab}(x) = \delta^{ab}\delta^3(x), \tag{4.2.34}$$

and is given explicitly by

$$\Delta^{ab}(x) = \delta^{ab}\frac{1}{2\pi i}\frac{1}{z}\tilde{\Delta}_\varphi \tag{4.2.35}$$

where

$$\tilde{\Delta}_\varphi = \frac{1}{2\pi}\left(\sum_{k=1}^{\infty}\frac{e^{ik\varphi}}{ik} + \varphi + \sum_{k=-\infty}^{-1}\frac{e^{ik\varphi}}{ik}\right), \tag{4.2.36}$$

which satisfies $\partial_\varphi\tilde{\Delta}_\varphi = \frac{1}{2\pi}\left(\sum_{k=-\infty}^{\infty}e^{ik\varphi}\right) = \delta(\varphi)$. The propagator can be shown to satisfy $\Delta^{ab}(x-y) = \Delta^{ba}(y-x)$. The two point function for ϕ can be determined, using (4.2.33), to be

$$\langle\phi^a(x)\phi^b(y)\rangle = -\pi i\hbar\Delta^{ab}(x-y). \tag{4.2.37}$$

This two-point function is multi-valued, as it includes the expression on the RHS of (4.2.36) as a factor. In order to make it single-valued, we ought to define it with a branch cut. We shall pick the branch cut to be from $r = 0$ to $(r = R, \varphi = \pi)$, where R is the radius of D. This effectively restricts φ in (4.2.36) to take values in $(-\pi, \pi)$. In this manner, we obtain a well-defined, single-valued two-point function.

Now, to compute the RHS of (4.2.30), we ought to expand each operator to linear order in ϕ

$$\langle g_{R_1}^{-1}(0, z_1)g_{R_1}(\pi, z_1) \otimes g_{R_2}^{-1}(\pi/2, z_2)g_{R_2}(3\pi/2, z_2)\rangle$$

$$= \langle(1 - \phi_a(0, z_1)T_{R_1}^a)(1 + \phi_b(\pi, z_1)T_{R_1}^b)$$

$$\otimes(1 - \phi_c(\pi/2, z_2)T_{R_2}^c)(1 + \phi_d(3\pi/2, z_2)T_{R_2}^d)\rangle + \cdots. \tag{4.2.38}$$

We only keep terms of quadratic or lower order in the fields, while taking self-contractions (i.e., correlators of operators with the same value of z) to be zero via regularization.

This regularization involves using the Feynman parametrization

$$\frac{1}{z} = \int_{t=0}^{\infty} \bar{z}\, e^{-\frac{z\bar{z}}{t}} \frac{\mathrm{d}t}{t^2}, \qquad (4.2.39)$$

whereby one can define a UV regulator by computing the integral over the region $t \geq \epsilon$, where ϵ is a positive real parameter that can be interpreted as the inverse of the UV cutoff. We shall take the $\epsilon \to 0$ limit at the end of calculations. Evaluating the integral over the region $t \geq \epsilon$, we have

$$\int_{t=\epsilon}^{\infty} \bar{z}\, e^{-\frac{z\bar{z}}{t}} \frac{\mathrm{d}t}{t^2} = \frac{1 - e^{-\frac{z\bar{z}}{\epsilon}}}{z}. \qquad (4.2.40)$$

The regularized propagator then takes the form

$$\Delta_{\mathrm{reg}}^{ab}(x - x') = \frac{\delta^{ab}}{2\pi i} \frac{\left(1 - e^{-\frac{|z-z'|^2}{\epsilon}}\right)}{z - z'} \widetilde{\Delta}_{\varphi - \varphi'}. \qquad (4.2.41)$$

In particular, in the limit where z approaches z', the regularized propagator is not divergent, unlike the unregularized one.

In addition, 1-point functions for ϕ can be shown to be zero using (4.2.33). Hence, we find that

$$\begin{aligned}
&\mathbb{1} + \langle \phi_a(0, z_1)\phi_c(\pi/2, z_2)\rangle T_{R_1}^a \otimes T_{R_2}^c - \langle \phi_a(\pi, z_1)\phi_c(\pi/2, z_2)\rangle T_{R_1}^a \otimes T_{R_2}^c \\
&\quad - \langle \phi_a(2\pi, z_1)\phi_c(3\pi/2, z_2)\rangle T_{R_1}^a \otimes T_{R_2}^c + \langle \phi_a(\pi, z_1)\phi_c(3\pi/2, z_2)\rangle T_{R_1}^a \\
&\quad \otimes T_{R_2}^c + O(\hbar^2) \\
&= \mathbb{1} - \frac{\hbar}{2}\delta_{ac}\frac{1}{z_1 - z_2}\widetilde{\Delta}_{-\frac{\pi}{2}} T_{R_1}^a \otimes T_{R_2}^c + \frac{\hbar}{2}\delta_{ac}\frac{1}{z_1 - z_2}\widetilde{\Delta}_{\frac{\pi}{2}} T_{R_1}^a \otimes T_{R_2}^c \\
&\quad + \frac{\hbar}{2}\delta_{ac}\frac{1}{z_1 - z_2}\widetilde{\Delta}_{\frac{\pi}{2}} T_{R_1}^a \otimes T_{R_2}^c - \frac{\hbar}{2}\delta_{ac}\frac{1}{z_1 - z_2}\widetilde{\Delta}_{-\frac{\pi}{2}} T_{R_1}^a \otimes T_{R_2}^c + O(\hbar^2) \\
&= \mathbb{1} + \frac{\hbar}{z_1 - z_2}(\widetilde{\Delta}_{\frac{\pi}{2}} - \widetilde{\Delta}_{-\frac{\pi}{2}}) T_{R_1}^a \otimes T_{R_2 a} + O(\hbar^2) \\
&= \mathbb{1} + \frac{\hbar}{z_1 - z_2} T_{R_1}^a \otimes T_{R_2 a} + O(\hbar^2). \qquad (4.2.42)
\end{aligned}$$

Here, we have used the fact that

$$\widetilde{\Delta}_{\frac{\pi}{2}} = \frac{1}{2\pi}\frac{\pi}{2} + \frac{1}{\pi}\left(1 - \frac{1}{3} + \frac{1}{5} - \frac{1}{7}\cdots\right) = \frac{1}{2}, \qquad (4.2.43)$$

and likewise $\widetilde{\Delta}_{-\frac{\pi}{2}} = -\frac{1}{2}$.

The result of (4.2.42) is in agreement with the order $\hbar$ correlation function for a pair of perpendicular Wilson lines in 4d Chern–Simons theory, that was computed in Section 1.3.2. This agreement between the 3d "chiral" WZW model correlation function and the corresponding 4d Chern–Simons theory correlator provides evidence for a holographic duality between the two theories.

4.2.2.1 *Non-perpendicular Wilson lines*

As explained in Chapter 1, in 4d Chern–Simons theory, for a pair of straight, crossed Wilson lines that are not perpendicular, their correlation function is the same as that of perpendicular Wilson lines, at linear order in $\hbar$. We would like to check that this is manifest in the dual computation using boundary local operators. This would demonstrate that topological invariance in the bulk is manifest along the boundary at linear order in $\hbar$.

As an example, consider two perpendicular Wilson lines, and rotate the vertical Wilson line clockwise by an angle, δ, as shown in Figure 4.4.

The four-point function we should compute is

$$\langle g_{R_1}^{-1}(0, z_1) g_{R_1}(\pi, z_1) \otimes g_{R_2}^{-1}(\pi/2 - \delta, z_2) g_{R_2}(3\pi/2 - \delta, z_2)\rangle. \tag{4.2.44}$$

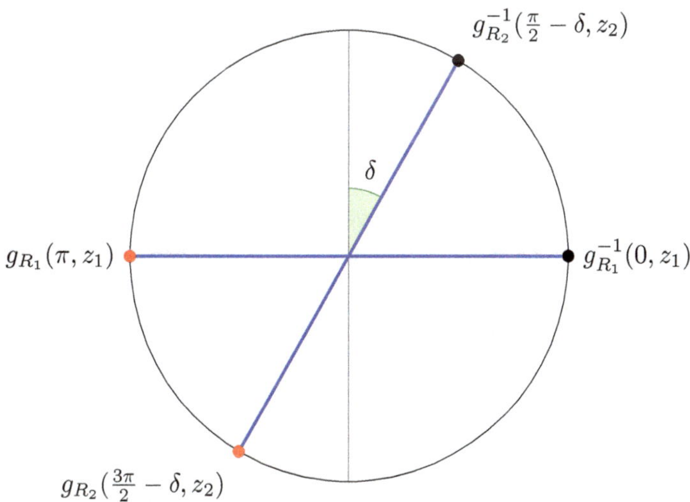

Figure 4.4.   Non-perpendicular Wilson lines on D.

Expanding each operator to linear order in ϕ as in (4.2.38), we find

$$
\mathbb{1} + \langle \phi_a(0, z_1)\phi_c(\pi/2 - \delta, z_2)\rangle T^a_{R_1} \otimes T^c_{R_2}
$$
$$
- \langle \phi_a(\pi, z_1)\phi_c(\pi/2 - \delta, z_2)\rangle T^a_{R_1} \otimes T^c_{R_2}
$$
$$
- \langle \phi_a(2\pi, z_1)\phi_c(3\pi/2 - \delta, z_2)\rangle T^a_{R_1} \otimes T^c_{R_2}
$$
$$
+ \langle \phi_a(\pi, z_1)\phi_c(3\pi/2 - \delta, z_2)\rangle T^a_{R_1} \otimes T^c_{R_2} + O(\hbar^2)
$$
$$
= \mathbb{1} - \frac{\hbar}{2}\delta_{ac}\frac{1}{z_1 - z_2}\widetilde{\Delta}_{-\frac{\pi}{2}+\delta}T^a_{R_1} \otimes T^c_{R_2} + \frac{\hbar}{2}\delta_{ac}\frac{1}{z_1 - z_2}\widetilde{\Delta}_{\frac{\pi}{2}+\delta}T^a_{R_1} \otimes T^c_{R_2}
$$
$$
+ \frac{\hbar}{2}\delta_{ac}\frac{1}{z_1 - z_2}\widetilde{\Delta}_{\frac{\pi}{2}+\delta}T^a_{R_1} \otimes T^c_{R_2}
$$
$$
- \frac{\hbar}{2}\delta_{ac}\frac{1}{z_1 - z_2}\widetilde{\Delta}_{-\frac{\pi}{2}+\delta}T^a_{R_1} \otimes T^c_{R_2} + O(\hbar^2)
$$
$$
= \mathbb{1} + \frac{\hbar}{z_1 - z_2}(\widetilde{\Delta}_{\frac{\pi}{2}+\delta} - \widetilde{\Delta}_{-\frac{\pi}{2}+\delta})T^a_{R_1} \otimes T_{R_2 a} + O(\hbar^2). \tag{4.2.45}
$$

Now, note that (4.2.36) can be rewritten as

$$
\widetilde{\Delta}_\varphi = \frac{\varphi}{2\pi} + \frac{1}{\pi}\sum_{k=1}^{\infty}\frac{\sin(k\varphi)}{k}. \tag{4.2.46}
$$

This implies that

$$
\widetilde{\Delta}_{\frac{\pi}{2}+\delta} = \frac{\frac{\pi}{2} + \delta}{2\pi} + \frac{1}{\pi}\sum_{k=1}^{\infty}\frac{\sin(k\frac{\pi}{2})\cos(k\delta) + \cos(k\frac{\pi}{2})\sin(k\delta)}{k} \tag{4.2.47}
$$

and

$$
\widetilde{\Delta}_{-\frac{\pi}{2}+\delta} = \frac{-\frac{\pi}{2} + \delta}{2\pi} - \frac{1}{\pi}\sum_{k=1}^{\infty}\frac{\sin(k\frac{\pi}{2})\cos(k\delta) - \cos(k\frac{\pi}{2})\sin(k\delta)}{k}. \tag{4.2.48}
$$

We find that the sums over k above can be separated into two types of sums, each having the form of a Fourier series, namely, the Fourier series for a square wave,

$$
\sum_{k=1}^{\infty}\frac{\sin(\frac{k\pi}{2})\cos(kx)}{k} = \frac{\pi}{4}\text{sign}(\cos(x)), \tag{4.2.49}
$$

and the Fourier series for a sawtooth wave,

$$
\sum_{k=1}^{\infty}\frac{\cos(\frac{k\pi}{2})\sin(kx)}{k} = \frac{-x}{2} + \frac{l\pi}{2}, \quad \pi\left(l - \frac{1}{2}\right) < x < \pi\left(l + \frac{1}{2}\right), \quad l \in \mathbb{Z},
$$
$$
\tag{4.2.50}
$$

for $x \in \mathbb{R}$. However, single-valuedness of the propagators involved in the computation (4.2.45) requires that $-\frac{\pi}{2} < \delta < \frac{\pi}{2}$, implying

$$\sum_{k=1}^{\infty} \frac{\sin(\frac{k\pi}{2})\cos(k\delta)}{k} = \frac{\pi}{4} \tag{4.2.51}$$

and

$$\sum_{k=1}^{\infty} \frac{\cos(\frac{k\pi}{2})\sin(k\delta)}{k} = -\frac{\delta}{2}. \tag{4.2.52}$$

From here, it follows that

$$\widetilde{\Delta}_{\frac{\pi}{2}+\delta} = \frac{1}{2},$$
$$\widetilde{\Delta}_{-\frac{\pi}{2}+\delta} = -\frac{1}{2} \tag{4.2.53}$$

for $-\frac{\pi}{2} < \delta < \frac{\pi}{2}$. As a result, (4.2.45) is in fact independent of the angle δ, and agrees precisely with the result for perpendicular Wilson lines. Hence, there is once again agreement with the results of Costello, Witten and Yamazaki [7], that were reviewed in Chapter 1.

Arbitrarily crossed Wilson lines

The preceding calculations can be generalized further to more general configurations of crossed Wilson lines, for which we expect to obtain the same result as (4.2.42) due to the topological invariance at linear order in $\hbar$. For instance, we can consider *both* Wilson lines rotated from perpendicularity, as shown in Figure 4.5. The corresponding four-point function is unaffected by the additional rotation, i.e., we find

$$\langle g_{R_1}^{-1}(0-\alpha, z_1)g_{R_1}(\pi-\alpha, z_1) \otimes g_{R_2}^{-1}(\pi/2-\delta, z_2)g_{R_2}(3\pi/2-\delta, z_2)\rangle$$

$$= \mathbb{1} + \frac{\hbar}{z_1 - z_2}(\widetilde{\Delta}_{\frac{\pi}{2}-\alpha+\delta} - \widetilde{\Delta}_{-\frac{\pi}{2}-\alpha+\delta})T_{R_1}^a \otimes T_{R_2 a} + O(\hbar^2)$$

$$= \mathbb{1} + \frac{\hbar}{z_1 - z_2}T_{R_1}^a \otimes T_{R_2 a} + O(\hbar^2) \tag{4.2.54}$$

(where $-\frac{\pi}{2} < -\alpha + \delta < \frac{\pi}{2}$ to ensure single-valued propagators) with the use of the identity (4.2.53).

A different generalization is that of perpendicular Wilson lines crossing at a point that is not the origin, $r = 0$, as shown in Figure 4.6. The four-point function in this case is also independent of the angles β and ρ shown

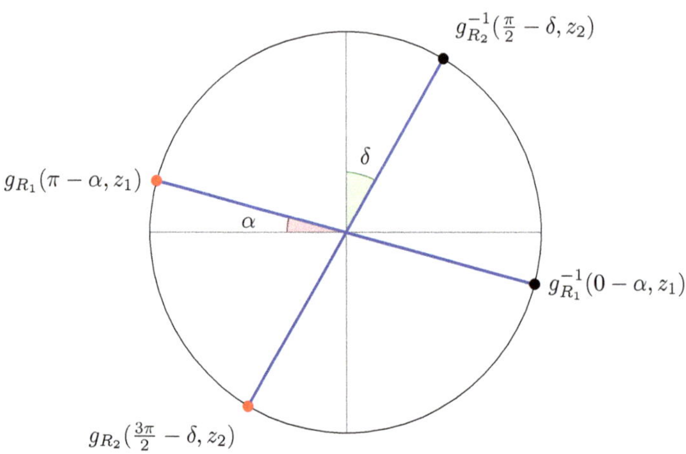

Figure 4.5. Crossed Wilson lines, both rotated from perpendicularity.

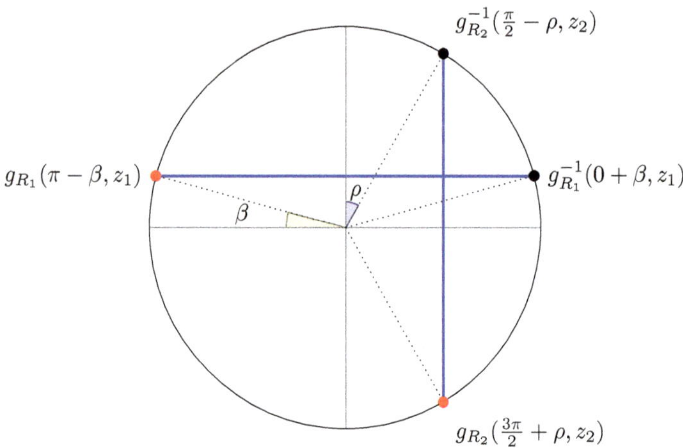

Figure 4.6. Perpendicular Wilson lines crossed away from the origin.

in the figure, i.e., we have

$$\langle g_{R_1}^{-1}(0 + \beta, z_1) g_{R_1}(\pi - \beta, z_1) \otimes g_{R_2}^{-1}(\pi/2 - \rho, z_2) g_{R_2}(3\pi/2 + \rho, z_2)\rangle$$

$$= \mathbb{1} + \frac{\hbar}{z_1 - z_2} \frac{1}{2} (\widetilde{\Delta}_{\frac{\pi}{2} + \beta - \rho} - \widetilde{\Delta}_{-\frac{\pi}{2} + \beta + \rho}$$

$$+ \widetilde{\Delta}_{\frac{\pi}{2} - \beta + \rho} - \widetilde{\Delta}_{-\frac{\pi}{2} - \beta - \rho}) T_{R_1}^a \otimes T_{R_2 a} + O(\hbar^2)$$

$$= \mathbb{1} + \frac{\hbar}{z_1 - z_2} T_{R_1}^a \otimes T_{R_2 a} + O(\hbar^2), \tag{4.2.55}$$

using the identity (4.2.53), where $-\frac{\pi}{2} < \beta + \rho < \frac{\pi}{2}$ and $-\frac{\pi}{2} < \beta - \rho < \frac{\pi}{2}$ to ensure single-valuedness of propagators. Note that the allowed ranges of $\beta + \rho$ and $\beta - \rho$ mean that the result is valid only when the Wilson lines are crossed.

Let us now study the most general configuration. We shall show that the four-point function corresponding to any *arbitrary* configuration of crossed Wilson lines has the same correlation function at order $\hbar$. Such a configuration, as depicted in Figure 4.7, is determined by four angles, namely α, β, γ, and ρ. The four-point function is then

$$\langle g_{R_1}^{-1}(0 + \beta - \alpha, z_1) g_{R_1}(\pi - \beta - \alpha, z_1)$$

$$\otimes g_{R_2}^{-1}(\pi/2 - \rho - \delta, z_2) g_{R_2}(3\pi/2 + \rho - \delta, z_2)\rangle$$

$$= \mathbb{1} + \frac{\hbar}{z_1 - z_2} \frac{1}{2} (\widetilde{\Delta}_{\frac{\pi}{2}+\beta-\rho-\alpha+\delta} - \widetilde{\Delta}_{-\frac{\pi}{2}+\beta+\rho-\alpha+\delta} + \widetilde{\Delta}_{\frac{\pi}{2}-\beta+\rho-\alpha+\delta}$$

$$- \widetilde{\Delta}_{-\frac{\pi}{2}-\beta-\rho-\alpha+\delta}) T_{R_1}^a \otimes T_{R_2 a} + O(\hbar^2). \tag{4.2.56}$$

Here, to ensure single-valuedness of propagators, we require $-\frac{3\pi}{2} < \beta - \widehat{r} - \alpha + \delta < \frac{\pi}{2}$, $-\frac{3\pi}{2} < -\beta + \widehat{r} - \alpha + \delta < \frac{\pi}{2}$, $-\frac{\pi}{2} < \beta + \widehat{r} - \alpha + \delta < \frac{3\pi}{2}$ and $-\frac{\pi}{2} < -\beta - \widehat{r} - \alpha + \delta < \frac{3\pi}{2}$.

However, to ensure that we are considering only crossed Wilson lines, we require the stronger conditions $-\frac{\pi}{2} < \beta - \widehat{r} - \alpha + \delta < \frac{\pi}{2}$, $-\frac{\pi}{2} < -\beta + \widehat{r} - \alpha + \delta < \frac{\pi}{2}$, $-\frac{\pi}{2} < \beta + \widehat{r} - \alpha + \delta < \frac{\pi}{2}$ and $-\frac{\pi}{2} < -\beta - \widehat{r} - \alpha + \delta < \frac{\pi}{2}$. These

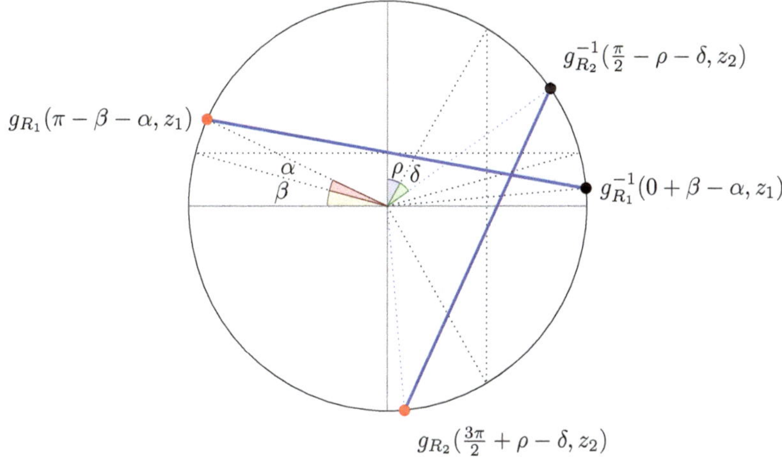

Figure 4.7. Arbitrarily inserted crossed Wilson lines.

conditions in turn allow us to use (4.2.53), whereby we find that (4.2.56) is

$$\mathbb{1} + \frac{\hbar}{z_1 - z_2} T^a_{R_1} \otimes T_{R_2 a} + O(\hbar^2). \tag{4.2.57}$$

It has thus been demonstrated that topological invariance along Σ of the bulk correlation function of two crossed Wilson lines (that realizes the rational R-matrix) is reflected in the four-point function of the boundary 3d "chiral" WZW model, up to linear order in $\hbar$.

4.2.3   *3d Toda theory*

We have so far considered 4d CS theory on a four-manifold with a single boundary along the topological plane. Picking the Dirichlet boundary condition where the gauge field component $\mathcal{A}_{\bar{z}}$ is set to vanish at the boundary, we obtained a 3d analogue of the chiral Wess–Zumino-Witten (WZW) model.

In this subsection, we shall generalize this construction by allowing Σ to have two boundaries, that is, we choose $\Sigma = S^1 \times I$. Moreover, we shall further impose an analogue of Nahm pole boundary conditions in addition to Dirichlet boundary conditions at each boundary. As we shall see, this leads to a three-dimensional analogue of 2d Toda field theory, known as 3d Toda theory. The original reference is the work of Ashwinkumar *et al.* [59], and the reader is encouraged to refer to this article for further details not contained here.

In what follows, we shall restrict ourselves to 4d Chern–Simons theory with gauge group $G = SL(N, \mathbb{C})$. To define the boundary conditions at the endpoints of the interval, we define the convenient coordinates

$$x^{\pm} = \tau \pm \frac{i}{2}\bar{z}, \tag{4.2.58}$$

where τ parametrizes S^1. The corresponding partial derivatives and complexified gauge fields are

$$\partial_{\pm} = \frac{1}{2}\left(\partial_\tau \mp 2i\partial_{\bar{z}}\right),$$

$$\mathcal{A}_{\pm} = \frac{1}{2}\left(\mathcal{A}_\tau \mp 2i\mathcal{A}_{\bar{z}}\right), \tag{4.2.59}$$

respectively.

At the endpoints of the interval, parametrized by $\sigma \in [0, \pi]$, Nahm-pole-type boundary conditions are imposed. These boundary conditions are defined as follows. Towards one end of the interval, namely $\sigma \to 0$, one imposes the behaviour

$$\mathcal{A} \to \frac{id\sigma}{\sigma}H + \frac{dx^+}{\sigma}T_+, \qquad (4.2.60)$$

where, for $\mathfrak{sl}(2) \subset \mathfrak{sl}(n)$, a homomorphism $\rho : \mathfrak{sl}(2) \to \mathfrak{sl}(N)$ has been picked such that the image, T_+, of the raising operator of $\mathfrak{sl}(2)$ is a maximal-length Jordan block. Explicitly, we have

$$T_+ = -i \begin{pmatrix} 0 & \mu_1 & 0 & \cdots & 0 \\ 0 & 0 & \mu_2 & \cdots & 0 \\ \vdots & \vdots & \vdots & \ddots & \vdots \\ 0 & 0 & 0 & \cdots & \mu_{N-1} \\ 0 & 0 & 0 & \cdots & 0 \end{pmatrix}, \qquad (4.2.61)$$

where μ_i are real and non-zero constants. Note that the boundary conditions (4.2.60) include

$$\mathcal{A}_- = 0 \qquad (4.2.62)$$

at $\sigma = 0$, which is a Dirichlet boundary condition.

At the other end of the interval where $\sigma \to \pi$, the behaviour

$$\mathcal{A} \to \frac{-id\sigma}{\sigma - \pi}H + \frac{dx^-}{\sigma - \pi}T_- \qquad (4.2.63)$$

is imposed, where for $\mathfrak{sl}(2) \subset \mathfrak{sl}(n)$ a homomorphism $\rho' : \mathfrak{sl}(2) \to \mathfrak{sl}(N)$ is picked such that the image, T_-, of the lowering operator is a maximal-length Jordan block. This is given explicitly by

$$T_- = -i \begin{pmatrix} 0 & 0 & \cdots & 0 & 0 \\ \nu_1 & 0 & \cdots & 0 & 0 \\ 0 & \nu_2 & \cdots & 0 & 0 \\ \vdots & \vdots & \ddots & \vdots & \vdots \\ 0 & 0 & \cdots & \nu_{N-1} & 0 \end{pmatrix}, \qquad (4.2.64)$$

where ν_i are real and non-zero constants. In particular, the boundary conditions (4.2.63) include the Dirichlet boundary condition

$$\mathcal{A}_+ = 0 \qquad (4.2.65)$$

at $\sigma = \pi$.

To derive 3d Toda theory, we shall first show that the Dirichlet boundary conditions (4.2.62) and (4.2.65) imply that the dynamics close to either endpoint of the interval can be described by a 3d "chiral" WZW model, which we shall refer to as a 3d WZW model in what follows for brevity.

Let us first consider the boundary at $\sigma = 0$ and the Dirichlet boundary condition (4.2.62). To prove the locality of 4d CS at $\sigma = 0$, we first vary the 4d CS action to give

$$\delta S_{4d\ CS} = \frac{1}{2\pi\hbar} \int_{I \times S^1 \times C} dz \wedge \mathrm{Tr}\left(\delta\mathcal{A} \wedge \mathcal{F} + d(\delta\mathcal{A} \wedge \mathcal{A})\right). \qquad (4.2.66)$$

Then, Stokes' theorem implies that the second term of (4.2.66) can be identified with a boundary term. Taking into account the Dirichlet boundary condition (4.2.62), this term vanishes.

To show the gauge invariance of 4d CS theory, we first extend the partial connection $\mathcal{A}$ to a full connection, which allows us to rewrite the 4d CS action as

$$S_{4d\ CS}[\mathcal{A}] = -\frac{1}{2\pi\hbar} \int_{I \times S^1 \times C} z\mathrm{Tr}\left(\mathcal{F} \wedge \mathcal{F}\right)$$

$$+ \frac{1}{2\pi\hbar} \int_{S^1 \times C} z\mathrm{Tr}\left(\mathcal{A} \wedge d\mathcal{A} + \frac{2}{3}\mathcal{A} \wedge \mathcal{A} \wedge \mathcal{A}\right). \qquad (4.2.67)$$

Utilizing the Dirichlet boundary condition $\mathcal{A}_- = 0$, and imposing the additional boundary condition $\mathcal{A}_z = 0$, one finds that the second term of (4.2.67) vanishes (note that $\mathcal{A}_z$ is not the complex conjugate of $\mathcal{A}_{\bar{z}}$ here). The remaining term is gauge invariant under large gauge transformations of the form

$$\mathcal{A} \to U\mathcal{A}U^{-1} - dUU^{-1}, \qquad (4.2.68)$$

so long as we restrict U to preserve the boundary conditions $\mathcal{A}_- = 0 = \mathcal{A}_z$. This is achieved by taking U to tend to the identity element of G at $\sigma = 0$.

Next, in order to derive the 3d WZW model at the boundary at $\sigma = 0$, we use the fact that the 4d CS action admits the alternate form

$$\frac{1}{2\pi\hbar} \int_{I \times S^1 \times C} dz \wedge d\sigma \wedge dx^- \wedge dx^+ \mathrm{Tr}\left(2\mathcal{A}_-\mathcal{F}_{+\sigma} - \mathcal{A}_\sigma\partial_-\mathcal{A}_+ + \mathcal{A}_+\partial_-\mathcal{A}_\sigma\right),$$

$$(4.2.69)$$

as $\sigma \to 0$, where $\mathcal{A}_-$ plays the role of a Lagrange multiplier. The Lagrange multiplier can be integrated out to produce the constraint $\mathcal{F}_{+\sigma} = 0$, which has the solutions

$$\mathcal{A}_\sigma = g^{-1}\partial_\sigma g \qquad (4.2.70a)$$

$$\mathcal{A}_+ = g^{-1}\partial_+ g, \qquad (4.2.70b)$$

where g is an $SL(N,\mathbb{C})$-valued field on $I \times S^1 \times C$. Changing variables from $\mathcal{A}_\sigma$ and $\mathcal{A}_+$ to g, (4.2.69) then becomes a 3d WZW model with the action

$$\lim_{\sigma \to 0} S_{\text{4d CS}}[\mathcal{A}] = S_{\text{3d WZW}}[g]$$

$$= \frac{1}{2\pi\hbar} \int_{S^1 \times C} dz \wedge dx^+ \wedge dx^- \, \mathrm{Tr}\left(\partial_+ g\, g^{-1} \partial_- g\, g^{-1}\right)$$

$$- \frac{1}{6\pi\hbar} \int_{I \times S^1 \times C} dz \wedge \mathrm{Tr}\left(dg\, g^{-1} \wedge dg\, g^{-1} \wedge dg\, g^{-1}\right).$$

$$(4.2.71)$$

Now, a gauge transformation (4.2.68) amounts to $g \to gU^{-1}$ in (4.2.70). As a result, one may change the value of g in the bulk without changing its value at $\sigma = 0$. This implies that the Wess–Zumino term in (4.2.71) does not depend on the choice of extension of the boundary value of g over the bulk manifold. Hence, one can divide out the volume of the gauge group to obtain a path integral of the form

$$\int D\tilde{g}\, e^{iS_{\text{3d WZW}}[\tilde{g}]}, \qquad (4.2.72)$$

where $\tilde{g}$ is an $SL(N,\mathbb{C})$-valued field on $S^1 \times C$.

The 3d WZW model action is invariant under the transformation

$$\tilde{g}(z, x^+, x^-) \to \widetilde{\Omega}(z, x^+)\tilde{g}\,\Omega(z, x^-), \qquad (4.2.73)$$

and the corresponding conserved currents take the form

$$J_+ = \tilde{g}^{-1}\,\partial_+\tilde{g} \qquad (4.2.74\text{a})$$

$$J_- = \partial_-\tilde{g}\,\tilde{g}^{-1}, \qquad (4.2.74\text{b})$$

and satisfy the current conservation equations

$$\partial_-(\tilde{g}^{-1}\partial_+\tilde{g}) = 0 \qquad (4.2.75)$$

and

$$\partial_+(\partial_-\tilde{g}\,\tilde{g}^{-1}) = 0. \qquad (4.2.76)$$

These equations are in fact just the equation of motion of the 3d WZW model. By comparing (4.2.70b) and (4.2.74a) we can identify $\mathcal{A}_+$ with J_+ on the boundary at $\sigma = 0$, and thus the remaining data of the boundary condition (4.2.60) also constrain J_+. This shall be crucial in the derivation of the 3d Toda theory action.

Let us now consider the other end of the interval, $\sigma = \pi$. Locality of 4d CS theory close to this boundary is guaranteed since the Dirichlet boundary condition (4.2.65) ensures that the boundary contribution to the variation (4.2.66) vanishes.

In order to show the gauge invariance of 4d CS theory, one should first extend the partial connection $\mathcal{A}$ to a full connection. Using the Dirichlet boundary condition (4.2.65), and imposing the additional boundary condition $\mathcal{A}_z = 0$, the second term of (4.2.67) vanishes. Here, we ought to restrict U such that the boundary conditions $\mathcal{A}_+ = 0 = \mathcal{A}_z$ are preserved, and this shall be achieved by taking U to tend to the identity element of G at $\sigma = \pi$.

Using the boundary condition (4.2.65), the 4d CS action can be reexpressed in the alternate form

$$\frac{1}{2\pi\hbar} \int_{I \times S^1 \times C} \mathrm{d}z \wedge \mathrm{d}\sigma \wedge \mathrm{d}x^- \wedge \mathrm{d}x^+ \, \mathrm{Tr}\left(2\mathcal{A}_+ \mathcal{F}_{-\sigma} + \mathcal{A}_\sigma \partial_+ \mathcal{A}_- - \mathcal{A}_- \partial_+ \mathcal{A}_\sigma\right),$$

$$(4.2.77)$$

as $\sigma \to \pi$. Here, $\mathcal{A}_+$ is a Lagrange multiplier, which can be integrated out to produce the constraint $\mathcal{F}_{\sigma-} = 0$. The solutions to this constraint are

$$\mathcal{A}_\sigma = -\partial_\sigma g \, g^{-1} \qquad\qquad (4.2.78a)$$

$$\mathcal{A}_- = -\partial_- g \, g^{-1}, \qquad\qquad (4.2.78b)$$

where g is an $SL(N, \mathbb{C})$-valued field on $I \times S^1 \times C$.[14]

Upon changing variables from $\mathcal{A}_\sigma$ and $\mathcal{A}_-$ to g, and taking note of the orientation of I, (4.2.77) becomes a 3d WZW model action:

$$\lim_{\sigma \to \pi} S_{\text{4d CS}}[\mathcal{A}] = S'_{\text{3d WZW}}[g]$$

$$= \frac{1}{2\pi\hbar} \int_{S^1 \times C} \mathrm{d}z \wedge \mathrm{d}x^+ \wedge \mathrm{d}x^- \, \mathrm{Tr}\left(\partial_+ g \, g^{-1} \partial_- g \, g^{-1}\right)$$

$$+ \frac{1}{6\pi\hbar} \int_{I \times S^1 \times C} \mathrm{d}z \wedge \mathrm{Tr}\left(\mathrm{d}g \, g^{-1} \wedge \mathrm{d}g \, g^{-1} \wedge \mathrm{d}g \, g^{-1}\right),$$

$$(4.2.79)$$

which differs from (4.2.71) by a minus sign multiplying the topological term.

Now, a gauge transformation of the form (4.2.68) amounts to $g \to Ug$ in (4.2.78a). This means that the value of g in the bulk can be modified without affecting its value at $\sigma = \pi$. Therefore, the Wess–Zumino term in (4.2.79) does not depend on the choice of extension of the boundary value

---

[14]We shall use g and $\tilde{g}$ to denote degrees of freedom associated with both endpoints of the interval, since these degrees of freedom shall eventually be identified, because the size of the interval can be taken to be vanishingly small due to the topological invariance along this direction.

of g over the bulk manifold. It then follows that one can divide out the volume of the gauge group to obtain a path integral of the form

$$\int D\tilde{g}\, e^{iS'_{\text{3d wzw}}[\tilde{g}]}, \qquad (4.2.80)$$

where $\tilde{g}$ is an $SL(N,\mathbb{C})$-valued field on $S^1 \times C$

With a convenient overall factor, the conserved currents for the $G \times G$ symmetry of (4.2.79) are

$$J'_+ = -\tilde{g}^{-1}\partial_+\tilde{g} \qquad (4.2.81\text{a})$$

$$J'_- = -\partial_-\tilde{g}\,\tilde{g}^{-1}, \qquad (4.2.81\text{b})$$

which satisfy $\partial_- J'_+ = 0$ and $\partial_+ J'_- = 0$, where the prime on J' indicates that these currents are localized to $\sigma \to \pi$. By comparing (4.2.78b) and (4.2.81b), we find that we can identify $\mathcal{A}_-$ with J'_- on the boundary at $\sigma = \pi$, meaning that the remaining data of the boundary condition (4.2.63) also constrain J'_-.

We shall now proceed to derive 3d Toda theory from the two 3d WZW models by implementing the constraints arising from the Nahm-pole-type boundary conditions. We shall use the fact that the topological invariance along the I direction allows us to identify the degrees of freedom associated with the two 3d WZW models.

Now, observe that T_+ in (4.2.60) (T_- in (4.2.63)) can be expressed as a sum of the positive (negative) simple roots with non-zero coefficients. Moreover, a current, J, can be projected onto Cartan directions, as well as roots of the Lie algebra, that is, we can write

$$J = \sum_{\alpha \in \Delta} (J^{-\alpha} R_{-\alpha} + J^{+\alpha} R_{+\alpha}) + \sum_i J^{0i} R_{0i}, \qquad (4.2.82)$$

where R_{0i} denotes the i-th Cartan generator for $i = 1, \cdots, N-1$, while Δ denotes the root space, with $R_{+\alpha}$ and $R_{-\alpha}$ denoting the generators associated with the positive and negative roots, respectively. Given that μ_i and ν_i are non-zero only for simple roots, corresponding to $R_{\pm i}$, it can be shown that the Nahm-pole-type boundary conditions (4.2.60) and (4.2.63) imply

$$J_+{}^{+i} = \mu_i, \qquad (4.2.83\text{a})$$

$$J_+{}^{0i} = 0 \qquad (4.2.83\text{b})$$

and

$$J'_-{}^{-i} = \nu_i, \tag{4.2.84a}$$

$$J'_-{}^{0i} = 0, \tag{4.2.84b}$$

respectively. Together, (4.2.83) and (4.2.84) are current constraints.

Using a Gauss decomposition, the 3d WZW field $\tilde{g}$ can be expressed as

$$\tilde{g} = \exp\left(i \sum_{\alpha \in \Delta} X_\alpha R_\alpha^+\right) \exp\left(i \sum_i \phi_i R_i^0\right) \exp\left(i \sum_{\alpha \in \Delta} Y_\alpha R_\alpha^-\right), \tag{4.2.85}$$

where X_α, Y_α and ϕ_i are scalar fields. From (4.2.85), $J_+ = \tilde{g}^{-1}\,\partial_+\tilde{g}$ may then be rewritten as

$$J_+ = \sum_i \left(i\partial_+ X_i e^{-C_{ij}\phi_j}\right) R_i^+$$

$$+ \left(-\sum_{i,k} C_{ik} Y_i Y_k \partial_+ X_k e^{-C_{kj}\phi_j} - i \sum_{i,j} C_{ij} Y_i \partial_+ \phi_j + i \sum_i \partial_+ Y_i \right) R_i^-$$

$$+ \sum_i \left(i\partial_+ \phi_i + 2i Y_i \partial_+ X_i e^{-C_{ij}\phi_j}\right) R_i^0 + \cdots ,$$

$$\tag{4.2.86}$$

where C_{ij} are elements of the Cartan matrix, $\exp\left(C_{ij}\phi_j\right) = \exp\left(\sum_j C_{ij}\phi_j\right)$, and where the ellipsis denotes terms involving non-simple roots. For $SL(N, \mathbb{C})$, the Cartan matrix is written as

$$[C_{ij}] = \begin{pmatrix} 2 & -1 & \cdots & 0 & 0 \\ -1 & 2 & \cdots & 0 & 0 \\ \vdots & \vdots & \ddots & \vdots & \vdots \\ 0 & 0 & \cdots & 2 & -1 \\ 0 & 0 & \cdots & -1 & 2 \end{pmatrix}. \tag{4.2.87}$$

Comparing (4.2.86) with (4.2.83) leads to

$$\partial_+ X_i e^{-C_{ij}\phi_j} = -i\mu_i \tag{4.2.88a}$$

$$\partial_+ \phi_i + 2 Y_i \partial_+ X_i e^{-C_{ij}\phi_j} = 0. \tag{4.2.88b}$$

Likewise, $J'_- = -\partial_- \tilde{g}\,\tilde{g}^{-1}$ can be recast as

$$
J'_- = \left(\sum_{i,k} C_{ik} X_i X_k \partial_- Y_k e^{-C_{kj}\phi_j} + i \sum_{i,j} C_{ij} X_i \partial_- \phi_j - i \sum_i \partial_- X_i \right) R_i^+
$$
$$
+ \sum_i \left(-i\partial_- Y_i e^{-C_{ij}\phi_j} \right) R_i^-
$$
$$
+ \sum_i \left(-i\partial_- \phi_i - 2i X_i \partial_- Y_i e^{-C_{ij}\phi_j} \right) R_i^0 + \cdots ,
$$

$$(4.2.89)$$

where the ellipsis denotes terms involving non-simple roots. Comparing (4.2.89) with (4.2.84) leads to

$$
\partial_- Y_i e^{-C_{ij}\phi_j} = i\nu_i \tag{4.2.90a}
$$

$$
\partial_- \phi_i + 2 X_i \partial_- Y_i e^{-C_{ij}\phi_j} = 0. \tag{4.2.90b}
$$

Combining (4.2.88) and (4.2.90) then furnishes the equation

$$
\partial_+ \partial_- \phi_i + 2\mu_i \nu_i e^{C_{ij}\phi_j} = 0, \tag{4.2.91}
$$

which resemble the equations of motion for 2d Toda theory. The action that has (4.2.91) as equations of motion is

$$
S_{\text{3d Toda}}[\phi]
$$
$$
= \frac{1}{2\pi\hbar} \int_{S^1 \times C} dz \wedge dx^+ \wedge dx^- \left(C_{ij} \partial_+ \phi^i \partial_- \phi^j - 4 \sum_i \mu_i \nu_i e^{C_{ij}\phi_j} \right),
$$

$$(4.2.92)$$

which has the form of a 3d analogue of analytically-continued 2d Toda theory. Therefore, the edge modes of 4d CS theory with Nahm-pole-type boundary conditions are 3d Toda fields.

Although we have derived 3d Toda field theory at the level of the equations of motion, one may alternatively also obtain 3d Toda theory from 4d CS theory via an off-shell procedure, i.e., working at the level of the action and path integral, rather than equations of motion. As before, the two constrained 3d WZW models can be identified with a single 3d WZW model due to the topological invariance along I, but now the Gauss decomposition (4.2.85) is substituted directly into the action. Subsequently, the implementation of the constraints (4.2.88) and (4.2.90) in the resulting action leads to the 3d Toda action (4.2.92).

It is expected that correlation functions of local operators in 3d Toda theory can be identified with correlation functions of Wilson lines in the bulk 4d Chern–Simons theory, although this has not been checked explicitly.

Note that if we pick $C = T^2$, parametrized by the coordinates θ^1 and θ^2, we can perform a dimensional reduction of 3d Toda theory on the circle in the θ^2 direction to obtain 2d Toda theory. Indeed, the dimensional reduction amounts to setting $\partial_{\bar{z}} \to \frac{1}{2}\partial_{\theta_1}$ and $dz \wedge d\bar{z} = -2id\theta_1 \wedge d\theta_2$, implying that $\partial_+ \to \partial_u = \frac{1}{2}(\partial_\tau - i\partial_{\theta_1})$ and $\partial_- \to \partial_{\bar{u}} = \frac{1}{2}(\partial_\tau + i\partial_{\theta_1})$. The 3d Toda action (4.2.92) then becomes

$$S_{\text{2d Toda}}[\phi] = \frac{iR}{2\hbar} \int_{C'} du \wedge d\bar{u} \left(C_{ij}\partial_u \phi^i \partial_{\bar{u}} \phi^j - 4 \sum_i \mu_i \nu_i e^{C_{ij}\phi_j} \right), \quad (4.2.93)$$

where the remaining two directions have been denoted by C', and R is the radius of the circle parametrized by θ^2. The action (4.2.93) takes the form of the usual 2d conformal Toda action, although the fields are analytically continued to be complex, since we started with 4d CS with complex gauge group.

4.2.3.1   *Three-dimensional W-algebras*

It is well known that 2d Toda field theory is underlaid by the rich algebraic structure of W-algebras. One can therefore expect that a similar infinite-dimensional symmetry algebra is present in 3d Toda theory. For example, at the classical level, one can show that the Poisson bracket

$$[\Theta\left(z',\xi'\right), \Theta(z,\xi)]_{\text{PB}}$$

$$= \left(\partial_\xi \Theta(z,\xi)\delta\left(\xi - \xi'\right) + 2\Theta(z,\xi)\delta'\left(\xi - \xi'\right) - \frac{1}{2}\delta'''\left(\xi - \xi'\right) \right) \delta\left(z - z'\right)$$

$$(4.2.94)$$

is satisfied for the case of $G = SL(2,\mathbb{C})$, where ξ is either x^+ or x^- (depending on whether x^- or x^+ is chosen to be the time direction, respectively), with

$$\Theta(z,\xi) = T(z,\xi) - \partial_\xi J_3(z,\xi), \quad (4.2.95)$$

where $T(z,\xi)$ is the analogue of the Segal-Sugawara energy-momentum tensor

$$T(z,\xi) = J_a(z,\xi)J^a(z,\xi), \quad (4.2.96)$$

with J_a for $a = 1, 2, 3$ denoting components of the $\mathfrak{sl}(2, \mathbb{C})$-valued current. This is just a 3d analogue of the 2d Virasoro algebra. Likewise, for higher-rank gauge groups, one can derive classical higher-spin 3d W-algebras, that can be canonically quantized to quantum 3d W-algebras.

Finally, we note that an important mathematical implication was obtained by embedding the 4d Chern–Simons theory on $S^1 \times I \times C$ with Nahm pole-type boundary conditions in a generalization of the supersymmetric gauge theory setup arising from the D4-NS5-brane system discussed in the previous chapter. This generalization involved defining the supersymmetric gauge theory on a manifold with corners, that is, $\mathbb{R}_+ \times S^1 \times I \times C$. The aforementioned mathematical result states that modules of 3d W-algebras defined on $S^1 \times C$ are modules for the quantized algebra of certain holomorphic functions on the Bogomolny moduli space of G-monopoles on $S^1 \times C$. Details of the derivation of this result can be obtained from Ref. [59].

Chapter 5

Toward Quantization and Dualities of Integrable Field Theories

We have so far studied integrable field theories in 4d Chern–Simons theory in Chapter 2, and this was purely at the classical level. We would like to be able to understand these field theories at the quantum level from 4d Chern–Simons theory. In particular, the problem of integrable quantization of non-ultralocal field theories is an important one, as the Maillet bracket (discussed in Chapter 2) of these theories is of a form that obstructs their discretization. This problem has not been resolved at present, and the hope is that 4d Chern–Simons theory could provide some insight in this direction.

In this chapter, we shall review some results where progress towards quantizing integrable field theories and deriving their dualities from the perspective of 4d Chern–Simons theory has been made. We first review how the 1-loop renormalization group flow of PCM-type integrable field theories can be derived from 4d Chern–Simons theory, from the work of Levine [73]. We then discuss discretizations of ultralocal integrable field theories from the perspective of 4d Chern–Simons theory, based on the work of Ashwinkumar, Sakamoto, and Yamazaki [75]. We then review the anomaly inflow mechanism for order surface defects in 4d Chern–Simons theory, based on unpublished results of Costello and Yamazaki. Finally, we review dualities that one can derive by dualizing defects in 4d Chern–Simons theory, based on work of Ashwinkumar, Sakamoto, and Yamazaki [75].

5.1 Renormalization Group Flow of Integrable Field Theories

The 1-loop renormalization group (RG) flow of integrable field theories of PCM type were shown to have a universal form by Levine [76], who

213

subsequently further interpreted this in terms of 4d Chern–Simons theory [73]. In what follows, for the sake of pedagogy, we shall review both of these derivations. We shall be interested in integrable field theories defined on a worldsheet denoted Σ, parametrized by Minkowski coordinates σ^+ and σ^-, i.e., we shall be concerned with relativistic 2d Lorentz-invariant field theories.

5.1.1   *2d integrable field theory derivation*

Consider an integrable field theory whose Lax connection has the form

$$\mathcal{L}_\pm(z) = \frac{\mathcal{J}_\pm}{1 \pm z}, \tag{5.1.1}$$

where $\mathcal{J}$ shall be referred to as the Lax current. A simple example of such a theory is the PCM, where $\mathcal{J}_\pm = g^{-1}\partial_\pm g$. In general, integrable deformations of the PCM have Lax connections of the form (5.1.1) with Lax currents that depend on the field g, but whose form is model-dependent.

The curvature of this Lax connection is given by

$$\begin{aligned}
F_{+-}(\mathcal{L}) &= \partial_+\mathcal{L}_- - \partial_-\mathcal{L}_+ + [\mathcal{L}_+, \mathcal{L}_-] \\
&= \frac{1}{1-z^2}F_{+-}(\mathcal{J}) + \frac{z}{1-z^2}\partial \cdot \mathcal{J},
\end{aligned} \tag{5.1.2}$$

where $F_{+-}(\mathcal{J}) = \partial_+\mathcal{J}_- - \partial_-\mathcal{J}_+ + [\mathcal{J}_+, \mathcal{J}_-]$ and $\partial \cdot \mathcal{J} = \partial_+\mathcal{J}_- + \partial_-\mathcal{J}_+$. Recall that the equations of motion of the theory ensure that this quantity vanishes, i.e., we have

$$\partial \cdot \mathcal{J} = 0, \quad F_{+-}(\mathcal{J}) = 0. \tag{5.1.3}$$

Now, note that certain combinations of the equations (5.1.3) may correspond to equations that hold off-shell, i.e., they are Bianchi identities. In the example of the PCM, we notice that $F_\pm(\mathcal{J}) = 0$ off-shell since $\mathcal{J}_\pm$ is just the Maurer-Cartan current $g^{-1}\partial_\pm g$. For other theories of PCM type, the Lax currents may solve one or several Bianchi identities of the form

$$P(g) \cdot F_{+-}(\mathcal{J}) \equiv O(g) \cdot (\partial \cdot \mathcal{J}), \tag{5.1.4}$$

where $O(g) : \mathfrak{g} \to \mathfrak{g}$ and $P(g) : \mathfrak{g} \to \mathfrak{g}$ are linear maps that depend on g.

The proposal of Levine is that, for all theories with Lax connections of the form (5.1.1) satisfying a certain assumption known as the Bianchi completeness assumption, the 1-loop divergence has a universal form.

Let us now review the derivation of this statement. Consider an integrable field theory with Lax connection given in terms of Lax currents and spectral parameter z, and expand the Lax currents around a background field $\overline{\mathcal{J}}$,

$$\mathcal{J}_\pm = \overline{\mathcal{J}}_\pm + a_\pm, \quad \overline{\mathcal{J}}_\pm = \mathcal{J}_\pm(\bar{g}), \tag{5.1.5}$$

with $\bar{g}$ denoting a background field for g that satisfies the equations of motion.

One now makes the Bianchi completeness assumption, which is the statement that to leading order in the background field expansion, some of the zero-curvature equations are equivalent to the Lax currents taking the correct form $\mathcal{J}_\pm = \mathcal{J}_\pm(g)$ in terms of the physical fields of the theory. This can be restated as follows. There is a subset of the zero-curvature equations of the Lax connection corresponding to the Bianchi identities, i.e.,

$$\text{Bianchi}_\alpha(g, \mathcal{J}) \equiv B_\alpha{}^I(g)\mathrm{zc}_I(\mathcal{J}) \equiv 0, \tag{5.1.6}$$

where $\mathrm{zc}_I(\mathcal{J}) = 0$ are the equations implied by the zero-curvature condition $F_\pm(\mathcal{L}) = 0$, while B_α^I are the components of a g-dependent matrix. The linearized versions of the Bianchi identities then correspond to

$$\text{Bianchi}_\alpha(\bar{g}) \cdot a \equiv B_\alpha{}^I(\bar{g})\mathrm{zc}_I(\mathcal{J}(\bar{g}) + a)|_{\mathcal{O}(a)} \equiv 0 \tag{5.1.7}$$

where $|_{\mathcal{O}(a)}$ indicates truncation at linear order in a. One can replace $B_\alpha^I(g)$ by $B_\alpha^I(\bar{g})$ to linear order since $\bar{g}$ is on-shell. The Bianchi completeness assumption means that (5.1.7) corresponds to the fluctuations $a_\pm$ of $\mathcal{J}_\pm$ being such that

$$\mathcal{J}_\pm(\bar{g}) + a_\pm = \mathcal{J}_\pm(\bar{g} + \phi), \tag{5.1.8}$$

that is, they arise from fluctuations ϕ of the field g.

Now, consider the 1-loop effective action of the integrable field theory on the worldsheet Σ, that takes the general form

$$\widehat{S}^{(1)}(\bar{g}) = -i \log \int \mathcal{D}\phi^i e^{i \int_\Sigma \phi^i \mathcal{O}_{ij}(\bar{g})\phi^j} = \frac{i}{2} \log \det \mathcal{O}(\bar{g}) \tag{5.1.9}$$

where $\phi^i \mathcal{O}_{ij}(\bar{g})\phi^j = \mathcal{L}(\bar{g} + \phi) - \mathcal{L}(\bar{g}) + \mathcal{O}\left(\phi^3\right)$ is the linearized Lagrangian up to quadratic order in the field ϕ. The linearized equations of motion are $\text{EOM}_i(\bar{g}) \cdot \phi \equiv \mathcal{O}_{ij}(\bar{g})\phi^j = 0$.

Now, one can make the simple but crucial observation that (5.1.9) has an alternate path integral representation,

$$2\widehat{S}^{(1)}(\bar{g}) = -i \log \int \mathcal{D}u^i \mathcal{D}\phi^i e^{i \int u^i \mathcal{O}_{ij}(\bar{g})\phi^j}$$

$$= -i \log \int \mathcal{D}u^i \mathcal{D}\phi^i e^{i \int u^i \, \mathrm{EOM}_i(\bar{g})\cdot\phi}, \qquad (5.1.10)$$

where we note the expected factor of 2 on the LHS, which occurs since

$$\log\det \begin{pmatrix} 0 & \mathcal{O}(\bar{g}) \\ \mathcal{O}(\bar{g})^T & 0 \end{pmatrix} = 2\log\det\mathcal{O}(\bar{g}) + \text{ constant }. \qquad (5.1.11)$$

One now employs the Bianchi completeness assumption, leading us to trade the integral over the fluctuations ϕ for the integral over the fluctuations $a_\pm$ of the Lax currents, while simultaneously enforcing the linearized Bianchi identities via delta-functionals inserted in the path integral:

$$2\widehat{S}^{(1)}(\bar{g}) = -i \log \int \mathcal{D}u^i \mathcal{D}a_\pm \delta\left(\mathrm{Bianchi}_\alpha(\bar{g}) \cdot a\right) e^{i \int_\Sigma u^i \, \mathrm{EOM}_i(\bar{g})\cdot a}$$

$$= -i \log \int \mathcal{D}u^i \mathcal{D}v^\alpha \mathcal{D}a_\pm \exp i \int_\Sigma \left(u^i \, \mathrm{EOM}_i(\bar{g}) + v^\alpha \, \mathrm{Bianchi}_\alpha(\bar{g})\right) \cdot a,$$

$$(5.1.12)$$

where v^α is an auxiliary field.

Since the equations of motion and Bianchi identities are related to the zero-curvature equations via a $\bar{g}$-dependent rotation, one then obtains

$$2\widehat{S}^{(1)}(\overline{\mathcal{J}}) = -i \log \int \mathcal{D}U^I \mathcal{D}a_\pm \exp i \int_\Sigma U^I \, \mathrm{zc}_I(\overline{\mathcal{J}}) \cdot a, \qquad (5.1.13)$$

where U^I is related to u and v via a change of variables. This expression does not depend explicitly on $\bar{g}$, but only on $\overline{\mathcal{J}}$. Thus, one observes that, barring extra gauge redundancies that have not been taken into account above, all theories satisfying the Bianchi completeness assumption have the same 1-loop on-shell effective action, given in terms of the Lax currents. For the specific case of Lax connections of the form (5.1.1), one obtains

$$2\widehat{S}^{(1)}(\overline{\mathcal{J}}) = -i \log \int \mathcal{D}u \mathcal{D}v \mathcal{D}a_\pm \exp i$$

$$\times \int_\Sigma \mathrm{Tr}\left[u\partial \cdot (\overline{\mathcal{J}} + a) + vF_{+-}(\overline{\mathcal{J}} + a)\right]\Big|_{\mathcal{O}(a)} \qquad (5.1.14)$$

by writing U in terms of variables $u \in \mathfrak{g}$ and $v \in \mathfrak{g}$.

One can now compute the 1-loop effective action using Feynman diagrams, where the Lax currents and their fluctuations are considered to be the fundamental fields. Working at quadratic order in the fluctuation fields $a_\pm, u, v$ in the effective action, and making a change of variables $(u, v) \to (x, y) = (u + v, u - v)$, one obtains

$$2\widehat{S}^{(1)} = -i \log \int \mathcal{D}x \mathcal{D}y \mathcal{D}a_\pm$$

$$\times \exp i \int_\Sigma \mathrm{Tr}\left[x\partial_+ a_- + y\partial_- a_+ + \frac{1}{2}(x - y)\left([\overline{\mathcal{J}}_+, a_-] + [a_+, \overline{\mathcal{J}}_-]\right)\right]$$

$$(5.1.15)$$

We now observe that the effective action has a Chern–Simons-like form, although x and y are scalars.

Using the Feynman rules that can be extracted from the effective action, one finds only a single divergent diagram shown in Figure 5.1, whose amplitude is

$$2\widehat{S}^{(1)} = \left(\int \frac{\mathrm{d}^2 l}{(2\pi)^2} \frac{1}{l_-(l + k)_+}\right) \frac{1}{4} f_{\alpha\beta\gamma} f_\delta^{\beta\gamma} \int_\Sigma \mathrm{d}^2\sigma \overline{\mathcal{J}}_+^\alpha \overline{\mathcal{J}}_-^\delta. \qquad (5.1.16)$$

Evaluating the integral gives

$$\widehat{S}^{(1)} = \frac{1}{4\pi} \int_\Sigma \mathrm{d}^2\sigma \widehat{L}^{(1)} = \frac{1}{4\pi} \int_\Sigma \mathrm{d}^2\sigma \log \frac{\Lambda}{\mu} \frac{1}{2} \mathsf{h}^\vee \mathrm{Tr}\left[\overline{\mathcal{J}}_+ \overline{\mathcal{J}}_-\right] + \text{finite} ,$$

$$(5.1.17)$$

where $\mathsf{h}^\vee$ is the dual Coxeter number, Λ is a UV cutoff, and μ the renormalization group scale. This further implies that the effective Lagrangian satisfies

$$\frac{d}{dt}\widehat{L}^{(1)} = -\frac{1}{2}\mathsf{h}^\vee \mathrm{Tr}\left[\overline{\mathcal{J}}_+ \overline{\mathcal{J}}_-\right] \quad (t \equiv \log\mu). \qquad (5.1.18)$$

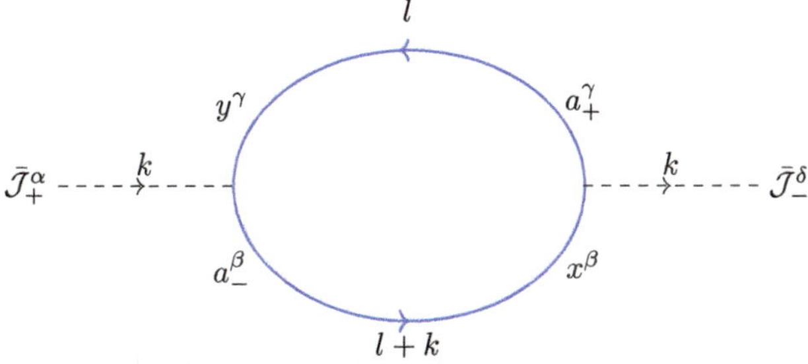

Figure 5.1.   Divergent 1-loop diagram.

This is the universal form for the 1-loop RG flow in integrable field theories of PCM-type. One can verify this in specific examples, such as the $\lambda-$ and $\eta-$deformed PCMs.

We have so far considered Lax operators with simple poles at ± 1 as in equation (5.1.1). One can generalize the above derivation to Lax connections with simple poles at more general locations on the spectral plane [76]. With the assumptions that the $+$ and $-$ components of $\mathcal{L}$ do not share poles, i.e., they correspond to two sets of poles P_+ and P_- with trivial intersection, and that there are no constant terms in $\mathcal{L}$, one can derive the more general result

$$\frac{d}{dt}\widehat{L}^{(1)} = -2\mathrm{h}^\vee \sum_{z_i \in P_+, w_j \in P_-} \frac{1}{(z_i - w_j)^2} \, \mathrm{Tr}\left[\overline{\mathcal{J}}_+^{(i)}\overline{\mathcal{J}}_-^{(j)}\right]. \tag{5.1.19}$$

5.1.2   *4d Chern–Simons derivation*

We shall now review the derivation of the universal 1-loop divergence from the perspective of 4d Chern–Simons theory. Let us start with the 4d Chern–Simons action with a convenient normalization given by

$$S_{4d} = \frac{i}{16\pi^2} \int_{\Sigma \times C} \varphi(z) \mathrm{d}z \wedge \mathrm{CS}(A). \tag{5.1.20}$$

The approach is once again via a background field expansion $A_m = \bar{A}_m + a_m$ where $m = \bar{z}, +, -$. The 4d action is expanded as

$$S_{4d}[\bar{A} + a] = S_{4d}[\bar{A}] + \int_{\Sigma \times C} \mathrm{Tr}\left[a^m \mathcal{O}_{mn}(\bar{A})a^n\right] + \dots, \tag{5.1.21}$$

where integration here means integration against the 4-form $\mathrm{d}z \wedge \mathrm{d}\bar{z} \wedge \mathrm{d}\sigma^+ \wedge \mathrm{d}\sigma^-$. The 1-loop effective action can then be computed as the determinant of $\mathcal{O}$,

$$\widehat{S}_{4d}^{(1)}[\bar{A}] = -i \log \int_{\mathrm{B.C.s}} \frac{\mathcal{D}a_m}{\mathrm{gauge}} e^{i \int_{\Sigma \times C} \mathrm{Tr}\left[a^m \mathcal{O}_{mn}(\bar{A})a^n\right]}$$
$$= \frac{i}{2} \log \det{}' \mathcal{O}(\bar{A}) \tag{5.1.22}$$

where 'B.C.s' here means that the boundary conditions on the gauge fields have been imposed, and $\det'$ is used to denote that the determinant is taken over gauge-inequivalent field configurations.

Just as in the 2d derivation, one can now rewrite the one-loop determinant in an alternate path integral form:

$$\widehat{S}_{4d}^{(1)} = -\frac{i}{2}\log\int_{\text{B.C.s}} \frac{\mathcal{D}a_m\mathcal{D}u^m}{\text{gauge}}\exp i\int_{\Sigma\times C}\text{Tr}\left[u^m\mathcal{O}_{mn}(\bar{A})a^n\right]$$

$$= -\frac{i}{2}\log\int_{\text{B.C.s}} \frac{\mathcal{D}a_m\mathcal{D}u^m}{\text{gauge}}\exp i \qquad (5.1.23)$$

$$\times\int_{\Sigma\times C}\varphi(z)\,\text{Tr}\left[u^{\bar{z}}F_{+-}(A) + u^-F_{\bar{z}+}(A) + u^+F_{\bar{z}-}(A)\right].$$

The fields u here play the role of Lagrange multipliers enforcing the equations of motion, and satisfy the same boundary conditions as the fluctuation fields a_m.

We now change variables using a formal gauge transformation

$$A_{\bar{z}} = -\partial_{\bar{z}}gg^{-1}, \quad A_{\pm} = g\mathcal{L}_{\pm}g^{-1} - \partial_{\pm}gg^{-1}. \qquad (5.1.24)$$

The integration is now over fluctuations of the new variables $(g, \mathcal{L}_{\pm})$, defined via

$$g = \bar{g}\gamma, \quad \mathcal{L}_{\pm} = \bar{\mathcal{L}}_{\pm} + \ell_{\pm}. \qquad (5.1.25)$$

Transforming the Lagrange multipliers as

$$u^m \to gu^mg^{-1} \qquad (5.1.26)$$

allows us to re-express the path integral as

$$\widehat{S}_{4d}^{(1)} = -\frac{i}{2}\log\int_{\text{B.C.s}} \frac{\mathcal{D}\gamma\mathcal{D}\ell_{\pm}\mathcal{D}u^m}{\text{gauge}}$$

$$\qquad (5.1.27)$$

$$\times\exp i\int_{\Sigma\times C}\varphi(z)\,\text{Tr}\left[u^{\bar{z}}F_{+-}(\mathcal{L}) + u^-\partial_{\bar{z}}\mathcal{L}_+ + u^+\partial_{\bar{z}}\mathcal{L}_-\right].$$

Integrating over the Lagrange multipliers u_+ and u_- then imposes the usual equations of motion

$$\varphi(z)\partial_{\bar{z}}\mathcal{L}_{\pm} = 0 \qquad (5.1.28)$$

via delta functional insertions in the path integral, i.e.,

$$\int\mathcal{D}\ell_{\pm}\mathcal{D}u^{\pm}\exp i\int_{\Sigma\times C}\varphi(z)\,\text{Tr}\left[u^-\partial_{\bar{z}}\mathcal{L}_+ + u^+\partial_{\bar{z}}\mathcal{L}_-\right]$$

$$\qquad (5.1.29)$$

$$= \int\mathcal{D}\ell_{\pm}\delta^{(4)}\left(\varphi(z)\partial_{\bar{z}}\mathcal{L}_+\right)\delta^{(4)}\left(\varphi(z)\partial_{\bar{z}}\mathcal{L}_-\right).$$

These equations of motion can be solved in the familiar manner to give

$$\mathcal{L}_{\pm} = \tilde{\mathcal{J}}_{\pm}^{0} + \sum_{i=1}^{k} \sum_{p=1}^{m_i} \frac{\tilde{\mathcal{J}}_{\pm}^{i,p}}{(z - a_i)^p}, \tag{5.1.30}$$

assuming we have a twist function of the form

$$\varphi(z) = K \frac{(z - a_1)^{m_1} \cdots (z - a_k)^{m_k}}{(z - b_1)^{n_1} \cdots (z - b_l)^{n_l}}, \tag{5.1.31}$$

where K is a constant.

The path integral over $\ell_{\pm}$ can be replaced by a path integral over fluctuations of $\tilde{\mathcal{J}}_{\pm}^{i,p}(\sigma)$, denoted $\tilde{\alpha}_{\pm}^{i,p}$:

$$\int \mathcal{D}\ell_{\pm} \delta^{(4)}\left(\varphi(z)\partial_{\bar{z}}\mathcal{L}_{+}\right) \delta^{(4)}\left(\varphi(z)\partial_{\bar{z}}\mathcal{L}_{-}\right) = \int \mathcal{D}\tilde{\alpha}_{\pm}^{i,p}. \tag{5.1.32}$$

where

$$\tilde{\mathcal{J}}^{i,p} = \overline{\mathcal{J}}^{i,p} + \tilde{\alpha}^{i,p}, \tag{5.1.33}$$

with $\overline{\mathcal{J}}^{i,p}$ denoting the background field. This change of variables incurs a divergent but constant Jacobian, which is neglected in the present analysis.

Now, denote P as the set of poles of the twist function $\varphi(z)$. The boundary conditions at the poles constrain the form of the Lax currents $\tilde{\mathcal{J}}$ as

$$\tilde{\mathcal{J}} = \tilde{\mathcal{J}}\left(\tilde{g}|_P\right) \tag{5.1.34}$$

where $\tilde{g}|_P$ denotes certain combinations of g and its derivatives evaluated at P. This motivates us to decompose the integral over the fluctuation field γ as

$$\mathcal{D}\gamma = \mathcal{D}\tilde{\gamma}|_P \, \mathcal{D}\gamma|_{\text{bulk}} \tag{5.1.35}$$

where γ_{bulk} denotes fluctuations away from P.

We however only need to integrate over $\tilde{\gamma}_P$ as the 4d gauge transformations imply the transformation $g \to \hat{u}g$, where $\hat{u}$ is set to be trivial at P in order to not modify the boundary conditions. We can thus gauge-fix and divide out the factor of $\mathcal{D}\gamma|_{\text{bulk}}$

$$\widehat{S}_{4d}^{(1)} \sim -\frac{i}{2} \log \int_{\text{B.C.s}} \frac{\mathcal{D}\gamma \mathcal{D}\tilde{\alpha}_{\pm}^{i,p} \mathcal{D}u^{\bar{z}}}{\text{gauge}} \exp i \int_{\Sigma \times C} \varphi(z) \operatorname{Tr}\left[u^{\bar{z}} F_{+-}(L(z, \tilde{\mathcal{J}}))\right]. \tag{5.1.36}$$

There is also the additional gauge invariance that arises from the change of variables (5.1.24), which is conventionally fixed by setting $\tilde{g}|_{\infty} = 1$.

Integration over $\tilde{\gamma}|_P$ imposes some Bianchi identities on $\tilde{\mathcal{J}}$ due to the boundary conditions, i.e.,

$$
\begin{aligned}
\int_{\text{B.C.s}} \mathcal{D}\tilde{\alpha}_\pm^{i,p}\, \mathcal{D}\tilde{\gamma}|_P &= \int \mathcal{D}\tilde{\alpha}^{i,p}\mathcal{D}\tilde{\gamma}\Big|_P\, \delta\left(\tilde{\mathcal{J}} - \tilde{\mathcal{J}}\left(\tilde{g}|_P\right)\right) \\
&= \int \mathcal{D}\tilde{\alpha}_\pm^{i,p} \prod_s \delta\left(B_s \cdot \tilde{\alpha}\right) \\
&= \int \mathcal{D}\tilde{\alpha}_\pm^{i,p}\mathcal{D}\tilde{v}^s \exp i \int_\Sigma \mathrm{d}^2\sigma\, \left(\tilde{v}^s\left(B_s \cdot \tilde{\alpha}\right)\right),
\end{aligned}
\tag{5.1.37}
$$

where $B_s \cdot \tilde{\alpha} = 0$ are linearized Bianchi identities and $\tilde{v}^s$ are 2d Lagrange multipliers.

In analogy to the 2d derivation, one assumes a Bianchi completeness identity, i.e., that the linearized Bianchi identities $B_s \cdot \tilde{\alpha} = 0$ that arise from the boundary conditions are a subset of the linearized zero-curvature equations $F_{+-}(\mathcal{L}(z, \overline{\mathcal{J}} + \tilde{\alpha})) = 0$. This then implies that the exponential factor in (5.1.37) can be discarded, since the zero curvature equations are already imposed via Lagrange multipliers in (5.1.36), and the effective action is given by

$$
\widehat{S}_{4d}^{(1)} \sim -\frac{i}{2} \log \int \mathcal{D}\tilde{\alpha}_\pm^{i,n}\mathcal{D}u^{\bar{z}}\ \exp i \int_{\Sigma \times C} \varphi(z)\, \mathrm{Tr}\left[u^{\bar{z}} F_{+-}(\mathcal{L}(z, \tilde{\mathcal{J}}))\right],
\tag{5.1.38}
$$

where the constant factor $\int \frac{\mathcal{D}\tilde{v}^s}{\text{gauge}}$ has also been discarded. Finally, noting that there is still a residual gauge invariance in the path integral due to only some modes of $u^{\bar{z}}$ appearing in the action, one can gauge fix to reduce the 4d path integral to the universal 2d path integral given in (5.1.13). This implies that the universal 1-loop divergence of a PCM-type integrable field theory is equivalent to the 1-loop divergence of 4d CS with suitable disorder surface defects.

5.2 Discretizations of Integrable Field Theories

In this section, we shall approach the problem of quantizing integrable field theories from the perspective of discretization, in the context of 4d Chern–Simons theory, following the work of Ashwinkumar, Sakamoto, and Yamazaki [75]. Such discretizations are widely-studied in the literature, and were pioneered by Faddeev and Reshetikhin, in their attempt to quantize the principal chiral model [78].

5.2.1 *Discretizations of surface defects*

From the perspective of 4d Chern–Simons theory, discretization of an integrable field theory can be interpreted in terms of discretization of order surface operators to line operators, such that the resulting system is an integrable lattice model of the type reviewed in Chapter 1. In what follows, we shall focus on Lorentzian integrable field theories, and hence Σ can be understood to be endowed with space and time coordinates, denoted by σ and τ respectively, with the corresponding lightcone coordinates denoted as $\sigma^+ = \tau + \sigma$ and $\sigma^- = \tau - \sigma$.

In general, one may discretize chiral and anti-chiral surface operators that take the form

$$\frac{1}{\hbar_{2d}} \int_{\Sigma \times \{z_\alpha^+\}} d\sigma^+ d\sigma^- \left(\mathscr{L}_\alpha \left(\phi^\alpha, \partial_- \phi^\alpha\right) |_+ + \mathcal{J}_a^\alpha A_-^a \right) \quad (\alpha = 1, \ldots n_+) \,,$$

$$\frac{1}{\hbar_{2d}} \int_{\Sigma \times \{z_{\bar\alpha}^-\}} d\sigma^+ d\sigma^- \left(\bar{\mathscr{L}}_{\bar\alpha} \left(\bar\phi^{\bar\alpha}, \partial_+ \bar\phi^{\bar\alpha}\right) |_- + \bar{\mathcal{J}}_a^{\bar\alpha} A_+^a \right) \quad (\bar\alpha = 1, \ldots n_-) \,.$$

$$(5.2.1)$$

Here, the notation $|_+$ and $|_-$ indicates quantities that transform as components of one-forms defined along the $+$ and $-$ lightcone directions on Σ, respectively. ϕ and $\bar\phi$ indicate fields supported on the chiral and anti-chiral surface defects, respectively, while $\mathcal{J}$ and $\bar{\mathcal{J}}$ indicate currents for the G-symmetry on each defect.[1]

We shall now review how to discretize one of these surface operators, which has the chiral form given in the first line of (5.2.1).

The essential reason we can discretize this defect with ease is that, given such a surface operator located at a point $z_+ \in \mathbb{CP}^1$, and with dependence on a set of fields ϕ^α, the partial derivatives of these fields are only with respect to σ^-. Thus, we can easily discretize it along the σ^+ direction. This discretization may be accomplished by replacing the integral over σ^+ in the defect action

$$\frac{1}{\hbar_{2d}} \int_{\Sigma \times \{z_+\}} d\sigma^+ d\sigma^- \mathscr{L} \left(\phi^\alpha, \partial_- \phi^\alpha; A_- |_{z_+}\right) \qquad (5.2.2)$$

by a Riemann sum over lattice points $\sigma^+ = \sigma_n^+ := \Delta n \ (n \in \mathbb{Z})$ that are equidistant, with lattice spacing Δ. The action that results is then a sum

---

[1] In this section, we shall denote the gauge group of 4d Chern–Simons theory using both G and $G_{\mathbb{C}}$.

over the 1d defect actions along the lightcone direction, which we denote as $\mathbb{R}_-$:

$$\frac{1}{\hbar_{1\mathrm{d}}} \sum_i \int_{\{\sigma_i^+\} \times \mathbb{R}_- \times \{z_+\}} \mathrm{d}\sigma^- \, \mathscr{L}\left(\phi_i^\alpha, \partial_- \phi_i^\alpha; A_{-,i}\right)$$

$$= \frac{1}{\hbar_{1\mathrm{d}}} \sum_{i=1}^N \int_{\{\sigma_i^+\} \times \mathbb{R}_- \times \{z_+\}} \mathrm{d}\sigma^- \left(\mathscr{L}\left(\phi_i^\alpha, \partial_- \phi_i^\alpha\right) + J_i^{\alpha,a} A_{-,a}\big|_{\sigma_i^+, z_+}\right),$$

$$(5.2.3)$$

where we defined

$$\phi_n^\alpha := \phi^\alpha\big|_{\sigma^+ = \sigma_n^+}, \quad A_{-,n} := A_-\big|_{\sigma^+ = \sigma_n^+}, \quad J_n^{\alpha,a} := \mathscr{J}^{\alpha,a}\big|_{\sigma^+ = \sigma_n^+}, \quad (5.2.4)$$

and the one-dimensional Planck constant, $\hbar_{1\mathrm{d}}$, by

$$\frac{1}{\hbar_{1\mathrm{d}}} = \frac{\Delta}{\hbar_{2\mathrm{d}}}. \tag{5.2.5}$$

An analogous procedure can be carried out for the anti-chiral surface operator located at a point z_- and with dependence on a set of fields denoted $\bar{\phi}$, i.e.,

$$\frac{1}{\hbar_{2\mathrm{d}}} \int_{\Sigma \times \{z_-\}} \mathrm{d}\sigma^+ \mathrm{d}\sigma^- \mathscr{L}\left(\bar{\phi}^{\bar{\alpha}}, \partial_+ \bar{\phi}^{\bar{\alpha}}; A_+\big|_{z_-}\right), \tag{5.2.6}$$

but we would now be discretizing in the σ^- direction instead into $\sigma^- = \sigma_n^- := n\Delta$. In this way, we arrive at another infinite set of line defects oriented along the σ^+ direction, i.e.,

$$\frac{1}{\hbar_{1\mathrm{d}}} \sum_{j=1}^N \int_{\mathbb{R}_+ \times \{\sigma_j^-\} \times \{z_-\}} \mathrm{d}\sigma^+ \left(\mathscr{L}\left(\bar{\phi}_j^{\bar{\alpha}}, \partial_+ \bar{\phi}_j^{\bar{\alpha}}\right) + J^a\left(\bar{\phi}_j^{\bar{\alpha}}\right) A_{+,a}\big|_{\sigma_j^-, z_-}\right),$$

$$(5.2.7)$$

where we defined $\bar{\phi}_l = \bar{\phi}_l\big|_{\sigma_j^-}$.

Observe that the currents supported by the surface defects have now become discrete modes, $J_{\pm,m}^{\alpha,a}$. To be precise, there is a pair of modes $J_{\pm,m}^{\alpha,a}$ supported on the null links $\sigma^+ = m\Delta$ and $\sigma^- = -m\Delta$ associated with each spatial lattice point $\sigma = m\Delta$. The discretized chiral and anti-chiral surface operators supporting these discrete modes are depicted in Figure 5.2. Prior to discretization, the Poisson brackets of the currents can be identified with a level 0 classical current algebra. As a result, the discrete modes of the currents satisfy the Poisson brackets

$$\{J_{\pm,m}^a, J_{\pm,n}^b\} = \frac{1}{\Delta} f^{ab}{}_c J_{\pm,n}^c \delta_{mn}, \tag{5.2.8}$$

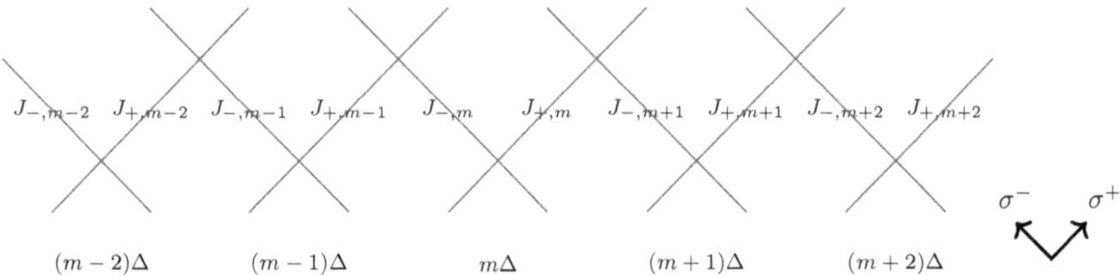

Figure 5.2.　The lightcone lattice where the discrete modes of the currents are supported along the null segments.

where the Dirac delta distribution of the classical current algebra at level 0 has been substituted by δ_{mn}/Δ.

We shall review many examples of surface defect discretizations in Sections 5.2.3 and 5.2.4.

One might worry that, when considering surface defects with fermionic degrees of freedom, the discretization procedure runs into problems associated with the doubling of the chiral fermions [79–82]. The present setup, however, does *not* involve discretizing the full 2d spacetime into lattices, but rather to discretize only *one* of the lightcone directions while keeping the orthogonal direction intact. The crucial point is that the chiral (or anti-chiral) theory has a kinetic term which does not involve any σ^+ (σ^-) derivatives and our discretization is only along the σ^+ (σ^-) direction. This guarantees that there is no doubling of chiral fermions and that our procedure is consistent.

We can thus discretize the 2d defects into a set of 1d defects without any obstruction, thereby discretizing the corresponding 2d integrable field theory. The 1d line defects are located at

$$\text{(chiral defect)} : \sigma^+ = m\Delta\,, \qquad m \in \mathbb{Z}, \tag{5.2.9}$$

$$\text{(anti-chiral defect)} : \sigma^- = n\Delta\,, \qquad n \in \mathbb{Z}. \tag{5.2.10}$$

In the literature, lightcone discretizations of integrable field theories of this sort has a long history (see e.g., [26, 83–86]), and this discretization plays a crucial role in the quantization of such theories in the approach of the quantum inverse scattering method of Faddeev and Reshetikhin [26]. The advantage of using 4d Chern–Simons theory in this context is that it allows us to systematically apply discretization to a very general class of integrable models.

When we have to deal with multiple chiral or anti-chiral defects, labeled by a pair of integers $n_+ > 1$ or $n_- > 1$, we need to be more careful in performing the discretization procedure. In this situation, all the discretized chiral (anti-chiral) defects are located on the same lightcone ray in the two-dimensional plane, which appears singular from the two-dimensional perspective. However, remember that the surface defects are separated along the spectral curve, C, and there are no actual overlaps or divergences. Moreover, 4d CS theory is topological along Σ, and thus we may freely move the 1d defects without affecting any physical quantities to obtain an integrable lattice model as in Figure 5.2. The example of $n_+ = 2$ and $n_- = 2$ is depicted in Figure 5.3. Moreover, we may additionally permute the chiral (or anti-chiral) defects without affecting any physics (see Figures 5.4 and 5.5 for examples).

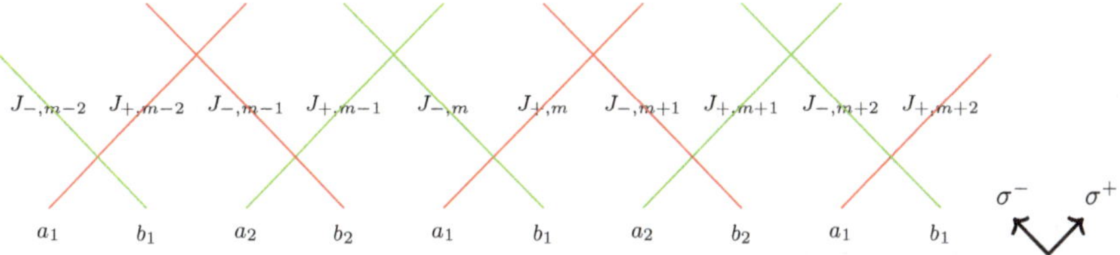

Figure 5.3. A discretization with two chiral and two anti-chiral defects ($n_+ = 2$, $n_- = 2$).

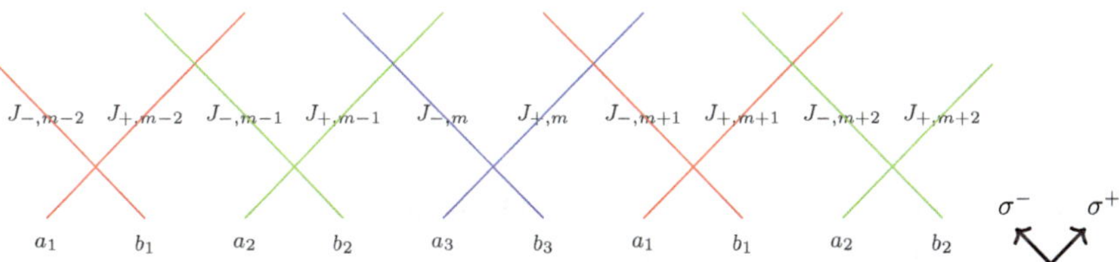

Figure 5.4. A possible discretization with three chiral and three anti-chiral defects ($n_+ = 3, n_- = 3$).

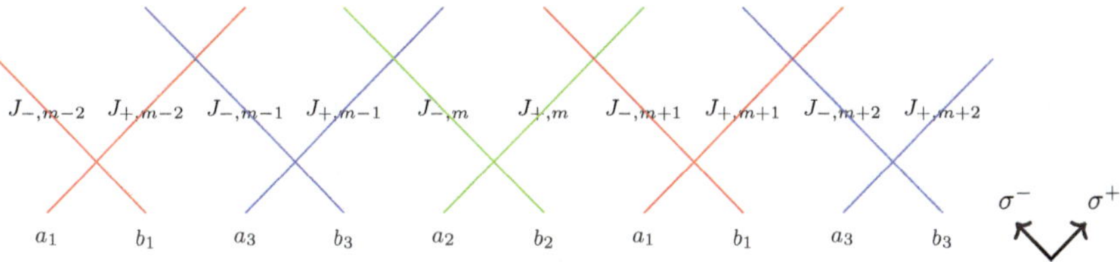

Figure 5.5. An alternate discretization with three chiral and three anti-chiral defects ($n_+ = 3, n_- = 3$).

In general, the path integral of 4d CS coupled to the 1d defects takes the form[2]

$$\int \mathcal{D}A \exp\left(\frac{i}{2\pi\hbar}\int_{\Sigma\times C} dz \wedge \mathrm{CS}(A)\right)$$

$$\times \prod_{\alpha=1}^{n_+}\prod_{i}^{N}\int \mathcal{D}\phi_i^\alpha \exp\left(\frac{i}{\hbar_{1d}}\int_{\{\sigma_i^+\}\times\mathbb{R}_-\times\{z_+\}} d\sigma^-\,\mathscr{L}\left(\phi_i^\alpha, \partial_-\phi_i^\alpha; A_-|_{\sigma_i^+,z_+}\right)\right)$$

$$\times \prod_{\bar\alpha=1}^{n_-}\prod_{j}^{N}\int \mathcal{D}\bar\phi_j^{\bar\alpha} \exp\left(\frac{i}{\hbar_{1d}}\int_{\mathbb{R}_+\times\{\sigma_j^-\}\times\{z_-\}} d\sigma^+\,\mathscr{L}\left(\bar\phi_j^{\bar\alpha}, \partial_+\bar\phi_j^{\bar\alpha}; A_+|_{\sigma_j^-,z_-}\right)\right).$$

$$(5.2.11)$$

As we shall observe from concrete examples in Sections 5.2.3 and 5.2.4, these 1d defects can be identified with gauge-invariant Wilson loops via a convenient quantization scheme, once we adopt suitable periodic or antiperiodic boundary conditions on the lattice. The path integral can thus be expressed as

$$\int \mathcal{D}A \exp\left(\frac{i}{2\pi\hbar}\int_{\Sigma\times C} dz \wedge \mathrm{CS}(A)\right)$$

$$\times \prod_{\alpha=1}^{n_+}\prod_{i=1}^{N}\mathrm{Tr}_{\rho_{+,i}}\mathcal{P}\exp\left(\frac{1}{\hbar_{1d}}\int_{\{\sigma_i^+\}\times S_-^1\times\{z_+\}} d\sigma^-\,A_-|_{\sigma_i^+,z_+}\right) \qquad (5.2.12)$$

$$\times \prod_{\bar\alpha=1}^{n_-}\prod_{j=1}^{N}\mathrm{Tr}_{\rho_{-,j}}\mathcal{P}\exp\left(\frac{1}{\hbar_{1d}}\int_{S_+^1\times\{\sigma_j^-\}\times\{z_-\}} d\sigma^+\,A_+|_{\sigma_j^-,z_-}\right).$$

Here, the traces are taken in representations $\rho_{+,i}$ and $\rho_{-,j}$, which depend on the matter content and action of the line defects. From Chapter 1, we know that this is precisely the configuration with which we realize integrable lattice models in 4d Chern–Simons theory. This is how one discretizes an integrable field theory to an integrable lattice model in 4d Chern–Simons theory.

5.2.2　*Quantization in terms of spin chains*

Once an integrable field theory has been discretized in 4d CS, one can go on to quantize it in a systematic manner.

---

[2]Since we are working in Lorentzian signature, the weight for the path integral is defined by $\exp(iS)$ for an action S.

Let us simplify the following presentation by specializing to the case of $n_+ = n_- = 1$. The Hilbert space, $\mathcal{H}$, of the discretized integrable field theory can be defined in terms of tensor products of modules for representations of $\mathfrak{g}$:

$$\mathcal{H} = \bigotimes_{i,j} (V_{+,i} \otimes V_{-,j}),\tag{5.2.13}$$

where we denote the representation space for $\rho_{+,i}$ ($\rho_{-,j}$) by $V_{+,i}$ ($V_{-,j}$). Here, the label i in (5.2.13) runs over all the integers, and $\mathcal{H}$ is infinite-dimensional. As in Figure 5.2, the chiral and anti-chiral defects correspond respectively to even and odd sites of a spin chain (that arise from the lattice at a fixed point in time), respectively, which leads to the natural labeling

$$\mathcal{H} = \bigotimes_n V_n,\tag{5.2.14}$$

with $V_{2m-1} := V_{+,i}$ and $V_{2m} := V_{-,j}$ and correspondingly $\rho_{2m-1} := \rho_{+,i}$ and $\rho_{2m} := \rho_{-,j}$.

Alternatively, we often impose the periodic boundary condition along the spatial direction, such that we have $V_{n+2N} = V_n$ for an integer N, so that we obtain a spin chain with a periodic boundary condition. Having defined the spin chain, we can go on to discuss the relationship between the quantum integrability of the field theory and that of the spin chain.

As we have discussed in Chapter 2, the integrability of a two-dimensional quantum field theory originates from a Wilson line for the Lax connection along the spatial direction. This Wilson line is included in addition to the Wilson lines discussed above that realize an integrable lattice model. Assuming that this auxiliary Wilson line is in the representation ρ_A with associated representation space V_A (where "A" stands for "auxiliary"), we now have an enlarged Hilbert space $\mathcal{H} \otimes V_A$.

This additional Wilson line, denoted W_A, gives rise to the configuration shown in Figure 5.6, where W_A crosses with the other Wilson lines along the lightcone directions. In Figure 5.6, and what follows, we pick the positions z_+ and z_- of the chiral and anti-chiral defects to be ν and $-\nu$, respectively. Since the interactions localized at the crossing of two Wilson lines in 4d CS generate an R-matrix (as explained in Chapter 1) we obtain a product of R-matrices along W_A:

$$\begin{aligned}T(z) = \cdots &R_{\rho_A,\rho_-}(z+\nu)R_{\rho_A,\rho_+}(z-\nu)\\ \times\,&R_{\rho_A,\rho_-}(z+\nu)R_{\rho_A,\rho_+}(z-\nu)\cdots.\end{aligned}\tag{5.2.15}$$

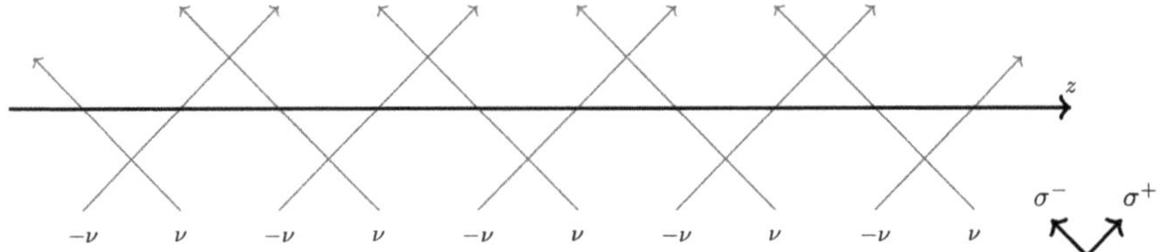

Figure 5.6. Wilson line configuration giving rise to inhomogeneous spin chain with alternating inhomogeneities.

(There are corrections to this formula from the framing anomaly, which, however, are not important for the understanding of what follows.) This product is known as the total monodromy matrix, and its matrix elements are operators that act on the Hilbert space, $\mathcal{H}$, defined in (5.2.13), while the matrix itself acts on the auxiliary space, V_A.

Now, let us consider two parallel Wilson lines of the type just introduced, denoted W_A and $W_{A'}$, along the spatial direction, and let us compactify the spatial direction so that we are taking the trace of each corresponding monodromy matrix along the auxiliary Hilbert space, V_A. We may now use the topological invariance along Σ to exchange the two Wilson lines, realizing the Yang–Baxter equation along the way. This implies that

$$[\mathrm{Tr}_A T(z), \mathrm{Tr}_{A'} T(z')] = 0 \,, \tag{5.2.16}$$

that is, the conserved charges of the integrable model (that are obtained by Taylor expanding $\mathrm{Tr}_A T(z)$) commute with each other, which implies quantum integrability. The conserved charges include the lightcone components of energy and momentum. These are unitary operators that can be expressed as[3]

$$U_+ = \mathrm{Tr}_A T(\nu)\,, \qquad U_-^\dagger = \mathrm{Tr}_A T(-\nu)\,, \tag{5.2.17}$$

and generate discrete shifts along the lightcone directions, i.e., U_+ generates shifts of the form $\sigma^+ \to \sigma^+ + \Delta$ and $U_-^\dagger$ generates shifts of the form $\sigma^- \to \sigma^- - \Delta$.

In order to further understand quantum integrability of the integrable field theory, we we would like to understand how the monodromy matrix for the spin chain is related to the monodromy operator before discretization. In

---

[3] Quantities such as $T(\nu)$ and $T(-\nu)$ that involve the R-matrix $R_{\rho\rho'}(0)$ are finite as long as they are normalized appropriately, e.g., using (1.3.19) in the case of $G = GL(N)$.

the literature on discretizations of integrable field theories, this relationship relies on the identifications

$$\mathcal{P} \exp\left[-\int_{m\Delta}^{(m-1)\Delta} d\sigma^- \, \mathcal{L}_-(z) \right] \sim R_{\rho_A, \rho_{2m-1}}(z, \nu),$$

$$\mathcal{P} \exp\left[-\int_{m\Delta}^{(m+1)\Delta} d\sigma^+ \, \mathcal{L}_+(z) \right] \sim R_{\rho_A, \rho_{2m}}(z, -\nu),$$

$$(5.2.18)$$

where $R_{\rho_1, \rho_2}(z_1, z_2)$ is the R-matrix in the tensor product $\rho_1 \otimes \rho_2$ satisfying the Yang–Baxter equation. Taking a product of these matrices, we obtain the following relation between the classical and quantum monodromy operators:

$$\mathcal{P} \exp\left[-\int_C \mathcal{L}(z) \right] \sim \prod_m R_{\rho_A, \rho_{2m-1}}(z, \nu) R_{\rho_A, \rho_{2m}}(z, -\nu) = T(z, \nu).$$

$$(5.2.19)$$

where the integration contour C is along the σ-direction. The monodromy here is invariant under smooth deformations of the contour, since the connection $\mathcal{L}$ is flat. When the σ-direction is compactified, C wraps the winding cycle parametrized by σ.

The identification (5.2.19) is what we obtain in 4d Chern–Simons theory, since the monodromy operators before and after discretization arise from the same operator, i.e, the Wilson line.

This identification suggests that one can quantize the Lax connection into a well-defined operator in the quantized Hilbert space of the 2d integrable field theory. Together with (5.2.16), this would imply the quantum integrability of the integrable field theory.

One can check the relation (5.2.18) in the semiclassical limit. To this end, let us expand the left-hand side of (5.2.18). One then obtains

$$\mathcal{P} \exp\left[-\int_{m\Delta}^{(m-1)\Delta} d\sigma^- \, \mathcal{L}_-(z) \right] = 1 + \Delta\hbar \, \frac{1}{z - \nu} J^a_{-,m} \rho_A(t_a) + \dots,$$

$$\mathcal{P} \exp\left[-\int_{m\Delta}^{(m+1)\Delta} d\sigma^+ \, \mathcal{L}_+(z) \right] = 1 + \Delta\hbar \, \frac{1}{z + \nu} J^a_{+,m} \rho_A(t_a) + \dots,$$

$$(5.2.20)$$

using the explicit expression of the Lax operator in terms of the currents discussed (and computed using 4d CS) in Chapter 2. Now, in quantizing the discretized integrable field theory, we ought to replace the Poisson brackets of the discretized currents in (5.2.8) by commutators, i.e., by making the replacement $\{\,,\,\} \to [\,,\,]$. This gives us the canonical commutator

$$[J^a_{\pm,m}, J^b_{\pm,n}] = \frac{1}{\Delta} f^{ab}{}_c J^c_{\pm,n} \delta_{mn}\,, \tag{5.2.21}$$

and hence each of the modes $J^a_{\pm,m}$ is in the adjoint representation of $\mathfrak{g}$. From (5.2.21), we know that the discretized current can be regarded as an operator acting on the auxiliary Hilbert space as

$$
\begin{aligned}
J_{-,m} &= J^a_{-,m}\rho_A(t_a) = \frac{1}{\Delta}\rho_{-,m}(t^a) \otimes \rho_A(t_a)\,,\\[6pt]
J_{+,m} &= J^a_{+,m}\rho_A(t_a) = \frac{1}{\Delta}\rho_{+,m}(t^a) \otimes \rho_A(t_a)\,.
\end{aligned}
\tag{5.2.22}
$$

By using this relation for the current, we obtain

$$
\begin{aligned}
\mathcal{P}\exp\!\left[-\int_{m\Delta}^{(m-1)\Delta} d\sigma^- \, \mathcal{L}_-(z)\right] &= 1 + \hbar\, r_{\rho_A,\rho_{-,m}}(z,\nu) + \dots\,,\\[6pt]
\mathcal{P}\exp\!\left[-\int_{m\Delta}^{(m+1)\Delta} d\sigma^+ \, \mathcal{L}_+(z)\right] &= 1 + \hbar\, r_{\rho_A,\rho_{+,m}}(z,-\nu) + \dots\,,
\end{aligned}
\tag{5.2.23}
$$

which coincides with the expansion of the right-hand side of (5.2.18) up to normalization.

We have so far discussed the discretization of an integrable field theory and then quantized the resulting lattice model. Let us now discuss the opposite direction, namely, the thermodynamic limit of the lattice model which leads us back to an integrable field theory.

One can go back to the field theory[4] by taking the thermodynamic limit

$$\Delta \to 0\,, \quad \hbar_{1d} : \text{ fixed}\,, \quad \hbar_{2d} \to 0 \tag{5.2.24}$$

or

$$\Delta \to 0\,, \quad \hbar_{1d} \to \infty\,, \quad \hbar_{2d} : \text{ finite}\,. \tag{5.2.25}$$

Since we are interested in the quantization of two-dimensional field theories, we shall be mainly interested in the double-scaling thermodynamic limit (5.2.25) where the 2d Planck constant is kept finite.

Let us now explain the relation between integrable structures before and after the thermodynamic limit (5.2.25) where $\Delta \to 0$, while also taking the semiclassical limit where $\hbar_{2d} = \hbar$ becomes infinitesimally small. The

---

[4]To go back to the original field theory before discretization, one needs to ensure $\xi \gg \Delta$, where ξ is the typical correlation length of the theory. This is ensured when the system is near criticality, while more care is needed for general massive theories.

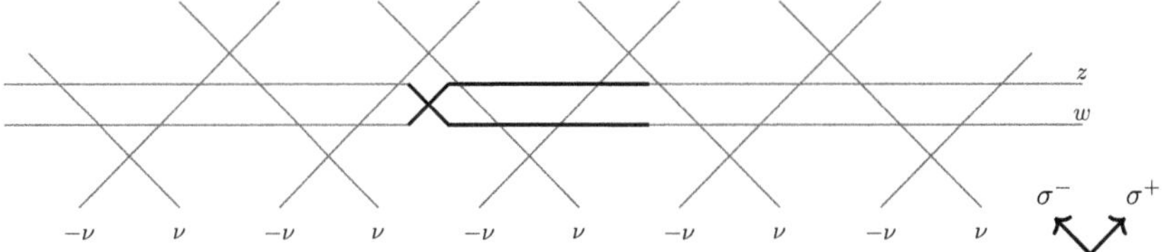

Figure 5.7.   The configuration that realizes the LHS of (5.2.26).

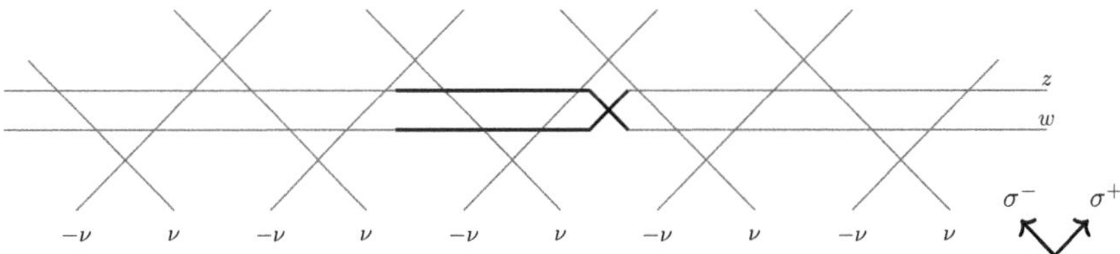

Figure 5.8.   The configuration that realizes the RHS of (5.2.26).

integrability of the one-dimensional quantum spin chain is guaranteed by the RLL relation, where "L" in this context denotes the Lax connection of the spin chain:

$$R_{12}(z - w)L_{n,1}(z)L_{n,2}(w) = L_{n,2}(w)L_{n,1}(z)R_{12}(z - w). \qquad (5.2.26)$$

This is a version of the Yang–Baxter equation, and this RLL relation can, in fact, be realized using Wilson lines in 4d Chern–Simons theory, as depicted in Figures 5.7 and 5.8.

We may rewrite the RLL relation as

$$[L_{n,1}(z), L_{n,2}(w)] = \big(1 - R_{12}(z - w)\big)L_{n,1}(z)L_{n,2}(w)$$
$$- L_{n,2}(w)L_{n,1}(z)\big(1 - R_{12}(z - w)\big). \qquad (5.2.27)$$

Up to normalizations that drop out of the RLL relation, the semiclassical limit of the R-matrix is given in terms of the classical r-matrix, r:

$$R(z) \longrightarrow 1 + \hbar\, r(z) + \cdots. \qquad (5.2.28)$$

Furthermore, we can generalize the relation (5.2.18) for the R-matrix to a relation for the L-operator:

$$L_{n,A}(z) \sim \mathcal{P}\exp\left[-\int_{n\Delta}^{(n+1)\Delta} dx\, \mathcal{L}(x; z)\right] = 1 - \Delta\mathcal{L}_{n,A}(z) + \cdots, \qquad (5.2.29)$$

so that we have

$$[L_{n,1}(z)\overset{\otimes}{,}L_{n,2}(w)] \longrightarrow \Delta^2\{\mathcal{L}_{n,1}(z),\mathcal{L}_{n,2}(w)\} \, , \tag{5.2.30}$$

in the semiclassical limit. By substituting the semiclassical expressions (5.2.28) and (5.2.30) into (5.2.27), we obtain

$$\{\mathcal{L}_{n,1}(z)\overset{\otimes}{,}\mathcal{L}_{n,2}(w)\} = \frac{\hbar}{\Delta}[r_{12}(z-w),\mathcal{L}_{n,1}(z)+\mathcal{L}_{n,2}(w)] \, , \tag{5.2.31}$$

or equivalently

$$\{\mathcal{L}_{n}(z)\overset{\otimes}{,}\mathcal{L}_{n}(w)\} = \frac{\hbar}{\Delta}[r_{12}(z-w),\mathcal{L}_{n}(z)\otimes 1 + 1 \otimes \mathcal{L}_{n}(w)] \, . \tag{5.2.32}$$

This is a discrete version of the ultralocal Poisson bracket algebra of the integrable field theories we considered previously in Chapter 2.

5.2.3   *Examples*

We now turn to concrete examples of discretizations of 2d surface defects. We will in turn discuss the coadjoint orbit defect, free fermion defect, free $\beta\gamma$ system defect, and finally the curved $\beta\gamma$ system defect.

We shall mostly restrict the discussion to the case of one chiral ($n_+ = 1$) and one anti-chiral ($n_- = 1$) defect. The generalization to multiple chiral and anti-chiral defects is straightforward. In addition, although we shall mostly focus on integrable field theories of rational type in explicit examples, our techniques also apply to trigonometric and elliptic integrable field theories.

5.2.3.1   *Coadjoint orbit defects*

We shall first discretize coadjoint orbit surface operators using the general procedure outlined in the previous section. For a chiral surface operator located at z_+, we shall discretize it along the σ^+ direction:

$$S_+^{\text{2d-coadj}} = \frac{1}{\hbar_{2d}} \int_{\Sigma \times \{z_+\}} \text{Tr}\left(\Lambda \mathcal{G}_{(+)}^{-1} D_- \mathcal{G}_{(+)}\right) \mathrm{d}\sigma^+\mathrm{d}\sigma^-$$

$$\to S_+^{\text{1d-coadj}} = \frac{1}{\hbar_{1d}} \sum_i \int_{\{\sigma_i^+\} \times \mathbb{R}_- \times \{z_+\}} \text{Tr}\left(\Lambda \mathcal{G}_{i(+)}^{-1} D_- \mathcal{G}_{i(+)}\right) \mathrm{d}\sigma^- \, ,$$

$$\tag{5.2.33}$$

where we have used $\mathcal{G}_{i(+)}$ and $\mathcal{G}_{i(+)}^{-1}$ to respectively denote $\mathcal{G}_{(+)}|_{\sigma_i^+}$ and $\mathcal{G}_{(+)}^{-1}|_{\sigma_i^+}$. We repeat the same procedure for an anti-chiral surface defect located at z_-, whereby we arrive at another infinite set of line defects

$$S_-^{\text{1d-coadj}} = \frac{1}{\hbar_{1d}} \sum_j \int_{\mathbb{R}_+ \times \{\sigma_j^-\} \times \{z_-\}} \text{Tr}\left(\Lambda \mathcal{G}_{j(-)}^{-1} D_+ \mathcal{G}_{j(-)}\right) d\sigma^+, \quad (5.2.34)$$

where $\mathcal{G}_{j(-)}$ and $\mathcal{G}_{j(-)}^{-1}$ respectively denote $\mathcal{G}_{(-)}|_{\sigma_j^-}$ and $\mathcal{G}_{(-)}^{-1}|_{\sigma_j^-}$.

Given that the defect actions are exponentiated in their respective path integrals, the path integral of 4d CS coupled to these defects takes the form

$$Z^{\text{1d-coadj}} = \int \mathcal{D}A \prod_{k=1}^{N} \mathcal{D}\mathcal{G}_{k(+)} \prod_{l=1}^{N} \mathcal{D}\mathcal{G}_{l(-)} \exp\left(\frac{i}{2\pi\hbar} \int_{\Sigma \times C} dz \wedge \text{CS}(A)\right)$$

$$\times \exp\left(i \sum_{i=1}^{N} \int_{S_-^1 \times \{z_+\}} \text{Tr}\left(\Lambda \mathcal{G}_{i(+)}^{-1} D_- \mathcal{G}_{i(+)}\right) d\sigma^-\right.$$

$$\left. + i \sum_{j=1}^{N} \int_{S_+^1 \times \{z_-\}} \text{Tr}\left(\Lambda \mathcal{G}_{j(-)}^{-1} D_+ \mathcal{G}_{j(-)}\right) d\sigma^+\right),$$

$$(5.2.35)$$

where we have adopted periodic boundary conditions to obtain S^1 for both lightcone directions.

We can re-express the line defects that appear in (5.2.35) as Wilson line operators, as long as Λ can be identified with the highest weight of an irreducible representation of the complex gauge group, $G_{\mathbb{C}}$.[5] This can be achieved via geometric quantization, or equivalently, via coherent state quantization; see, e.g., [87] for a helpful review. The crucial point here is that the partition function of a quantum mechanical system can be written schematically as

$$\text{Tr}_{\mathcal{H}} \mathcal{T} \exp\left(\int dt\, H\right), \quad (5.2.37)$$

---

[5] If one is interested in the compact form of $G_{\mathbb{C}}$, a reality condition such as

$$\overline{\mathcal{G}_{(\pm)}} = \mathcal{G}_{(\pm)}, \qquad \overline{A|_{\Sigma}} = A|_{\Sigma}, \quad (5.2.36)$$

can be imposed, where $A|_{\Sigma}$ denotes the components of the gauge field along Σ. As a result, the surface defect fields and gauge fields along Σ are valued in the real subgroup and real subalgebra of $G_{\mathbb{C}}$ and $\mathfrak{g}_{\mathbb{C}}$, respectively.

and for the quantum mechanical systems obtained via discretization, the Hilbert space $\mathcal{H}$ is a representation of $G_{\mathbb{C}}$, the time direction, t, is the support of the Wilson line, H is the Hamiltonian that becomes identified with the gauge field, A, in the aforementioned representation, and the time ordering T becomes path ordering. This idea shall apply in the discretizations of other surface operators we will encounter below as well.

Thus, the partition function of the coupled 4d–2d system of 4d CS theory and the coadjoint orbit surface operators can be expressed as

$$Z^{\text{1d-coadj}} = \int \mathcal{D}A \exp\left(\frac{i}{2\pi\hbar}\int_{\Sigma \times C} dz \wedge \mathrm{CS}(A)\right)$$

$$\times \prod_{i=1}^{N} \mathrm{Tr}_{\rho_-} \mathcal{P} \exp\left(i\int_{S^1} d\sigma^- \mathcal{A}^{(z_+)}_{-,i}\right)$$

$$\times \prod_{j=1}^{N} \mathrm{Tr}_{\rho_+} \mathcal{P} \exp\left(i\int_{S^1} d\sigma^+ \mathcal{A}^{(z_-)}_{+,j}\right), \qquad (5.2.38)$$

where $\mathcal{A}^{(z_+)}_{-,i} = A_-(\sigma,z)|_{\sigma^+=\sigma_i^+,\,z=z_+}$ and $\mathcal{A}^{(z_-)}_{+,i} = A_+(\sigma,z)|_{\sigma^-=\sigma_i^-,\,z=z_-}$. A discretization of the Faddeev–Reshetikhin model can thus be understood in terms of discretizations of coadjoint orbit surface defects to Wilson loops followed by a regularization where the number of Wilson loops in both lightcone directions, N, is taken to be finite. The representations of these Wilson loops are determined by the highest weight that corresponds to Λ.

Let us now discuss generalizations of this discretization of the Faddeev–Reshetikhin model. Recall that the Faddeev–Reshetikhin model was derived by coupling 4d CS theory to two coadjoint orbit defects, each supporting fields valued in the complexified Lie group $G_{\mathbb{C}}$, denoted $\mathcal{G}_{(+)}$ and $\mathcal{G}_{(-)}$ [30], as reviewed in Chapter 2. The 4d–2d action is given by

$$S[A, \mathcal{G}_{(\pm)}] = \frac{1}{2\pi\hbar}\int_{\Sigma \times C} dz \wedge \mathrm{CS}(A)$$

$$+ \frac{1}{\hbar_{2d}}\int_{\Sigma \times \{z_+\}} \mathrm{Tr}\left(\Lambda \mathcal{G}_{(+)}^{-1} D_- \mathcal{G}_{(+)}\right) d\sigma^+ \wedge d\sigma^-$$

$$+ \frac{1}{\hbar_{2d}}\int_{\Sigma \times \{z_-\}} \mathrm{Tr}\left(\Lambda \cdot \mathcal{G}_{(-)}^{-1} D_+ \mathcal{G}_{(-)}\right) d\sigma^+ \wedge d\sigma^-, \qquad (5.2.39)$$

where Λ is an element of the real Lie subalgebra of the complex Lie algebra $\mathfrak{g}_\mathbb{C}$. In the case $G_\mathbb{C} = SL(2,\mathbb{C})$, we take $\Lambda = -i\sigma^3/2$, where $\sigma^3 = \begin{pmatrix} 1 & 0 \\ 0 & -1 \end{pmatrix}$. Caudrelier *et al.* [88] have shown that the 2d Zakharov-Mikhailov action, an ultralocal integrable field theory that includes the Faddeev–Reshetikhin model as a special case, can also be derived from 4d CS theory coupled to order surface operators:

$$\frac{1}{2\pi\hbar} \int_{\mathbb{R}^2 \times C} dz \wedge \mathrm{CS}(A) + S_{\text{defect}}\big(A, \{\phi_m\}_{m=1}^{N_1}, \{\chi_n\}_{n=1}^{N_2}\big), \qquad (5.2.40)$$

where

$$S_{\text{defect}}\big(A, \{\phi_m\}_{m=1}^{N_1}, \{\chi_n\}_{n=1}^{N_2}\big)$$

$$:= -\frac{1}{\hbar_{2d}} \sum_{m=1}^{N_1} \int_{\mathbb{R}^2 \times \{a_m\}} \mathrm{Tr}\big(\phi_m^{-1} D_+ \phi_m U_m^{(0)}\big) d\sigma^+ d\sigma^-$$

$$-\frac{1}{\hbar_{2d}} \sum_{n=1}^{N_2} \int_{\mathbb{R}^2 \times \{b_n\}} \mathrm{Tr}\big(\chi_n^{-1} D_- \chi_n V_n^{(0)}\big) d\sigma^+ d\sigma^-, \qquad (5.2.41)$$

where ϕ_m and χ_n are fields valued in $G_\mathbb{C}$, while $U_m^{(0)}$ and $V_n^{(0)}$ are constant non-dynamical elements valued in the associated Lie algebra, $\mathfrak{g}_\mathbb{C}$.

The 4d-2d coupled system (5.2.40) is analogous in form to the system with action (5.2.39). The main difference is that $U_m^{(0)}$ and $V_n^{(0)}$ are, in general, not the same element of $\mathfrak{g}_\mathbb{C}$, and also that there can be multiple copies of the surface operators. For example, when $N_1 = N_2 = 1$, discretization would lead to two sets of Wilson lines in *different* representations, with $U_m^{(0)}$ and $V_n^{(0)}$ determining highest weights. For general values of N_1 and N_2, the discretization can be identified with a more general null lattice where there is a repeating pattern for every N_1 and N_2 Wilson lines for each lightcone coordinate, respectively. An example of this for $N_1 = 2$ and $N_2 = 2$ is shown in Figure 5.3.

5.2.3.2 *Fermionic defects*

Let us next turn to the discretization of defects described by chiral/anti-chiral free-fermion surface defects. These defects can be utilized in 4d CS to realize the 2d massless Thirring model and 2d chiral Gross-Neveu models, as reviewed in Chapter 2. Discretizing these surface defects would thus lead to discretizations of these models from the viewpoint of 4d CS theory.

The path integral for 4d CS coupled to the discretized surface defects reads

$$Z_f^{\mathrm{Th}}(z_{L,R}) = \int \mathcal{D}A \mathcal{D}\psi_L \mathcal{D}\psi_R \exp\left(iS_f[A,\psi_L,\psi_R]\right)$$

$$= \int \mathcal{D}A \prod_n Z_{L,n}[\mathcal{A}_{-,n}^L] \prod_m Z_{R,m}[\mathcal{A}_{+,m}^R] \exp\left(iS_{\mathrm{CS}}[A]\right),$$

$$(5.2.42)$$

where $Z_{L,n}[\mathcal{A}_{-,n}]$ and $Z_{R,m}[\mathcal{A}_{+,m}]$ are

$$Z_{L,n}[\mathcal{A}_{-,n}] = \int \mathcal{D}\psi_{n,L} \exp\left(-\int_{S_-^1 \times \{z_+\}} d\sigma^- \psi_{L,n}^\dagger (\partial_- + \mathcal{A}_{-,n})\psi_{L,n}\right),$$

$$(5.2.43)$$

$$Z_{R,m}[\mathcal{A}_{+,m}] = \int \mathcal{D}\psi_{m,R} \exp\left(-\int_{S_+^1 \times \{z_-\}} d\sigma^+ \psi_{R,m}^\dagger (\partial_+ + \mathcal{A}_{+,m})\psi_{R,m}\right).$$

$$(5.2.44)$$

Here we have chosen anti-periodic boundary conditions in both lightcone directions.

Using well-known results on the equivalence between fermionic quantum mechanical systems and Wilson loops that are reviewed in Appendix C, the partition functions $Z_{L,n}[\mathcal{A}_{-,n}]$ and $Z_{R,m}[\mathcal{A}_{+,m}]$ can be written as

$$Z_{L,n}[\mathcal{A}_{-,n}] = W_{L,n}[\mathcal{A}_{-,n}] = \mathrm{Tr}_{\mathcal{H}_n} \mathcal{P} \exp\left(-\oint_{S^1} d\sigma^- \mathcal{A}_{-,n}^a \widehat{\rho}(t_a)\right),$$

$$(5.2.45)$$

$$Z_{R,m}[\mathcal{A}_{+,m}] = W_{R,m}[\mathcal{A}_{+,m}] = \mathrm{Tr}_{\mathcal{H}_m} \mathcal{P} \exp\left(-\oint_{S^1} d\sigma^+ \mathcal{A}_{+,m}^a \widehat{\rho}(t_a)\right),$$

$$(5.2.46)$$

where $\widehat{\rho}$ corresponds to the Schwinger representation of $G_{\mathbb{C}} = \mathfrak{gl}(N,\mathbb{C})$. As explained in Appendix C, these Wilson loops are in the direct sum of antisymmetric representations.

Hence, the lattice discretization of the partition function of the massless Thirring model is given by the expectation value of the product of Wilson loops

$$Z_f^{\mathrm{2d\text{-}Th}}(z_\pm) \to \int \mathcal{D}A \prod_n W_{L,n}[\mathcal{A}_-] \prod_m W_{R,m}[\mathcal{A}_+] \exp\left(i\,S_{\mathrm{CS}}[A]\right).$$

$$(5.2.47)$$

When Σ is compactified into a torus, the expectation value (5.2.47) is equivalent to the partition function of the rational vertex model on a torus, that is, each Wilson loop intersection corresponds to an element of the rational R-matrix. This lattice discretization via Wilson loops in the lightcone directions corresponds to the lightcone discretization of the work of Destri and de Vega [83–85].

As mentioned above, the Wilson loops we have considered so far are each in a reducible representation of $GL(N, \mathbb{C})$. Indeed, a given state $|\Psi\rangle$ in the the representation space has a finite expansion involving the fermionic creation operators $\widehat{\psi}_j^\dagger$ acting on a vacuum, of the form

$$|\Psi\rangle = \Psi_0 |0\rangle_f + \Psi^j \widehat{\psi}_j^\dagger |0\rangle_f + \cdots + \frac{1}{N!} \Psi^{j_1 \cdots j_N} \widehat{\psi}_{j_1}^\dagger \cdots \widehat{\psi}_{j_N}^\dagger |0\rangle_f , \qquad (5.2.48)$$

where the indices denoted j are valued in $1, \ldots, N$. Here, each coefficient $\Psi^{j_1 \cdots j_k}$ transforms in an antisymmetric product of the fundamental representation of $GL(N, \mathbb{C})$, i.e., the k-th antisymmetric representation of $GL(N, \mathbb{C})$.

As reviewed in Appendix C, if one imposes appropriate constraints, one can obtain a Wilson loop operator in the k-th antisymmetric representation. The quantum mechanical description of this operator is the constrained 1d fermion system

$$W_f^{(l)}[A] = \int \mathcal{D}\psi^\dagger \mathcal{D}\psi \mathcal{D}\tilde{A}^f \exp\left(i\, S_f^{(l)} \right) \exp\left(-ik_{\text{eff}} \int dt\, \tilde{A}_t^f \right) ,$$

$$S_f^{(l)} = \int dt\, \psi_i^\dagger i(\delta_j^i \partial_t + A_t^a \rho(t_a)^i{}_j - i\delta_j^i \tilde{A}_t^f)\psi^j , \qquad (5.2.49)$$

where $\tilde{A}^f$ is an auxiliary gauge field that can be integrated out to impose the necessary constraints, where k_{eff} is the integer k shifted by $-N/2$ due to Weyl ordering, i.e.,

$$k_{\text{eff}} = k - \frac{N}{2} , \qquad (5.2.50)$$

where ψ^j is a fermionic field in the fundamental representation of $GL(N, \mathbb{C})$, and where $\psi_j^\dagger$ is the complex conjugate of ψ^j.

The line operator defined by (5.2.49) can be obtained from a constrained chiral surface operator via the discretization procedure described previously. Therefore, one can define surface operators associated with a fixed antisymmetric representation, which can be used to realize integrable field theories in 4d Chern–Simons theory.

The preceding discussion can be generalized further to the case of arbitrary irreducible representations. As described in Appendix C, one can use fermionic quantum mechanics to describe a Wilson line associated with an irreducible representation, $\tilde{\rho}$, of $GL(N, \mathbb{C})$, characterized by the following Young tableau with k_I boxes in the I-th column ($1 \leq I \leq K$):

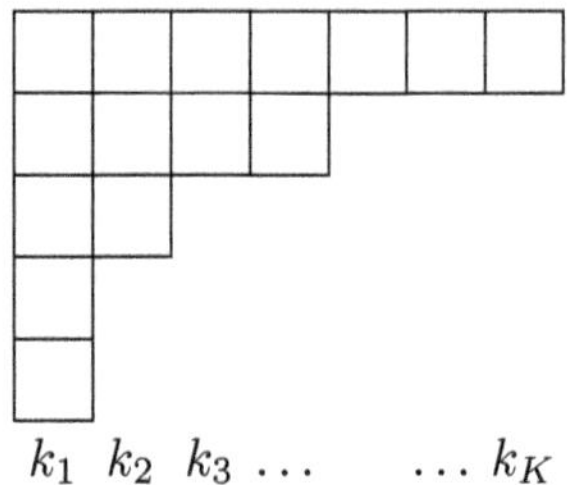

$$k_1 \;\; k_2 \;\; k_3 \ldots \qquad \ldots k_K$$

By taking the thermodynamic limit of a collection of such quantum mechanical systems, one can realize a constrained chiral or antichiral surface operator in 4d CS. For example, the path integral description of the constrained antichiral surface operator is

$$\int \mathcal{D}\psi^\dagger \mathcal{D}\psi \mathcal{D}\tilde{A}^f \, \exp\left(iS_f\right) \exp\left(-i \int_\Sigma \mathrm{d}\sigma^+ \mathrm{d}\sigma^- \sum_{I=1}^{K} k_{\mathrm{eff},I} \, (\tilde{A}^f_+)_{II}\right),$$

$$(5.2.51)$$

$$S^f = \int_\Sigma \mathrm{d}\sigma^+ \mathrm{d}\sigma^- \sum_{I,J=1}^{K} \sum_{i,j=1}^{N} \psi_i^{\dagger I} (i\delta^{IJ}\delta_j^i \partial_+ + \delta^{IJ} A_+^a \rho(t_a)^i{}_j + \delta^{ij}(\tilde{A}^f_+)_{IJ})\psi_J^j,$$

$$(5.2.52)$$

where the auxiliary gauge field $\tilde{A}^f$ is valued in the Borel subalgebra of $\mathfrak{u}(K)$, and has nonzero components $(\tilde{A}^f)_{IJ}$ with $I \geq J$, and where $k_{\mathrm{eff},I}$ is the integer k_I shifted by $-N/2$. Constrained chiral and antichiral surface operators play an important role in realizing dualities between integrable field theories in 4d Chern–Simons theory, as we shall see in Section 5.4.1.

5.2.3.3   *Free $\beta\gamma$ defects*

For the chiral free $\beta\gamma$ defect, the discretization procedure leads us to a 1d quantum mechanical system with action of the form

$$S_+^{\mathrm{1d}\text{-}\beta\gamma} = \int_{\mathbb{R}_- \times \{z_+\}} \mathrm{d}\sigma^- \beta^i \bar{\partial}_A \gamma_i = \int_{\mathbb{R}_- \times \{z_+\}} \mathrm{d}\sigma^- \left(\beta^i \partial_- \gamma_i + A_-^a \beta^i \rho(t_a)_i{}^j \gamma_j\right).$$

$$(5.2.53)$$

Likewise, the anti-chiral free $\beta\gamma$ system can be discretized to 1d quantum mechanical systems of the form

$$S_{-}^{\text{1d-}\beta\gamma} = \int_{\mathbb{R}_+ \times \{z_-\}} \mathrm{d}\sigma^+ \bar{\beta}^i \partial_A \bar{\gamma}_i = \int_{\mathbb{R}_+ \times \{z_-\}} \mathrm{d}\sigma^+ \left(\bar{\beta}^i \partial_+ \bar{\gamma}_i + A_+^a \bar{\beta}^i \rho(t_a)_i{}^j \bar{\gamma}_j \right).$$

$$(5.2.54)$$

Here, the fields β and γ are in general complex-valued, but one may specify an integration cycle for the path integral, such that β^i and γ_i are both real, or complex conjugates of each other, that is, $\beta_i = z_i^*$ and $\gamma^i = z^i$. In the latter case, the associated integrable field theory can be described as a bosonic version of the massless Thirring model.

As explained in Appendix C, the path integrals of these quantum mechanical systems correspond to Wilson loops in the (infinite) direct sum of symmetric representations. In analogy to the case of free fermions, a Wilson loop in a fixed irreducible symmetric representation corresponds to a constrained version of (5.2.53) or (5.2.54), which can be obtained from a constrained surface defect via discretization.

Moreover, such constrained quantum mechanical systems can be generalized to include additional bosonic degrees of freedom and flavor symmetry, in order to realize Wilson loops in arbitrary irreducible representations of the gauge group. For example, consider the irreducible representation $\tilde{\rho}$ specified by the Young tableau with l_I boxes in the I-th row ($1 \leq I \leq L$):

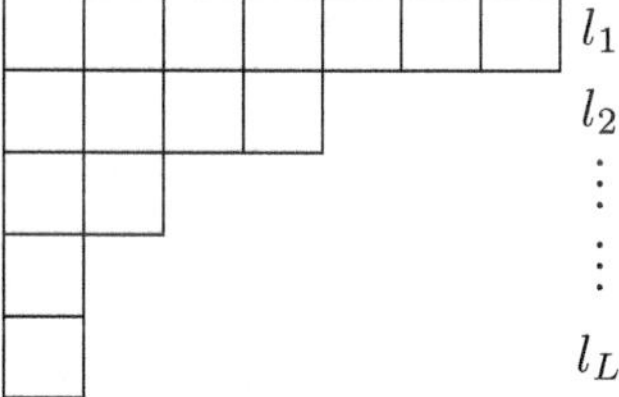

A collection of Wilson loops in this fixed irreducible representation can be obtained by discretizing a constrained surface defect. For example, Wilson loops of this type supported along the σ^+ direction can be obtained by discretizing the following chiral surface operator:

$$\int \mathcal{D}z \mathcal{D}z^* \mathcal{D}\tilde{A}^b \exp\left(iS^b\right) \exp\left(-i \int_\Sigma \mathrm{d}\sigma^+ \mathrm{d}\sigma^- \sum_{I=1}^L l_{\text{eff},I} \, (\tilde{A}_-^b)_{II}\right), \quad (5.2.55)$$

$$S^b = \int_\Sigma \mathrm{d}\sigma^+ \mathrm{d}\sigma^- \sum_{\alpha,\beta=1}^L \sum_{i,j=1}^N z_i[I]^*$$
$$\times \left(\delta^{IJ}\delta^i_j \partial_- + \delta^{IJ} A_-^a \rho(t_a)^i{}_j + \delta^{ij}(\tilde{A}_-^b)_{IJ}\right) z^j[J], \quad (5.2.56)$$

where the auxiliary gauge field $(\tilde{A}^b_-)_{IJ}$ is valued in the Borel subalgebra $\mathfrak{b}_L$ of $\mathfrak{u}(L)$ and has non-zero components $(\tilde{A}^b_-)_{IJ}$ with $I \geq J$. Here, the flavor indices are $I, J, \ldots$, and count the number of rows of the Young tableau of interest (this differs from the case of fermions, where the flavor index counted the number of columns). In addition, $l_{\mathrm{eff},I}$ measures the size of each row, since $l_{\mathrm{eff},I} = l_I + \frac{N}{2}$. The surface operator (5.2.55) and its antichiral counterpart shall play an important role in the study of dualities of integrable field theories in Section 5.4.1.

5.2.3.4 *Curved $\beta\gamma$ defects*

Let us now turn to the coupling of 4d CS theory to curved $\beta\gamma$ defects that were reviewed in Chapter 2, and the discretization of these defects.

Applying the discretization procedure to a chiral curved $\beta\gamma$ defect, we obtain a line defect with the action

$$\int_{S^1} \left(\beta^i \partial_- \gamma_i + A_{a,-} \beta^i \rho^a_i(\gamma) \right) \mathrm{d}\sigma^- \tag{5.2.57}$$

after adopting periodic boundary conditions for the lightcone direction. This is a one-dimensional gauged quantum mechanical system, where the field γ_i corresponds to a local coordinate on a curved target space. The Hamiltonian of this system is equal to $A^a_- \beta^i \rho_{a,i}(\gamma)$.

As explained in Appendix C, one can quantize the action of such a line defect such that the composite field $\sigma_a = \beta^i \rho_{a,i}(\gamma)$ acts on the resulting Hilbert space in an infinite-dimensional representation of $\mathfrak{g}$, known as a Verma module. For example, in the case of $\mathfrak{g} = \mathfrak{sl}(2)$ and $\mathbb{CP}^1$ target space described in Appendix C, the quantization of the line defect is such that $\sigma_a = \beta^i \rho_{a,i}(\gamma)$ satisfy the $\mathfrak{sl}(2)$ Lie algebra, while the Casimir is $\sigma_a \sigma^a = a(a+1)$, where $a \in \mathbb{C}$ is a parameter that accounts for the ambiguity in operator ordering in defining the σ^a.

From the transformation law given in Appendix C for σ_F, used when moving between charts, we further observe that a can be interpreted as the deformation parameter for affine deformations of $T^*\mathbb{CP}^1$, which are encapsulated in $H^1\left(\mathbb{CP}^1, \mathcal{O}(-2)\right)$. By Čech-Dolbeault isomorphism, this is equivalent to $H^{1,1}(\mathbb{CP}^1)$, implying that instead of a Bohr-Sommerfeld quantization condition where periods of the Kähler form are quantized in terms of integers, these periods may be valued in any complex number.

The results for $\mathbb{CP}^1$ generalize to the case of flag manifold target spaces $G_{\mathbb{C}}/B$ for gauge groups, $G_{\mathbb{C}}$, of higher rank, where the corresponding line operators are associated with Verma modules for $\mathfrak{g}$. The affine deformation parameters in this case are valued in $H^{1,1}(G_{\mathbb{C}}/B) = \mathfrak{h}$.

The discretization of integrable field theories associated with curved $\beta\gamma$ defects thus leads to integrable lattice models involving R-matrices associated with infinite-dimensional representations of $\mathfrak{g}$. There is more to say about discretizations of curved $\beta\gamma$ systems using the language of vertex algebras, as we shall see in the following subsection.

5.2.4 *Discretizations of vertex algebra defects*

We have so far considered defect theories that are free theories with explicit Lagrangians, and we have first discussed the discretization at the Lagrangian level. Having a Lagrangian is actually not a necessity for a surface defect to be discretized, and we can couple the 4d CS theory to more general non-Lagrangian chiral or anti-chiral defects, each with a global symmetry, G, and we will still obtain an integrable model.

As examples of more general defects, we discuss chiral (anti-chiral) defects described algebraically by a Vertex Algebra (VA). This is an algebraic framework for describing the chiral half of a CFT (see Appendix D for review), and in this setup we do not necessarily have a Lagrangian description.[6]

For the coupling to the 4d CS theory, the defect needs to have a global G-symmetry. The algebraic counterpart of this statement is that the VA contains the current algebra of G as a subalgebra, and in the following we will assume this condition throughout. In fact, as we shall show explicitly, all the chiral/anti-chiral defects that we have studied thus far support affine Kac–Moody algebras.

When we discretize the integrable field theory, the question is then how to discretize the VA defect. From the discussion of discretization above, we expect that this amounts to some kind of dimensional reduction of the 2d defect theory from 2d to 1d.[7]

We can be more explicit in this procedure. Suppose that we have an operator $\mathcal{O}$ of a VA defined on $\Sigma = \mathbb{C}$, with complex coordinate w, with the mode expansion

$$\mathcal{O}(w) = \sum_{n=0}^{\infty} \mathcal{O}_n w^{-\deg(\mathcal{O})+n} , \tag{5.2.58}$$

---

[6]A vertex algebra becomes a Vertex Operator Algebra (VOA) when we include the stress-energy tensor. The axioms of the vertex algebra, however, are sufficient for the purposes of this section.

[7]This can be regarded as a field theory version of T-duality [42, 100].

where we denoted the conformal weight of the operator $\mathcal{O}$ as $\deg(\mathcal{O})$. When we discuss dimensional reduction we first need to apply a conformal transformation from a plane to a cylinder and extract the zero mode, which is given by

$$o(\mathcal{O}) := \mathcal{O}_{\deg(\mathcal{O})-1}. \tag{5.2.59}$$

We thus conclude that the dimensionally-reduced algebra should be spanned by the zero modes $o(\mathcal{O})$.

Interestingly, the discussion above regarding the dimensional reduction of VA's is known in the mathematical literature, where the dimensionally-reduced algebra is known as Zhu's algebra [101, 104] associated with the VA; see Appendix D and, e.g., [102] and [103, Section 4] for an introduction to Zhu's algebra. Notably, the product of the VA induces an associative $\star$-product in the Zhu's algebra.

Current algebra

We can discuss the dimensional reduction of the current algebra, which we assumed to be a subalgebra of the defect VA. The basic OPE of an affine VOA $V_k(\mathfrak{g})$ (at a non-critical level) is given by

$$J_a(w)J_b(0) \sim \frac{k\delta_{ab}}{w^2} + \frac{f_{ab}{}^c J_c(0)}{w} + (J_a J_b), \tag{5.2.60}$$

where δ_{ab} is the Killing form on $\mathfrak{g}$. Denoting $[J_a] \in A(V)$ as j_a and using (D.1.2), we can compute products in the Zhu algebra,

$$
\begin{aligned}
j_a \star j_b &= \mathrm{Res}_w \left(\frac{(\hbar_\star w + 1)}{w} \left(\frac{f_{ab}{}^c J_c(0)}{w} + (J_a J_b) \right) \right) \\
&= (j_a j_b) + \hbar_\star f_{ab}{}^c j_c,
\end{aligned}
\tag{5.2.61}
$$

where $(j_a j_b) = [(J_a J_b)]$ denotes conformal normal ordering. Moreover, one can derive the commutation relation of the Lie algebra $\mathfrak{g}$. This follows since there are two expressions for the star product

$$
\begin{aligned}
a * b &= \mathrm{Res}_w \left(Y(a, w) \frac{(\hbar_\star w + 1)^{\deg(a)}}{w} b \right) \\
&= \mathrm{Res}_w \left(Y(b, w) \frac{(\hbar_\star w + 1)^{\deg(b)-1}}{w} a \right),
\end{aligned}
\tag{5.2.62}
$$

and using these, it can be shown that [104]

$$a * b - b * a \equiv \mathrm{Res}_w \left(Y(a, w) \frac{(\hbar_\star w + 1)^{\deg(a)}}{w} b \right)$$

$$- \mathrm{Res}_w \left(Y(a, w) \frac{(\hbar_\star w + 1)^{\deg(a)-1}}{w} b \right)$$

$$= \mathrm{Res}_w \left(Y(a, w)(\hbar_\star w + 1)^{\deg(a)-1} b \right), \qquad (5.2.63)$$

whereby we find that (5.2.60) implies

$$j_a \star j_b - j_b \star j_a = f_{ab}{}^c j_c . \qquad (5.2.64)$$

In fact, we can recover the whole of the universal enveloping algebra $U(\mathfrak{g})$: we can define a basis of $U(\mathfrak{g})$ using the $\star$-product to define higher order products of j_a (with fixed ordering), and using the associativity of the $\star$-product to define commutators between such higher-order products. The latter form a linearly independent basis of $U(\mathfrak{g})$, according to the Poincaré–Birkhoff–Witt (PBW) theorem. We therefore conclude that Zhu's algebra for the current algebra is the universal enveloping algebra, $U(\mathfrak{g})$.

This analysis suggests that an order chiral/anti-chiral surface operator that supports an affine VOA ought to be discretizable to a Wilson line in a representation of $U(\mathfrak{g})$.

We can summarize the discussion above using the following diagram:

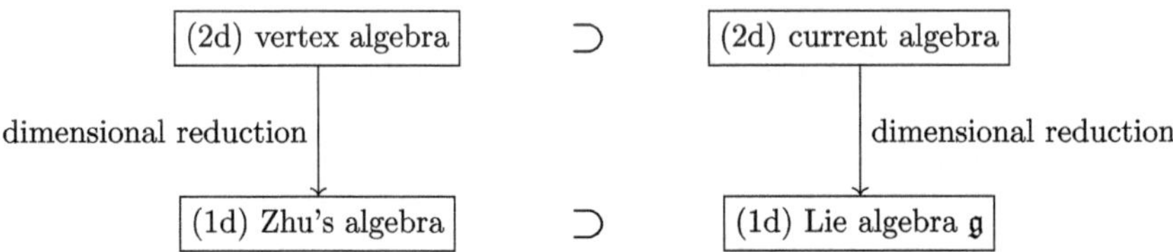

Free $\beta\gamma$ vertex algebra

As a simple example, let us revisit the example of the free $\beta\gamma$ system in Section 5.2.3.3 in the algebraic framework.

The free $\beta\gamma$ VA is described by the standard OPE between $\beta^i(w)$ and $\gamma_i(w')$:

$$\beta^i(w)\gamma_j(w') \sim -\frac{\delta^i_j}{w - w'} . \qquad (5.2.65)$$

For free $\beta\gamma$ systems, the currents that appear in the surface operator action have the Wakimoto free field realization $\mathcal{J}^a = \beta_i \rho(t^a)^i{}_j \gamma^j$. This gives rise to a current algebra, and in particular to the familiar current algebra that arises from symplectic bosons when β and γ are complex conjugates of each other. Let us recall this construction.

Given the surface operator of the form

$$\frac{1}{2\pi} \int_\Sigma \mathrm{d}^2 w (\beta_i \partial_{\overline{w}} \gamma^i + A^a_{\overline{w}} \beta_i \rho(t_a)^i{}_j \gamma^j) \,, \tag{5.2.66}$$

the free-field OPE implies the following OPE for the normal-ordered currents $\mathcal{J}_a = (\beta_i \rho(t_a)^i{}_j \gamma^j)$ of the G-symmetry:

$$\mathcal{J}_a(w)\mathcal{J}_b(w') \sim -\frac{\rho(t_a)_{ij}\rho(t_b)^{ji}}{(w-w')^2} + \frac{\rho(t_a)_{ij}\gamma^j(w)\beta^k(w')\rho(t_b)_{ki}}{(w-w')}$$

$$-\frac{\beta^k(w)\rho(t_a)_{ki}\rho(t_b)^i{}_j\psi^j(w')}{(w-w')}$$

$$\sim -\frac{\rho(t_a)_{ij}\rho(t_b)^{ji}}{(w-w')^2} + \frac{f_{ab}{}^c \mathcal{J}_c(w')}{w-w'} \,, \tag{5.2.67}$$

where the parenthesis indicates conformal normal-ordering. Here we find a second-order pole in the OPE arising due to double contractions. The corresponding commutator can be computed to be an affine Kac–Moody algebra,

$$[\mathcal{J}_{an}, \mathcal{J}_{bm}] = f_{ab}{}^c \mathcal{J}^c_{n+m} - cn\delta_{ab}\delta_{n+m,0} \,, \tag{5.2.68}$$

where the coefficient c is determined by

$$\rho(t_a)_{ij}\rho(t_b)^{ji} = c\delta_{ab} \,. \tag{5.2.69}$$

and c depends on the choice of gauge group and choice of representation, ρ. For example, if γ is in the fundamental representation of $SU(N)$, then $c = 1$.

Since the c-dependent term is a double pole, it does not contribute to the Lie algebra commutator one obtains from the zero modes of the affine Kac–Moody algebra. However, as we shall see in Section 5.3, the value of c determines the quantum gauge anomaly of chiral surface operators that support affine Kac–Moody algebras.

Free fermion vertex algebra

Given the surface operator of the form

$$\frac{1}{2\pi} \int_\Sigma \mathrm{d}^2 w (\psi_i^* \partial_{\overline{w}} \psi^i + A_{\overline{w}}^a \psi_i^* \rho(t_a)^i{}_j \psi^j) , \tag{5.2.70}$$

we have the standard free-field OPE

$$\psi_i^*(w)\psi^j(w') \sim \frac{\delta_i^j}{(w - w')} , \tag{5.2.71}$$

which implies the following OPE for the normal-ordered currents $\mathcal{J}_a = (\psi_i^* \rho(t_a)^i{}_j \psi^j)$ of the G-symmetry:

$$\mathcal{J}_a(w)\mathcal{J}_b(w') \sim \frac{\rho(t_a)_{ij}\rho(t_b)^{ji}}{(w - w')^2} + \frac{\rho(t_a)_{ij}\psi^j(w)\psi_k^*(w')\rho(t_b)^k{}_i}{(w - w')}$$

$$- \frac{\psi_k^*(w)\rho(t_a)^k{}_i \rho(t_b)^i{}_j \psi^j(w')}{(w - w')}$$

$$\sim \frac{\rho(t_a)_{ij}\rho(t_b)^{ji}}{(w - w')^2} + \frac{f_{ab}{}^c \mathcal{J}_c(w')}{w - w'} . \tag{5.2.72}$$

Here, as in the free $\beta\gamma$ VA, we find a second-order pole in the OPE arising due to double contractions, but with an opposite sign. This is the Wakimoto free-field realization of an affine Kac–Moody algebra in terms of fermions. The corresponding commutator can be computed to be

$$[\mathcal{J}_{an}, \mathcal{J}_{bm}] = f_{ab}{}^c \mathcal{J}_{c\,n+m} + cn\delta_{ab}\delta_{n+m,0} , \tag{5.2.73}$$

where c is defined by

$$\rho(t_a)_{ij}\rho(t_b)^{ji} = c\,\delta_{ab} . \tag{5.2.74}$$

Once again, since the c-dependent term is a double pole, it will not contribute to the Lie algebra commutator one obtains from the zero modes of the affine Kac–Moody algebra.

Coadjoint orbit defect

In order to describe the coadjoint orbit defect

$$\int_\Sigma \mathrm{Tr}\left(\Lambda \cdot \mathcal{G}^{-1} D_{\overline{w}} \mathcal{G}\right) \mathrm{d}^2 w \tag{5.2.75}$$

in terms of a VA, we need to take into account the fact that the degrees of freedom of the defect are valued in the flag variety, $G_\mathbb{C}/B$. In fact, we need a sheaf of VAs defined over each coordinate chart of $G_\mathbb{C}/B$. To facilitate this analysis, we can show that the coadjoint orbit defect action for $G_\mathbb{C}/B$ (with no reality condition imposed) is equivalent to the curved $\beta\gamma$ defect action with flag manifold target space, and with holomorphic vector fields taking the form of twisted differential operators. This is expected since coadjoint orbits are affine deformations of cotangent bundles of flag manifolds, $T^*(G_\mathbb{C}/B)$.

The relationship between the defects can be derived as follows.[8] Consider the coadjoint orbit action

$$\int_\Sigma \mathrm{d}w\mathrm{d}\overline{w}\ \mathrm{Tr}\Lambda(G^{-1}\partial_{\overline{w}}G + G^{-1}A_{\overline{w}}^a t_a G)\,. \tag{5.2.76}$$

We then consider a Gauss decomposition of $G = SL(2,\mathbb{C})$, such that

$$G = \begin{pmatrix} 1 & 0 \\ \gamma & 1 \end{pmatrix}\begin{pmatrix} e^{i\phi} & 0 \\ 0 & e^{-i\phi} \end{pmatrix}\begin{pmatrix} 1 & \beta \\ 0 & 1 \end{pmatrix} = \begin{pmatrix} e^{i\phi} & e^{i\phi}\gamma \\ e^{i\phi}\beta & e^{i\phi}\beta\gamma + e^{-i\phi} \end{pmatrix}\,, \tag{5.2.77}$$

where β, γ and ϕ are maps from Σ to $\mathbb{C}$. We also fix the weight Λ to be

$$\Lambda = a\begin{pmatrix} 1 & 0 \\ 0 & -1 \end{pmatrix} = 2at_H\,, \tag{5.2.78}$$

where $a \in \mathbb{C}$.

We can then show that

$$\mathrm{Tr}(\Lambda G^{-1}\partial_{\overline{w}}G) = -2ae^{2i\phi}\beta\partial_{\overline{w}}\gamma + 2ai\partial_{\overline{w}}\phi\,,$$

$$\mathrm{Tr}(\Lambda G^{-1}T_H G) = 2ae^{2i\phi}\beta\gamma + a\,,$$

$$\mathrm{Tr}(\Lambda G^{-1}T_E G) = 2ae^{2i\phi}\beta\gamma^2 + 2a\gamma\ ,\ \mathrm{Tr}(\Lambda G^{-1}T_F G) = -2ae^{2i\phi}\beta\,, \tag{5.2.79}$$

using the Chevalley basis for $\mathfrak{sl}_2$. The term proportional to $\partial_{\overline{w}}\phi$ does not contribute to the action due to Stokes' theorem. Redefining β as $-2ae^{2i\phi}\beta \to \beta$, we find that the action (5.2.76) becomes

$$\int_\Sigma \mathrm{d}w\mathrm{d}\overline{w}(\beta\partial_{\overline{w}}\gamma + A_{\overline{w}}^H(-\beta\gamma + a) + A_{\overline{w}}^F\beta + A_{\overline{w}}^E(-2\beta\gamma^2 + 2a\gamma))\,, \tag{5.2.80}$$

---

[8]We shall consider 2d (chiral) defects, but the derivation works for 1d line defects as well.

which is a curved $\beta\gamma$ defect with target space $\mathbb{CP}^1$, with the generators of the $SL(2,\mathbb{C})$ isometry taking the form of twisted differential operators, as discussed in Section 5.2.3.4 and Appendix C. Moreover, we can generalize this computation for coadjoint orbit defects associated with any flag manifold.

As such, the analysis of VAs associated with curved $\beta\gamma$ defects, reviewed below, encompasses that of coadjoint orbit defects, as it involves the appropriate chiralization of sheaves of twisted differential operators.

5.2.4.1 *Sheaves of vertex algebras and curved $\beta\gamma$ defects*

Let us now turn to the example of the curved $\beta\gamma$ system. To describe this example, we need to generalize our discussion of a VA to "a sheaf of VA". We shall explain this concept in the specific example of the curved $\beta\gamma$ system.

The target space of the curved $\beta\gamma$-system is a curved manifold, X, which can be covered by a set of local coordinate charts, each of which can be regarded as a copy of flat space. The VA counterpart of this statement is that the free $\beta\gamma$ systems associated to each coordinate chart will be "glued together" appropriately, so that one finds an algebra defined globally on X. This is what is meant by a sheaf of VA.

Sheaves of $\beta\gamma$ systems can be formulated mathematically in the language of chiral differential operators (CDOs). Chiral differential operators were first studied by Malikov *et al.* [105], with the aim of defining sheaves of VAs on curved manifolds. They provide a mathematically rigorous definition of the chiral part of a conformal field theory with curved target space. In what follows, we shall quantize the curved $\beta\gamma$ defects by coupling a generalization of chiral differential operators, namely *twisted* chiral differential operators, to the 4d CS theory. This embedding of twisted chiral differential operators into the 4d CS theory is novel, and the twist in question is with respect to the 4d CS gauge field.[9] We shall also describe discretization in terms of the relationship between a sheaf

---

[9]A different embedding of twisted chiral differential operators in 4d CS theory was studied by Khan [19]. In that setup, the underlying $\beta\gamma$-system is an example of a holomorphic surface defect along the holomorphic surface C discussed in Chapter 1, while in this chapter, we consider such systems along Σ.

of twisted chiral differential operators and a sheaf of twisted differential operators.

In what follows, we shall focus on the chiral defect defined in (2.1.52) for $\Sigma = \mathbb{C}$. For ease of analysis, we shall employ the gauge $A_{\bar{w}} = 0$ at the location of the defect, z_+. The equations of motion then imply that

$$\partial_{\bar{w}}\gamma_i = 0\,,$$
$$\partial_{\bar{w}}\beta^i = 0\,. \tag{5.2.81}$$

In addition, in holomorphic gauge $A_{\bar{z}} = 0$, the 4d CS equation of motion implies that

$$\partial_{\bar{w}}A_w = 0\,. \tag{5.2.82}$$

Thus, we may perform the Laurent expansions

$$\gamma_i(w) = \sum_{m\in\mathbb{Z}} \frac{\gamma_{im}}{w^m}\,, \tag{5.2.83}$$

$$\beta^i(w) = \sum_{m\in\mathbb{Z}} \frac{\beta^i_m}{w^{m+1}}\,, \tag{5.2.84}$$

$$A_w(w) = \sum_{m\in\mathbb{Z}} \frac{A_m}{w^{m+1}}\,, \tag{5.2.85}$$

which all converge in an annulus around $w = 0$. In addition, the standard OPE between $\beta^i(w)$ and $\gamma_i(w)$ holds, i.e.,

$$\beta^i(w)\gamma_j(w') \sim -\frac{\delta^i_j}{w - w'}\,, \tag{5.2.86}$$

while the OPE of A_w with β^i and γ_i is non-singular. We shall further specify the gauge for the Cartan component of the gauge field (5.2.85), denoted A^h_w, by gauging away the non-singular terms in its Laurent expansion by a w-dependent gauge transformation. Such a gauge transformation would not affect the gauge that we have previously chosen locally for $A_{\bar{w}}$ and $A_{\bar{z}}$. The annihilation operators, i.e., A^h_m for $m > 0$, then annihilate any state in the Hilbert space, and are thus effectively equal to zero as well since we are interested in non-trivial Hilbert spaces. We are thus left with

$$A^h_w(w) = \frac{A^h_0}{w} \tag{5.2.87}$$

for the Cartan component of $A^h_w(w)$.

Now, the OPE of conserved currents that arise from the G-symmetry of a 2d curved $\beta\gamma$ system with target space X is known to have an anomalous term [105, 106]:

$$
\begin{aligned}
J_V(w)J_W(w') \sim\ & -\frac{\partial_j V^i(w)\partial_i W^j(w')}{(w-w')^2} - \frac{(V^i\partial_i W^j - W^i\partial_i V^j)\beta_j}{w-w'} \\
\sim\ & -\frac{\partial_j V^i \partial_i W^j(w)}{(w-w')^2} - \frac{(V^i\partial_i W^j - W^i\partial_i V^j)\beta_j}{w-w'} \\
& -\underbrace{\frac{(\partial_k\partial_j V^i)(\partial_i W^j \partial\gamma^k)}{w-w'}}_{\text{anomalous term}},
\end{aligned}
\tag{5.2.88}
$$

where $J_V = -V^i\beta_i$, and where V and W are holomorphic Killing vector fields on X that generate the G symmetry. The first term does not contribute to the Lie algebra commutator as it contains a double pole. As shown in Appendix D, the algebra of zero modes is

$$
[J_{V0}, J_{W0}] = J_{[V,W]0} + \tilde{J}_0.
\tag{5.2.89}
$$

However, this is not in the form we want, as the second term is the anomaly that spoils the closure of the Lie algebra. It is related to the obstruction to gluing $\beta\gamma$ systems contained in open sets of the target space, as reviewed in Appendix D, and can be understood as a Lie algebra extension by $\tilde{J}_0$.

We will need to define a consistent global sheaf of chiral differential operators on the target space. We review how this is accomplished for Kähler manifolds, denoted X, in Appendix D. The crucial point is that such a definition is only possible when the first Pontryagin class of X vanishes, i.e., $p_1(X) = 0$, and this can be interpreted as the vanishing of an anomaly of the $\beta\gamma$ system [106].[10]

Having defined a global sheaf of chiral differential operators, we can go on to study its global sections. Let us focus on the example of $X = \mathbb{CP}^1$, with charts parametrized by γ and γ', where $\gamma = 1/\gamma'$. As explained in

---

[10]There is another anomaly vanishing condition, which requires $c_1(\Sigma)c_1(X) = 0$. This condition shall not play a major role in our analysis, since we shall only consider Σ such that $c_1(\Sigma) = 0$.

[106], the global sections in this case take the following form[11]

$$
\begin{aligned}
J_+(w) &= -(\beta(\gamma\gamma)) + 2\partial\gamma & &= \beta', \\
J_-(w) &= \beta & &= -(\beta'(\gamma'\gamma')) + 2\partial\gamma', \\
J_3(w) &= -(\gamma\beta) & &= (\gamma'\beta'),
\end{aligned}
\tag{5.2.95}
$$

where the parentheses indicate normal-ordering. One can show that these are globally well-defined via the CFT automorphisms

$$
\gamma' = \frac{1}{\gamma}, \qquad \beta' = -(\beta(\gamma\gamma)) + 2\partial\gamma
\tag{5.2.96}
$$

that are used to glue fields across open sets (the term proportional to $\partial\gamma$ is necessary in the gluing law to ensure that the $\beta(w)\beta(w')$ OPE is preserved

---

[11]The sign flip in the sign of J_3 follows from a rearrangement lemma, which states

$$
((AB)(CD)) = (A(C(DB))) + (A(C([B,D]))) + (A(([B,C])D)) + (A([CD,B]))
$$
$$
+ (((([CD,A])B) + ([(AB),(CD)]).
\tag{5.2.90}
$$

We shall use the standard identity

$$
([A,B]) = \sum_{n>0} \frac{(-1)^{n+1}}{n!} \partial^n \{AB\}_n(w),
\tag{5.2.91}
$$

where OPEs are defined as

$$
A(z)B(w) \equiv \sum_{n=1}^{N} \frac{\{AB\}_n(w)}{(z-w)^n}.
\tag{5.2.92}
$$

Substituting $A = \frac{1}{\gamma}$, $B = \gamma$, $C = \beta$ and $D = \gamma$ into (5.2.90), we find

$$
\left(\frac{1}{\gamma}\gamma\right)(\beta\gamma) = \left(\frac{1}{\gamma}(\beta(\gamma\gamma))\right) + 0 + 0 - \frac{\partial\gamma}{\gamma} - \frac{\partial\gamma}{\gamma} + 0,
\tag{5.2.93}
$$

where the second term vanishes since γ has a regular OPE with itself and hence $([\gamma,\gamma]) = 0$, the third term vanishes since the $\gamma - \beta$ OPE has no field-dependent singular terms and hence $([\gamma,\beta]) = 0$, and the last term vanishes since the OPE of $\beta\gamma$ with 1 is regular. Thus, we find that

$$
(\gamma\beta) = \left(\frac{1}{\gamma}(\beta(\gamma\gamma))\right) - 2\frac{\partial\gamma}{\gamma}.
\tag{5.2.94}
$$

This implies the gluing law of J_3.

in both charts). Their OPEs can be shown to be

$$J_+(w)J_-(w') \sim \frac{2J_3}{w-w'} - \frac{2}{(w-w')^2},$$

$$J_3(w)J_3(w') \sim -\frac{1}{(w-w')^2},$$

$$J_3(w)J_+(w') \sim \frac{J_+(w')}{w-w'},$$

$$J_3(w)J_-(w') \sim -\frac{J_-(w')}{w-w'},$$

(5.2.97)

which implies that they satisfy an $SL(2)$ current algebra at critical level. The zero mode algebra of this current algebra is indeed an ordinary Lie algebra.

However, one can show that the rescaled Sugawara energy-momentum tensor in this case is zero, i.e.,

$$S = \left((J_3 J_3) + \frac{1}{2}(J_+ J_-) + \frac{1}{2}(J_- J_+) \right) = 0. \qquad (5.2.98)$$

Hence, we do not expect to obtain non-trivial representations of $\mathfrak{g}$ from the zero-mode algebra of the current algebra. This suggests that one should consider further quantum corrections that can modify the form of the global sections (5.2.95). Indeed, given that we are dealing with a gauged $\beta\gamma$-system, it is more natural to consider sheaves of *twisted* CDOs, first studied by Arakawa *et al.* in [107]. The global sections in this case obtain further corrections from the presence of gauge fields. As we shall review below, sheaves of twisted CDOs are the natural chiralization of sheaves of twisted differential operators, and are thus useful to describe VAs associated with gauged, curved $\beta\gamma$-defects.

For twisted CDOs, the $\beta\gamma$ OPE remains the same, but the CFT automorphisms used to glue fields across open sets take the form

$$\tilde{\gamma} = \frac{1}{\gamma},$$

$$\tilde{A}_w^h = A_w^h,$$

(5.2.99)

$$\tilde{\beta} = -(\beta(\gamma\gamma)) + 2\partial\gamma - (\gamma A_w^h).$$

In the present context, we have identified the twist operator with A_w^h, the Cartan component of the w-component of the 4d CS gauge field. There are still currents that generate a critical level affine Kac–Moody algebra, but they take the form

$$
\begin{aligned}
J_+(w) &= -(\beta(\gamma\gamma)) + 2\partial\gamma - (\gamma A_w^h) &&= \tilde{\beta}\,, \\
J_-(w) &= \beta &&= -(\tilde{\beta}(\tilde{\gamma}\tilde{\gamma})) + 2\partial\tilde{\gamma} - (\tilde{\gamma}\tilde{A}_w^h)\,, \\
J_3(w) &= -(\gamma\beta) - \frac{1}{2}A_w^h &&= (\tilde{\gamma}\tilde{\beta}) + \frac{1}{2}\tilde{A}_w^h\,.
\end{aligned}
$$

$$(5.2.100)$$

The rescaled energy-momentum tensor in this case is nonzero, i.e.,

$$
S(w) = \left((J_3 J_3) + \frac{1}{2}(J_+ J_-) + \frac{1}{2}(J_- J_+) \right) = \frac{1}{4}(A_w^h(w))^2 - \frac{1}{2}\partial_w A_w^h(w)\,.
$$

$$(5.2.101)$$

From the transformation law for β in (5.2.99) and the discussion in Section 5.2.3.4, we deduce that A_w^h ought to be related to the deformation parameter for affine deformations of $T^*\mathbb{CP}^1$.

Let us now return to the discretization of the $\beta\gamma$ system. On a cylinder with holomorphic coordinate $\tilde{w} = t + i\theta$, the action of the chiral $\beta\gamma$ system is

$$
\int_{\mathbb{R}\times S^1} \beta^i \bar{D}_{\bar{\tilde{w}}}\, \gamma_i \, d\tilde{w} d\bar{\tilde{w}} := \int_{\mathbb{R}\times S^1} \left(\beta^i \partial_{\bar{\tilde{w}}} \gamma_i + A_{a,\bar{\tilde{w}}} \beta^i \rho_{a,i}(\gamma) \right) d\tilde{w} d\bar{\tilde{w}}\,,
$$

$$(5.2.102)$$

and

$$
\partial_{\bar{\tilde{w}}} = \frac{1}{2}(\partial_t - i\partial_\theta)\,.
$$

$$(5.2.103)$$

Dimensionally reducing along the circular direction parametrized by θ then gives the 1d action

$$
\frac{1}{2}\int_{\mathbb{R}} \beta^i \partial_t^A \gamma_i\,,
$$

$$(5.2.104)$$

if we identify $A_{\bar{\tilde{w}}}$ with A_t. As explained earlier, the algebraic counterpart of the 1d theory ought to be given by the sheaf of Zhu's algebra associated with the sheaf of the vertex algebra of the surface defect, which in this case is the sheaf of twisted chiral differential operators.

Now, a result of [107, Theorem 4.7] states that the Zhu algebra of a sheaf of twisted chiral differential operators is equivalent to a sheaf of twisted differential operators. A technical point in the proof of [107] is the requirement that only regular singularities are permitted in the Laurent expansion of $A_w^h(w)$, which plays the role of the central character. This is necessary (from [107, Theorem 5.2]) because a module over a sheaf of twisted chiral differential operators with a central character $\chi(z)$ can be identified with a module over a sheaf of twisted differential operators, defined with respect to a weight given by χ_0, when $\chi(z)$ has a regular singularity, but is zero otherwise. Indeed, from (5.2.87), the form of A_w^h satisfies this requirement in the gauge that we are in. The advantage of arriving at twisted differential operators by starting from twisted chiral differential operators is that we know which $\beta\gamma$ surface defects have vanishing anomalies and can be discretized consistently.

Let us try to understand the relationship between twisted chiral differential operators and twisted differential operators in greater detail, for the example of $X = \mathbb{CP}^1$. An immediate observation that can be made is that if we set $A_w^h(w) = 2a/w$ for a nonzero complex number, a, and perform the Laurent expansion

$$S(w) = \sum_{n=0}^{\infty} \frac{S_n}{w^{n+2}} \tag{5.2.105}$$

of the rescaled energy-momentum tensor (5.2.101), we arrive at $S_0 = a(a + 1)$, which is precisely the expression for the Casimir found in (C.3.7).

In addition, since A_w^h has regular OPEs with β, γ and itself, $S(w)$ has a regular OPE with itself as well as the currents that generate the critical level affine Kac–Moody algebra. Thus, the modes obtained from the Laurent expansion (5.2.105) commute with the Laurent modes of the currents that appear in the expansion

$$J^a(w) = \sum_{n=-\infty}^{\infty} \frac{J_n^a}{w^{n+1}} \, . \tag{5.2.106}$$

In other words, we have

$$[S_n, J_m^a] = 0 \,, \qquad [S_n, S_m] = 0 \,. \tag{5.2.107}$$

Note that the standard relation

$$S_n = \frac{1}{2} \sum_a \sum_m : J_m^a J_{n-m}^a : \tag{5.2.108}$$

among the Laurent modes also holds. In particular, the zero mode S_0 is

$$S_0 = \frac{1}{2} \sum_a \sum_m : J_m^a J_{-m}^a : \, . \tag{5.2.109}$$

Define a highest weight state $|\eta\rangle$ via

$$J_0^3 |\eta\rangle = \eta |\eta\rangle \,, \tag{5.2.110}$$

and

$$J_n^a |\eta\rangle = 0 , \qquad n > 0 . \tag{5.2.111}$$

Here, $|\eta\rangle$ is also an eigenstate for S_0, since $[S_0, J_0^3] = 0$. Moreover, defining a highest weight module in the standard manner, i.e.,

$$J_0^- J_0^- \ldots J_0^- |\eta\rangle , \tag{5.2.112}$$

we find that S_0 indeed behaves as a Casimir operator since $[S_0, J_0^-] = 0$, and since S_0 acts effectively as $\frac{1}{2} \sum_a \, : J_0^a J_0^a :$ on $|\eta\rangle$ due to normal ordering. Thus, we find that we can define non-trivial modules of conformal dimension zero that take the form of highest weight modules for the Lie algebra, $\mathfrak{g}$.

We have so far focused on twisted chiral differential operators on $\mathbb{CP}^1$. Generalization of the above techniques to $X = \mathbb{CP}^n$ for $n > 1$ is not possible, since the curved $\beta\gamma$ system is anomalous for $X = \mathbb{CP}^{n>1}$ which has non-vanishing $p_1(X)$. This is consistent with the result that sigma models on all $\mathbb{CP}^n$ for $n > 1$ are known to be non-integrable at the quantum level [108, 111].

Generalization to the case of flag manifolds

The above construction can be straightforwardly generalized to the case of flag manifolds of the form $G_\mathbb{C}/B$. Here, we can define global sections that generate the relevant affine Lie algebra, $\widehat{\mathfrak{g}}_{-h^\vee}$, at critical level, since $p_1(G_\mathbb{C}/B) = 0$. We may then obtain representations of $\mathfrak{g}$ as representations of the Zhu algebra of the globally-defined sheaf of twisted CDOs on $G_\mathbb{C}/B$, described in terms of twisted differential operators. It is natural to twist the ordinary sheaf of CDOs in this case with respect to the Cartan elements of the 4d CS field A_w. These Cartan elements determine the deformation parameters for affine deformations of $T^*G_\mathbb{C}/B$, which correspond to $H^{1,1}(G_\mathbb{C}/B) = \mathfrak{h}$.

One can also discuss twisted differential operators on even more general manifolds, such as partial flag varieties $G_\mathbb{C}/P$, where P is a parabolic subgroup of $G_\mathbb{C}$. The associated representations of $G_\mathbb{C}$ are, in general, parabolic Verma modules. However, sigma models on some examples of such manifolds are known to be non-integrable at the quantum level (for example, as mentioned above, $\mathbb{CP}^2$ [108, 111]). Hence, the thermodynamic limit of line operators defined as 1d gauged $\beta\gamma$ systems with such target spaces ought to give rise to anomalous surface operators, even if the line operators themselves are not anomalous (for example, recall from [7] that a Wilson line in any representation of $SL(N, \mathbb{C})$ is quantizable).

5.2.5 *Thermodynamic limit as polarization of D-branes in string theory*

We shall now review a string-theoretic interpretation of the process whereby a large number of Wilson lines gives rise to an order surface operator in 4d CS theory. We shall in turn review this interpretation for Wilson lines realized using bosonic and fermionic quantum mechanics, Wilson lines associated with finite-dimensional irreducible representations, and Wilson lines associated with infinite-dimensional representations.

We shall utilize the embedding of the 4d CS theory with gauge group $GL(N, \mathbb{C})$ in type IIB string theory derived in [46] utilizing a stack of N D5-branes in an Ω-background, that was reviewed in Chapter 3.[12]

As part of the analysis, one identifies the string-theoretic description of order surface operators in 4d Chern-Simons theory. In particular, the derivation of bosonic surface defects in 4d CS theory from Ω-deformed field theories is analogous to the derivation of 4d CS theory from the Ω-deformed topological-holomorphic 6d gauge theory reviewed in Chapter 3, and further details can be found in the relevant literature [112, 113].

Let us consider the configuration of D5- and D3-branes given in (3.2.88). We shall identify the 0 and 1 directions with the analytic continuation to Euclidean signature of the lightcone coordinates utilized in previous sections.

Now, if we place a large number of $D3_b$ or $D3_f$ branes at various points along, say, the '1' direction, these branes can polarize via the Myers effect [118, 119] into a single $D5_b$ or $D5_f$ brane, wrapping the '1' and '6' directions. This is related to how a finite number of fundamental strings polarize to D3-branes realizing Wilson lines in fixed symmetric representations in 4d $\mathcal{N} = 4$ SYM [89, 90], which is itself the low-energy worldvolume theory of a stack of D3-branes. Indeed, S-duality of said fundamental strings results in D1-branes, and two T-dualities takes the D1-D3 system to a D3-D5 system, which is precisely our starting point for the realization of line operators in 4d CS theory. The D3-branes we are interested in can thus be understood to polarize to D5-branes via the Myers effect.

The version of the Myers effect that is relevant is related to that of [120], where D-branes polarize into supertubes. For example, for $D3_b^+$, if we

---

[12] As explained in Chapter 3, this is equivalent via T-duality to the type IIA string theory embedding of the 4d CS theory using the D4-NS5 brane system, derived in [52, 53] and reviewed in Chapter 3, and in principle we could employ the latter to describe the thermodynamic limit and surface operators in 4d CS as well.

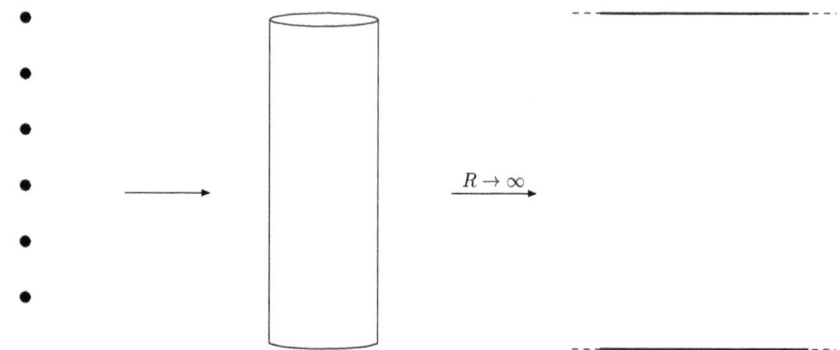

Figure 5.9.　The polarization of a large number of D3-branes (depicted as points) to a D5-brane that realizes the thermodynamic limit of Wilson lines in 4d CS theory. For example, for $\mathrm{D3}_b^+$, the vertical direction is '1' while the horizontal direction is '6'.

modify the '6' direction, which is the fiber direction of the cotangent bundle $T^*\Sigma$ (with the topology of $\mathbb{R}$), to be a circle with radius R, a large number of D3-branes placed at points along the '1' direction can polarize into a D5-brane with supertube topology. The radius of the supertube depends on the strength of the magnetic field generated by the D3-branes, and considering an infinite number of them (since we are interested in the thermodynamic limit) allows for polarization into a D5-brane where $R \to \infty$, as in Figure 5.9.

The resulting configuration is shown in the following table:

	Σ		$\mathbb{R}^2_\epsilon$		C		$N\Sigma \subset T^*\Sigma$		$\mathbb{R}^2_{-\epsilon}$	
	0	1	2	3	4	5	6	7	8	9
D5	×	×	×	×	×	×				
$\mathbf{D5}_b^+$	×	×	×	×			×	×		
$\mathbf{D5}_b^-$	×	×	×	×			×	×		
$\mathbf{D5}_f^+$	×	×					×	×	×	×
$\mathbf{D5}_f^-$	×	×					×	×	×	×

$$(5.2.113)$$

Let us now try to understand the D5-$\mathrm{D5}_f^+$ system. The $\mathrm{D5}_f^+$ constraint is

$$\epsilon_R = i\Gamma_{016789}\epsilon_L . \qquad (5.2.114)$$

Combining it with the constraint coming from the main D5-brane, we find

$$i\Gamma_{012345}\epsilon = i\Gamma_{016789}\epsilon , \qquad (5.2.115)$$

where we have denoted ϵ_L as ϵ. Using

$$i\Gamma_{0123456789}\epsilon = \epsilon \,, \tag{5.2.116}$$

we arrive at

$$i\Gamma_{012345}\epsilon = \Gamma_{2345}\epsilon \,, \tag{5.2.117}$$

or

$$i\Gamma_{01}\epsilon = \epsilon \,. \tag{5.2.118}$$

Likewise, for the anti-D5$_f$-brane denoted D5$_f^-$ (obtained via polarization of the anti-D3$_f$-brane, D3$_f^-$), we will get

$$-i\Gamma_{01}\epsilon = \epsilon \,. \tag{5.2.119}$$

The conditions (5.2.118) and (5.2.119) indicate that the preserved supercharges at the two intersections of the D5-D5$_f^+$ and D5-D5$_f^-$ systems have opposite charges with regard to rotations along the 01-plane.[13] Moreover, using the supersymmetry algebra, these conditions can be used to show that at the intersections, the theories are either holomorphic or anti-holomorphic, i.e., either the $\partial_{\bar{w}}$ or ∂_w operator is trivial in Q-cohomology. This is analogous to how the condition (3.2.20) can be used to show that the twist of the D5-brane worldvolume theory that localizes it to 4d CS theory has anti-holomorphic dependence (on C) that is trivial with regard to the cohomology of the supercharge used in the twist [46]. Notably, the conditions (5.2.118) and (5.2.119) are not compatible with the supercharge that respects topological invariance along Σ and that results in 4d CS theory. However, this is expected since the surface operators have holomorphic or antiholomorphic dependence on Σ.

We can perform an analogous analysis for the bosonic case using

$$(\Gamma_{23} + \Gamma_{89})\epsilon = 0 \,, \tag{5.2.120}$$

which is one of the constraints of (3.2.15). The D5$_b^+$-brane constraint is

$$\epsilon_R = i\Gamma_{012367}\epsilon_L \,. \tag{5.2.121}$$

Combining it with the constraint coming from the main D5-brane, we obtain

$$i\Gamma_{012345}\epsilon = i\Gamma_{012367}\epsilon \,. \tag{5.2.122}$$

---

[13]Essentially the same D-brane setup involving a D5-brane intersecting at different hyperplanes with a D5-brane and an anti-D5-brane was considered in [121], where the effective description of the system as a Gross-Neveu model was also derived.

Using

$$i\Gamma_{0123456789}\epsilon = \epsilon\,, \tag{5.2.123}$$

we arrive at

$$i\Gamma_{012345}\epsilon = \Gamma_{4589}\epsilon\,. \tag{5.2.124}$$

Using (5.2.120) we find

$$i\Gamma_{012345}\epsilon = -\Gamma_{4523}\epsilon\,, \tag{5.2.125}$$

which is equivalent to

$$i\Gamma_{01}\epsilon = -\epsilon\,. \tag{5.2.126}$$

Likewise, an analogous analysis involving the anti-D5$_b$-brane, D5$_b^-$, will give

$$i\Gamma_{01}\epsilon = \epsilon\,. \tag{5.2.127}$$

Thus, we find that the bosonic surface operators realized using D5$_b^+$ and D5$_b^-$ are holomorphic and anti-holomorphic theories.

Here, the D5$_b$-D5 intersection is a four-dimensional worldvolume with topology $\mathbb{C} \times \mathbb{R}_\epsilon^2$, where strings stretched between the branes give rise to a 4d $\mathcal{N} = 2$ hypermultiplet subject to the Ω-deformation of Kapustin's topological-holomorphic twist [122], which further localizes to a 2d theory of bosons [112, 113], which can be identified with a free $\beta\gamma$ surface operator in 4d CS theory:

$$\int_\Sigma \mathrm{Tr}_{\mathbb{C}^N}\left(\varphi\,\partial_{\bar{w}}^A\tilde{\varphi}\right), \tag{5.2.128}$$

where β and γ can be identified with φ and $\tilde{\varphi}$, respectively. The $\bar{w}$ derivative is a result of the dependence of the twist on the holomorphic structure of Σ.

The D5$_f$-D5 brane system is T-dual to the D4-D6 brane system that can be described by a surface operator supporting N chiral free fermions [115, 123, 124], and thus the strings stretched between the D5$_f$- and D5-branes give rise to a 2d theory of chiral fermions with the action

$$\int_\Sigma \mathrm{Tr}_{\mathbb{C}^N}\left(\psi\,\partial_{\bar{w}}^A\tilde{\psi}\right). \tag{5.2.129}$$

This is a free-fermion surface operator in 4d CS theory.

We shall now review the string-theoretic interpretation of the thermodynamic limit for Wilson lines in 4d CS theory associated with fixed, finite-dimensional irreducible representations of the gauge group. Let us first focus on Wilson lines realized by quantum mechanics with bosonic degrees of freedom. This bosonic quantum mechanics can be equivalently described in terms of coadjoint orbit line defects or free $\beta\gamma$ systems with constraints; an example of this equivalence was first shown by Diakonov and Petrov [125] and their derivation was generalized in Appendix H of Ashwinkumar, Sakamoto and Yamazaki [75].

Let us consider the brane configuration in (3.2.101). The polarization process described above also takes place when considering a large number of *stacks* of D3-branes in the configuration of (3.2.101). To be precise, we can separate the D3-branes in the initial configuration along the '8' or '9' direction, consider a large number of copies of such a configuration, whereby the polarization process results in a D5-brane at the location of each D3-brane in the configuration prior to polarization, and then bring together the resulting D5-branes to form a stack again. The polarization process results in a surface operator associated with a fixed, irreducible representation of $GL(N, \mathbb{C})$, which can be described either via a coadjoint orbit surface defect or a free $\beta\gamma$ system with constraints. The brane configuration that results from the polarization process is:

	$\overbrace{\qquad}^{\Sigma}$		$\overbrace{\qquad}^{\mathbb{R}^2_\epsilon}$		$\overbrace{\qquad}^{C}$		$\overbrace{\quad}^{N\Sigma \subset T^*\Sigma}$		$\overbrace{\;}^{\mathbb{R}^2_{-\epsilon}}$	
	0	**1**	**2**	**3**	**4**	**5**	**6**	**7**	**8**	**9**
$\mathbf{D5}_i$	×	×	×	×	×	×				
$\mathbf{D5}^{l_\alpha}_\alpha$	×	×	×	×			×	×		

$$(5.2.130)$$

Analogous analysis holds for the thermodynamic limit of fermionic quantum mechanics associated with irreducible finite-dimensional representations, with the resulting surface defect corresponding to the following brane configuration:

	$\overbrace{\qquad}^{\Sigma}$		$\overbrace{\qquad}^{\mathbb{R}^2_\epsilon}$		$\overbrace{\qquad}^{C}$		$\overbrace{\quad}^{N\Sigma \subset T^*\Sigma}$		$\overbrace{\;}^{\mathbb{R}^2_{-\epsilon}}$	
	0	**1**	**2**	**3**	**4**	**5**	**6**	**7**	**8**	**9**
$\mathbf{D5}_i$	×	×	×	×	×	×				
$\mathbf{D5}^{k_\alpha}_\alpha$	×						×		×	×

$$(5.2.131)$$

One can also interpret the thermodynamic limit of line operators of 4d CS theory associated with infinite-dimensional representations in string theory. These line operators were previously obtained via discretization of curved $\beta\gamma$ surface operators. Consider the brane configuration realizing such line operators given in (3.2.107). Having a large number of stacks of D3-branes in this configuration results in polarization of these branes into D5-branes in the following brane configuration, which are expected to realize curved $\beta\gamma$ surface operators:

	Σ		$\mathbb{R}^2_\epsilon$		C		$N\Sigma \subset T^*\Sigma$		$\mathbb{R}^2_{-\epsilon}$	
	0	**1**	**2**	**3**	**4**	**5**	**6**	**7**	**8**	**9**
D5	×	×	×	×	×	×	A			
D5	×	×	×	×			×	×		
NS5	×		×	×			B	×	×	×

$$(5.2.132)$$

This brane configuration is expected to be described by a 4d $\mathcal{N} = 2$ theory, that localizes to the path integral of a curved $\beta\gamma$ surface operator. Indeed, identifying the '7' direction with a circle of large radius, T-duality along this direction leads us to a system of D4-branes suspended between D6-branes and NS5-branes, which at low energy should be described by a 4d $\mathcal{N} = 2$ theory that is related to the 3d $\mathcal{N} = 4$ sigma model on $T^*(G_\mathbb{C}/B)$ via dimensional reduction.

5.3 Anomaly Inflow Mechanism for Order Defects in 4d Chern–Simons Theory

In order to engineer a large class of 2d integrable field theories, we use surface defects that are chiral or anti-chiral, meaning that there is only dependence of the surface defect on one of the complex coordinates on Σ. Such a 2d defect with classical gauge symmetry often suffers from a quantum gauge anomaly, which we ought to address if we wish to quantize the related integrable field theory.

The mechanism that ensures that such anomalies cancel takes the form of 4d–2d anomaly inflow , which, as we shall see, requires a quantum correction to the meromorphic one-form, ω, that enters the definition of 4d CS theory.

We shall first elucidate how gauge anomalies arise for any surface operator with degrees of freedom that give rise to an affine Kac–Moody algebra. Subsequently we demonstrate how they are canceled by coupling to the bulk 4d CS theory.[14]

---

[14]This result was originally derived by Costello and Yamazaki, but was unpublished. See also [75, 131, 132].

Let us consider a defect that supports an affine Kac–Moody algebra, which includes all the order surface defects encountered previously. The naive gauge anomaly arising from such a system has been computed in Appendix B. In what follows, we shall only consider infinitesimal gauge transformations, but the anomaly inflow mechanism even works for finite gauge transformations (at least for chiral free-fermion systems); see Appendix G of Ref. [75] for more details.

Now, the computation in Appendix B would lead us to deduce that the gauge variation of the effective action of a chiral defect is given by

$$\delta W_+ = \frac{k_+}{2\pi} \int_\Sigma \mathrm{d}^2 w \, \partial_w \epsilon_a(w, \bar{w}) A^a_{\bar{w}}(w, \bar{w}) \,, \tag{5.3.1}$$

where k_+ is the level of the associated affine Kac–Moody symmetry. Analogously, the gauge variation of the effective action with an anti-chiral defect, with Kac–Moody symmetry with level k_-, is given by

$$\delta W_- = \frac{k_-}{2\pi} \int_\Sigma \mathrm{d}^2 w \, \partial_{\bar{w}} \epsilon_a(w, \bar{w}) A^a_w(w, \bar{w}) \,. \tag{5.3.2}$$

However, it is important to impose the Wess–Zumino consistency condition [133] to obtain the correct form of the anomaly for either defect. This can be restated as the requirement that, when employing the BRST gauge-fixing scheme, the BRST operator Q_{BRST} is nilpotent, i.e., $Q_{\mathrm{BRST}}^2 = 0$ when acting on the effective action, where the BRST transformations are

$$\begin{aligned} Q_{\mathrm{BRST}} A^a_\mu &= -D_\mu c^a \,, \\ Q_{\mathrm{BRST}} c^a &= \frac{1}{2} f^{abc} c_b c_c \,, \end{aligned} \tag{5.3.3}$$

for $\mu = w, \bar{w}$. In order to achieve this, we add a local counterterm to the effective action that is proportional to $\int_\Sigma \mathrm{d}^2 w \mathrm{Tr}(A_\mu A^\mu)$, and whose BRST variation is proportional to

$$\int_\Sigma \mathrm{d}^2 w \, \mathrm{Tr}(\partial_\mu c A^\mu) \,. \tag{5.3.4}$$

Choosing the coefficient of the counterterm such that this variation becomes

$$-\frac{k_+}{4\pi} \int_\Sigma \mathrm{d}^2 w \mathrm{Tr}(\partial_w c A_{\bar{w}} + \partial_{\bar{w}} c A_w) \,, \tag{5.3.5}$$

the anomaly which takes the form $\frac{k_+}{2\pi} \int_\Sigma \mathrm{d}^2 w \, \mathrm{Tr}(\partial_w c(w, \bar{w}) A_{\bar{w}}(w, \bar{w}))$ is then modified to

$$-\frac{k_+}{4\pi} \int_\Sigma \mathrm{d}^2 w \mathrm{Tr}(-\partial_w c A_{\bar{w}} + \partial_{\bar{w}} c A_w) = \frac{k_+}{4\pi} \int_\Sigma \mathrm{Tr}(\mathrm{d}c \wedge A) \,. \tag{5.3.6}$$

This can be shown to be invariant under the transformation (5.3.3), since

$$Q_{\mathrm{BRST}} \int_{\Sigma} (\epsilon^{\mu\nu} \partial_{\mu} c_a A_{\nu}^a) d^2 w$$

$$= \int_{\Sigma} (\epsilon^{\mu\nu} \partial_{\mu} c_a D_{\nu} c^a) d^2 w + \int_{\Sigma} \epsilon^{\mu\nu} \partial_{\mu} \left(\frac{1}{2} f^{abc} c_b c_c \right) A_{a\nu} d^2 w$$

$$= \int_{\Sigma} (\epsilon^{\mu\nu} \partial_{\mu} c_a \partial_{\nu} c^a) d^2 w + \int_{\Sigma} (\epsilon^{\mu\nu} \partial_{\mu} c_a f^{abc} A_{\nu b} c_c) d^2 w$$

$$+ \int_{\Sigma} \epsilon^{\mu\nu} \frac{1}{2} f^{abc} \partial_{\mu} c_b c_c A_{a\nu} d^2 w + \int_{\Sigma} \epsilon^{\mu\nu} \frac{1}{2} f^{abc} c_b \partial_{\mu} c_c A_{a\nu} d^2 w, \quad (5.3.7)$$

where the first term vanishes due to integration by parts, while the remaining terms cancel upon using the Jacobi identity. Thus, the effective action which includes the counterterm is invariant under Q_{BRST}^2, and the Wess–Zumino consistency condition is satisfied.

The Q_{BRST}-invariant gauge anomaly for the chiral defect is thus

$$\delta W_+ = \frac{k_+}{4\pi} \int_{\Sigma} \mathrm{Tr}(\mathrm{d}\epsilon \wedge A). \tag{5.3.8}$$

Analogously, upon adding the aforementioned local counterterm with appropriate coefficient, the gauge variation of the effective action for an anti-chiral defect with Kac–Moody symmetry with level k_- is given by

$$\delta W_- = -\frac{k_-}{4\pi} \int_{\Sigma} \mathrm{Tr}(\mathrm{d}\epsilon \wedge A). \tag{5.3.9}$$

Let us now elucidate the anomaly inflow mechanism whereby the aforementioned anomalies can be canceled. We shall be slightly more general, and consider several copies each of the chiral and anti-chiral surface defects, at positions $z_{+,\alpha}$ and $z_{-,\beta}$, and levels $k_{+,\alpha}$ and $k_{-,\beta}$, where $\alpha = 1, \ldots, n_+$ and $\beta = 1, \ldots, n_-$.

Now, under the infinitesimal gauge transformation $\delta A = -\mathrm{d}\epsilon - [A, \epsilon]$, the 4d CS action transforms as

$$\delta S_{\mathrm{CS}}[A] = \frac{1}{2\pi\hbar} \int_{\Sigma \times \mathbb{CP}^1} \mathrm{d}\omega \wedge \mathrm{Tr}\,(\mathrm{d}\epsilon \wedge A), \tag{5.3.10}$$

where we have picked $C = \mathbb{CP}^1$. The upshot is that this transformation implies that the gauge anomalies of the surface defects can be canceled out by considering a quantum correction that results in a shift in ω,

$$\omega \mapsto \omega_{\text{eff}} = \omega - \frac{\hbar}{4\pi i} \left(\sum_{\alpha=1}^{n_+} \frac{k_{+,\alpha}}{z - z_{+,\alpha}} - \sum_{\beta=1}^{n_-} \frac{k_{-,\beta}}{z - z_{-,\beta}} \right) dz. \tag{5.3.11}$$

It is straightforward to verify that the variation (5.3.10) is localized at the simple poles $z = z_{+,\alpha}$, $z_{-,\beta}$ of ω_{eff} as

$$\frac{1}{2\pi\hbar} \int_{\Sigma \times \mathbb{CP}^1} d\omega_{\text{eff}} \wedge \text{Tr}\,(d\epsilon \wedge A): \tag{5.3.12}$$

$$:= -\frac{1}{4\pi} \sum_{\alpha=1}^{n_+} k_{+,\alpha} \int_{\Sigma \times \{z_{+,\alpha}\}} \text{Tr}\,(d\epsilon \wedge A)$$

$$+ \frac{1}{4\pi} \sum_{\beta=1}^{n_-} k_{-,\beta} \int_{\Sigma \times \{z_{-,\beta}\}} \text{Tr}\,(d\epsilon \wedge A), \tag{5.3.13}$$

and that these terms cancel with the chiral (and anti-chiral) anomalies which take the form given in (5.3.8) and (5.3.9).

Note that the absence of an extra pole at infinity requires

$$\sum_{\alpha=1}^{n_+} k_{+,\alpha} = \sum_{\beta=1}^{n_-} k_{-,\beta}, \tag{5.3.14}$$

and this condition on the levels is necessary to ensure gauge anomaly cancelation. For example, given a single chiral and a single anti-chiral free-fermion defect, the number of fermionic fields on each of these defects must be the same to satisfy this condition. If the condition (5.3.14) is not satisfied, it is expected that the classical integrability of the system will be broken due to the quantum effects of the theory.

The results of Ref. [35], reviewed earlier in Chapter 2, imply that the poles in ω_{eff} give rise to non-ultralocal terms appearing in the Poisson algebra of the Lax connection of the integrable field theory. Thus, the anomaly inflow mechanism suggests that the 2d integrable field theories associated with chiral surface defects that each support an affine Kac–Moody algebra are non-ultralocal at the quantum level.

In fact, in the study of the massless Thirring model in Ref. [134], a Schwinger term appears in the current algebra at the quantum level, such that the quantum algebra formed by the Lax connection for the massless

Thirring model becomes non-ultralocal due to quantum corrections, corroborating the observation above. It is expected that this phenomenon occurs for other 2d integrable field theories arising from other chiral and anti-chiral surface defects.

The one-loop shift in the meromorphic one-form ω can in fact be interpreted in terms of the emergence of disorder surface defects in 4d Chern–Simons theory. To illustrate this, we consider the simplest example of one chiral and one anti-chiral defect. For the anomaly cancelation condition (5.3.14) to be satisfied, the levels for the two defects ought to be the same— we denote this level as k.

The one-loop corrected one-form given by

$$\omega_{\text{eff}} = \omega - \frac{\hbar}{4\pi i}\left(\frac{k}{z - z_+} - \frac{k}{z - z_-}\right)\mathrm{d}z\,, \tag{5.3.15}$$

can be reexpressed as

$$\omega_{\text{eff}} = \frac{(z - \zeta_+)(z - \zeta_-)}{(z - z_+)(z - z_-)}\mathrm{d}z\,. \tag{5.3.16}$$

We thus find that ω_{eff} has the following poles and zeroes:

$$\mathfrak{p} = \{z_\pm, \infty\}\,, \qquad \mathfrak{z} = \{\zeta_\pm\}\,, \tag{5.3.17}$$

where $z = \infty$ is a double pole, $z = z_\pm$ are simple poles, and $z = \zeta_\pm$ are simple zeroes given explicitly by

$$\zeta_\pm = \frac{1}{2}\left(z_+ + z_- \pm \sqrt{(z_+ - z_-)\left(z_+ - z_- - \frac{i\hbar k}{\pi}\right)}\right)\,. \tag{5.3.18}$$

If we take the semiclassical limit $\hbar \to 0$, the poles $z_\pm$ and zeroes $\zeta_\pm$ cancel each other. From a different perspective, we say that the quantum correction to ω "pair-creates" a pair of poles and zeroes of the one-form.

As reviewed in Chapter 2, when one studies 4d CS theory with a meromorphic one-form with poles and zeroes as in (5.3.16), one ought to impose appropriate boundary conditions on the gauge fields to ensure that the requirement of having elliptic boundary conditions is satisfied. By elliptic boundary conditions, we mean boundary conditions that ensure that the theory has a unique propagator.

At the poles $z_\pm$, for some choice of gauge, we shall impose Dirichlet boundary conditions where chiral and anti-chiral components of the gauge fields are required to vanish:

$$A_+|_{z=z_-} = 0,$$
$$A_-|_{z=z_+} = 0,$$

(5.3.19)

where we now utilize lightcone notation for later reference. At the simple zeroes $\zeta_\pm$ we impose singular boundary conditions where the chiral and anti-chiral components of the gauge fields have poles:

$$A_+ \sim \mathcal{O}\left(\frac{1}{z - \zeta_+}\right),$$
$$A_- \sim \mathcal{O}\left(\frac{1}{z - \zeta_-}\right).$$

(5.3.20)

The implications of the emergence of disorder defects shall be explored further in Section 5.4.2.

5.4 Dualities of Integrable Field Theories

In this section, we review how 4d Chern–Simons theory can be used to derive various dualities of integrable field theories by employing dualities of the surface defects that realize them, based on the work of Ashwinkumar *et al.* [75].

5.4.1 *Dualities from discretizations of defects*

In Section 5.2, the discretizations of various integrable field theories were investigated from the perspective of 4d Chern–Simons theory, which amounted to discretizations of 2d order surface defects. Crucially, the line operators that arise from such discretizations were dualized into Wilson lines in 4d Chern–Simons theory.

This implies a rather strong statement at the level of surface defects, that can be obtained by taking the thermodynamic limit of Wilson lines in a particular representation. Such a Wilson line has multiple descriptions, and thus can give rise to different surface defects in the thermodynamic limit. The upshot is that given a particular lattice model realized by 4d CS whereby the thermodynamic limit gives us well-defined chiral and anti-chiral surface operators, one can have several 2d integrable field theory descriptions of the limit. These are expected to be dual theories, and in principle, given that many different defects are equivalent, we obtain an enormous duality web of integrable field theories.

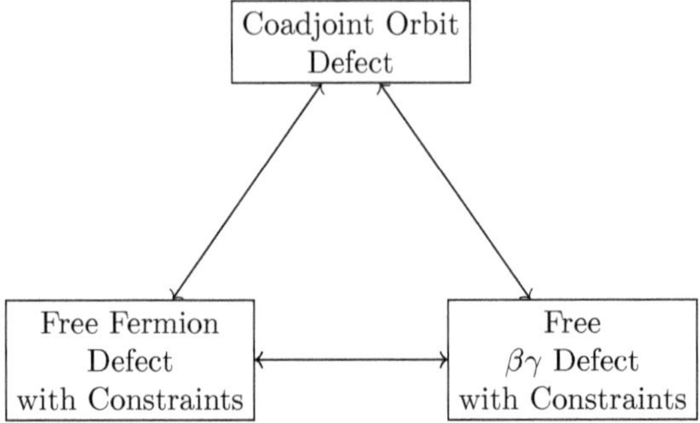

Figure 5.10. Seed triality among 1d line operators.

To elucidate these dualities, let us now describe a triality of integrable field theories that arises from the triality of the three line defects shown in Figure 5.10.

To be precise, we shall consider an irreducible representation, $\mathcal{R}$, of $GL(N, \mathbb{C})$ which is characterized by a Young tableau with k_I boxes in the I-th column $(1 \leq I \leq K)$ and l_α boxes in the α-th row $(1 \leq \alpha \leq L)$, respectively:

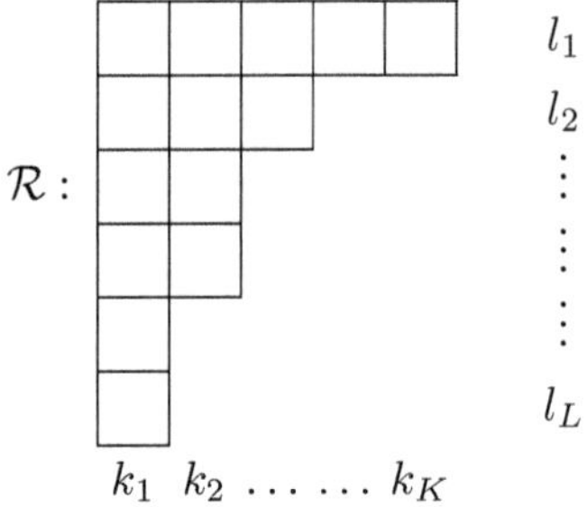

Here, we have $N \geq k_1 \geq k_2 \geq \cdots \geq k_K$ and $l_1 \geq l_2 \geq \cdots \geq l_L$. We know from Section 5.2 that the Wilson loop operator associated with the representation $\mathcal{R}$ can be described using either a 1d fermion defect, a 1d $\beta\gamma$ defect, or a 1d coadjoint orbit defect. The actions for the first two of these line operators are:

$$W_{\mathcal{R}}^f[A] = \int \mathcal{D}\psi \mathcal{D}\psi^\dagger \mathcal{D}\tilde{A}^f \, \exp\left(iS^f\right) \exp\left(-i \int \mathrm{d}t \sum_{I=1}^{K} k_{\mathrm{eff},I} \, (\tilde{A}_t^f)_{II}\right),$$

$$(5.4.1)$$

$$S^f = \int \mathrm{d}t \sum_{I,J=1}^{K} \sum_{i,j=1}^{N} \psi_i^{I\dagger} (i\delta^{IJ}\delta_j^i \partial_t + \delta^{IJ} A_t^a \rho(t_a)^i{}_j + \delta^{ij}(\tilde{A}_t^f)_{IJ}) \psi_J^j \,,$$

$$(5.4.2)$$

and

$$W_{\mathcal{R}}^b[A] = \int \mathcal{D}z \mathcal{D}z^* \mathcal{D}\tilde{A}^b \, \exp\left(iS^b\right) \exp\left(-i \int \mathrm{d}t \sum_{\alpha=1}^{L} l_{\mathrm{eff},\alpha} \, (\tilde{A}_t^b)_{\alpha\alpha}\right),$$

$$(5.4.3)$$

$$S^b = \int \mathrm{d}t \sum_{\alpha,\beta=1}^{L} \sum_{i,j=1}^{N} z_i[\alpha]^* \left(\delta^{\alpha\beta}\delta_j^i \partial_t + \delta^{\alpha\beta} A_t^a \rho(t_a)^i{}_j + \delta^{ij}(\tilde{A}_t^b)_{\alpha\beta}\right) z^j[\beta] \,,$$

$$(5.4.4)$$

where the only non-zero components of $(\tilde{A}_t^f)_{IJ}$ and $(\tilde{A}_t^b)_{\alpha\beta}$ are the $I \geq J$ and $\alpha \geq \beta$ components, respectively. The (quantum corrected) constants $k_{\mathrm{eff},I}$, $l_{\mathrm{eff},\alpha}$ are

$$k_{\mathrm{eff},I} = k_I - \frac{N}{2}, \qquad l_{\mathrm{eff},\alpha} = l_\alpha + \frac{N}{2}. \qquad (5.4.5)$$

In Ref. [75], it was further shown that the partition functions for these two line operators were equal.

We may take the thermodynamic limit of either description of the Wilson line in the representation $\mathcal{R}$. Given the two descriptions in terms of bosonic or fermionic degrees of freedom, one ends up with two descriptions in the thermodynamic limit as surface operators with bosonic or fermionic degrees of freedom. For example, for a collection of Wilson lines supported along the σ^+ direction, the two descriptions of the surface operators one obtains in the thermodynamic limit are

$$Z^f[\psi, \psi^\dagger, A] = \int \mathcal{D}\psi \mathcal{D}\psi^\dagger \mathcal{D}\tilde{A}^f$$

$$\times \exp\left(iS^f\right) \exp\left(-i \int \mathrm{d}\sigma^+ \mathrm{d}\sigma^- \sum_{I=1}^{K} k_{\mathrm{eff},I} \, (\tilde{A}_+^f)_{II}\right),$$

$$(5.4.6)$$

$$S^f = \int \mathrm{d}\sigma^+ \mathrm{d}\sigma^- \sum_{I,J=1}^{K} \sum_{i,j=1}^{N} \psi_i^{I\dagger} (i\delta^{IJ}\delta_j^i \partial_+ + \delta^{IJ} A_+^a \rho(t_a)^i{}_j + \delta^{ij}(\tilde{A}_+^f)_{IJ}) \psi_J^j \,,$$

$$(5.4.7)$$

and

$$Z^b[z, z^*, A] = \int \mathcal{D}z \mathcal{D}z^* \mathcal{D}\tilde{A}^b \, \exp\left(iS^b\right)$$

$$\times \exp\left(-i \int \mathrm{d}\sigma^+ \mathrm{d}\sigma^- \sum_{\alpha=1}^{L} l_{\mathrm{eff},I}\left(\tilde{A}^b_+\right)_{\alpha\alpha}\right), \tag{5.4.8}$$

$$S^b = \int \mathrm{d}\sigma^+ \mathrm{d}\sigma^- \sum_{\alpha,\beta=1}^{L} \sum_{i,j=1}^{N} z_i[\alpha]^*$$

$$\times \left(\delta^{\alpha\beta}\delta^i_j\partial_+ + \delta^{\alpha\beta}A^a_+\rho(t_a)^i{}_j + \delta^{ij}(\tilde{A}^b_+)_{\alpha\beta}\right) z^j[\beta]. \tag{5.4.9}$$

Hence, a version of bosonization holds between these surface operators, given its equivalent descriptions in terms of bosonic or fermionic degrees of freedom.

Moreover, by further considering the thermodynamic limit of a coadjoint orbit line defect in the representation $\mathcal{R}$, we obtain a third description of the same surface defect.

Let the aforementioned surface operator be located at z_+ on the holomorphic plane $\mathbb{CP}^1$ of 4d CS. Then, considering another surface operator of opposite chirality at position $z_- \in \mathbb{CP}^1$, with the same dual descriptions, would lead us to dual integrable field theories that are related via bosonization. Suppressing the projection conditions, the constrained bosonic and fermionic defects would lead to dual integrable field theories that have the actions

$$S^f = \int \mathrm{d}^2\sigma \left(\sum_{I=1}^{K} i\psi_i^{I\dagger}\partial_+\psi_I^i + \sum_{I'=1}^{K'} i\psi_i^{I'\dagger}\partial_-\psi_{I'}^i \right.$$

$$\left. + \frac{i}{4}\frac{\hbar}{z_+ - z_-} \sum_{I=1}^{K} \sum_{J'=1}^{K'} \left(\psi_i^{I\dagger}(t^a)^i{}_j\psi_I^j\right)\left(\psi_k^{J'\dagger}(t_a)^k{}_l\psi_{J'}^l\right) \right), \tag{5.4.10}$$

and

$$S^b = \int \mathrm{d}^2\sigma \left(\sum_{\alpha=1}^{L} z_i[\alpha]^*\partial_+z^i[\alpha] + \sum_{\alpha'=1}^{L'} z_i[\alpha']^*\partial_-z^i[\alpha'] \right.$$

$$\left. + \frac{i}{4}\frac{\hbar}{z_+ - z_-} \sum_{\alpha=1}^{L} \sum_{\beta'=1}^{L'} \left(z_i[\alpha]^*(t^a)^i{}_j z^j[\alpha]\right)\left(z_k[\beta']^*(t_a)^k{}_l z^l[\beta']\right) \right), \tag{5.4.11}$$

where $\int_\Sigma \mathrm{d}^2\sigma = \int_\Sigma \mathrm{d}\tau \wedge \mathrm{d}\sigma = -\frac{1}{2}\int_\Sigma \mathrm{d}\sigma^+ \wedge \mathrm{d}\sigma^-$. The chiral and anti-chiral surface operators considered here can each be in an arbitrary irreducible representation of $GL(N,\mathbb{C})$, which we denote $\mathcal{R}$ and $\mathcal{R}'$, respectively.

This is a bosonization duality between constrained fermionic and constrained bosonic massless Thirring models, that are each associated with a pair of fixed irreducible representations $\mathcal{R}$ and $\mathcal{R}'$ of $GL(N,\mathbb{C})$. Since the surface defects employed to show this duality are related via discretization to coadjoint orbit defects associated with representations $\mathcal{R}$ and $\mathcal{R}'$, we obtain a further duality to a generalized Faddeev–Reshetikhin model associated with the highest weights of $\mathcal{R}$ and $\mathcal{R}'$, denoted respectively as $\Lambda_\mathcal{R}$ and $\Lambda_{\mathcal{R}'}$, with the action

$$S_{\mathrm{FR}}\left[g_{(\pm)}\right] = \int_\Sigma d^2\sigma\,\mathrm{Tr}\left(\Lambda_\mathcal{R} g_{(+)}^{-1}\partial_- g_{(+)} + \Lambda_{\mathcal{R}'} g_{(-)}^{-1}\partial_+ g_{(-)}\right.$$
$$\left. - \frac{i}{4}\frac{\hbar}{z_+ - z_-}g_{(+)}\Lambda_\mathcal{R} g_{(+)}^{-1}g_{(-)}\Lambda_{\mathcal{R}'}g_{(-)}^{-1}\right). \tag{5.4.12}$$

We thus arrive at a triality relationship between the constrained bosonic massless Thirring model, its fermionic counterpart, and the generalized Faddeev–Reshetikhin model, depicted in Figure 5.11.

Although we have observed evidence for bosonization dualities via the discretization of surface defects, one would like to know if the duality continues to hold without discretizing the field theories. In other words,

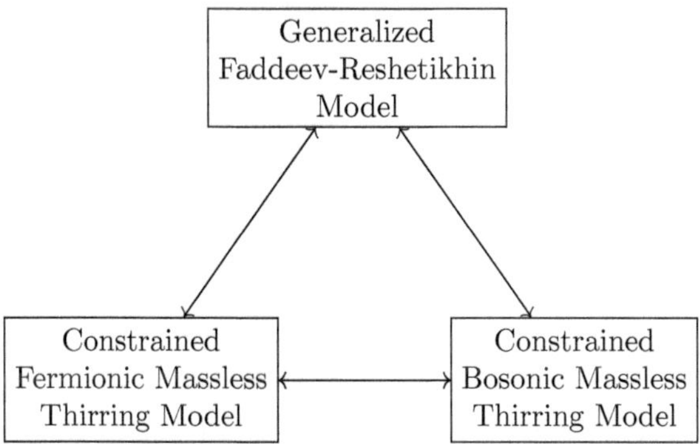

Figure 5.11.   An example of a triality of integrable field theories generated from the seed triality of line defects shown in Figure 5.10.

we would like to know if the 2d defects themselves are dual to each other prior to discretization. To achieve this, it would be desirable to show that the levels of the affine Kac–Moody algebras generated by each defect are equal.

To this end, recall that the coadjoint orbit defects can be interpreted as curved $\beta\gamma$ defects, at least classically. The latter give rise to affine Kac–Moody algebras at the critical level, i.e., $k = -\mathsf{h}^\vee$ (see (5.2.97)). Moreover, as explained in Appendix H of Ref. [75], the coadjoint orbit defect can be understood to arise from a reduced chiral WZW defect, where the conserved current of the latter obeys $\mathrm{Tr}(J^2) = \text{constant}$. As we shall elucidate further below, the latter behavior also arises from constrained 2d boson and fermion defects, which also suggests that the level of the associated affine Kac–Moody algebra is critical.

In particular, the constraints on the 2d boson and fermion defects fix the particle number in these systems, which then implies that the rescaled Sugawara energy-momentum tensor built out of the currents ought to be constant. The upshot is that this rescaled energy-momentum tensor, $\tilde{T}(z) = (J^a J_a)$, is proportional to the identity operator on Fock space. This implies that the OPE between $\tilde{T}(z)$ and any other operator, including $J^a(z)$, ought to be regular. This in turn means that the modes of $\tilde{T}(z)$ commute with each other, as well as with all the modes of the affine Kac–Moody current. This can be understood to arise from the $k \to -\mathsf{h}^\vee$ limit of the conventional Sugawara energy-momentum tensor, defined as

$$T(z) = \frac{1}{2(k + \mathsf{h}^\vee)}(J^a J_a)\,, \tag{5.4.13}$$

whose modes satisfy

$$[L_n, J_m^a] = -m J_{n+m}^a\,. \tag{5.4.14}$$

If we denote the modes of $\tilde{T}(z)$ as $\tilde{L}_n$, we would have

$$\left[\tilde{L}_n, J_m^a\right] = -2(k + \mathsf{h}^\vee) m J_{n+m}^a\,, \tag{5.4.15}$$

and in the critical level limit, the RHS of (5.4.15) goes to zero. This suggests that all three defects that were used to derive the triality in this section can be understood as constrained systems, where the constraints fix the level of the associated affine Kac–Moody algebra to be critical.

5.4.2 *Dualities from bosonizations of defects*

We have so far described dualities of integrable field theories arising from dualizing defects in the context of discretization. However, one does not have to be restricted to surface defect dualizations that arise from discretization. Indeed, one can show further equivalences between, e.g., free-fermion defects and certain bosonic defects, such as a defect defined as the path integral over a gauged WZW model, which we shall refer to as a WZW defect.

In what follows, we shall first recall two bosonization dualities for the multi-flavor massless Thirring model

$$S_{\mathrm{Th}}[\psi] = \frac{1}{\hbar_{2\mathrm{d}}} \int_{\Sigma} \mathrm{d}^2\sigma$$

$$\times \left(\sum_{I=1}^{N_F} \left(\psi_{L,I}^\dagger i\partial_- \psi_{L,I} + \psi_{R,I}^\dagger i\partial_+ \psi_{R,I} \right) - \frac{1}{4}\frac{i\hbar}{z_+ - z_-} \mathcal{J}_+^a \mathcal{J}_{a-} \right),$$

$$(5.4.16)$$

where N_F is the number of flavors, which has the Lax operator

$$\mathscr{L} = \frac{\hbar}{4}\frac{\mathcal{J}_+^a \rho\left(t_a\right)}{z - z_+}\mathrm{d}\sigma^+ - \frac{\hbar}{4}\frac{\mathcal{J}_-^a \rho\left(t_a\right)}{z - z_-}\mathrm{d}\sigma^-,$$

$$(5.4.17)$$

$$\mathcal{J}_+^a = \frac{1}{\hbar_{2\mathrm{d}}} \sum_{I=1}^{N_F} \psi_{L,I}^\dagger \rho\left(t^a\right) \psi_{L,I}, \quad \mathcal{J}_-^a = \frac{1}{\hbar_{2\mathrm{d}}} \sum_{I=1}^{N_F} \psi_{R,I}^\dagger \rho\left(t^a\right) \psi_{R,I},$$

based on the work of Polyakov and Wiegmann [135, 136]. We shall then explain how these bosonization dualities can be derived from 4d Chern–Simons theory, via dualization of surface defects.

5.4.2.1 *2d derivation of bosonization dualities*

To derive the bosonization dualities, we first rewrite the action (5.4.16) by introducing a Lie algebra-valued auxiliary field, B, as follows:

$$\tilde{S}_f^{\mathrm{Th}} = \frac{1}{\hbar_{2d}} \int_{\Sigma} \mathrm{d}^2\sigma \left(\sum_{I=1}^{N_F} \psi_{L,I}^\dagger i\partial_- \psi_{L,I} + \sum_{I=1}^{N_F} \psi_{R,I}^\dagger i\partial_+ \psi_{R,I} + iB_-^a \mathcal{J}_{+,a} \right.$$

$$\left. + iB_+^a \mathcal{J}_{-,a} + \hbar_{2d}G^{-1}\mathrm{Tr}\left(B_+ B_-\right) \right)$$

$$= S_+^f[B_-, \psi_L] + S_-^f[B_+, \psi_R] + \int_{\Sigma} \mathrm{d}^2\sigma\, G^{-1}\mathrm{Tr}(B_+ B_-),$$

$$(5.4.18)$$

where G is defined as

$$G = \frac{1}{4}\frac{i\hbar}{z_+ - z_-}. \tag{5.4.19}$$

From [135, 136],[15] it is known that the partition functions for the actions $S_+^f[B_-, \psi_L]$ and $S_-^f[B_+, \psi_R]$ can be respectively evaluated to be

$$\widehat{Z}_+^f[B_-] = e^{-iN_F S_{\mathrm{wzw}}[u_{(+)}]}, \qquad \widehat{Z}_-^f[B_+] = e^{-iN_F S_{\mathrm{wzw}}[u_{(-)}^{-1}]}, \tag{5.4.20}$$

where the two independent lightcone components of the auxiliary field B are now parametrized as

$$B_+ = u_{(-)}^{-1}\partial_+ u_{(-)}, \qquad B_- = u_{(+)}^{-1}\partial_- u_{(+)}, \tag{5.4.21}$$

where $u_{(\pm)} \in G$, and where $S_{\mathrm{wzw}}[g]$ is the WZW action

$$S_{\mathrm{wzw}}[g] = -\frac{1}{2\pi}\int_\Sigma \mathrm{d}^2\sigma\, \mathrm{Tr}\left(g^{-1}\partial_+ g g^{-1}\partial_- g\right) + S_{\mathrm{wz}}[g]$$

$$S_{\mathrm{wz}}[g] = \frac{1}{12\pi}\int_{\Sigma\times\mathbb{R}_+} \mathrm{Tr}\left(g^{-1}\mathrm{d}g \wedge g^{-1}\mathrm{d}g \wedge g^{-1}\mathrm{d}g\right). \tag{5.4.22}$$

Thus, the partition function Z^{Th} of the multi-flavor massless Thirring model becomes

$$Z^{\mathrm{Th}} = \int \mathcal{D}\psi \mathcal{D}\psi^\dagger\, e^{iS_f^{\mathrm{Th}}[\psi,\psi^\dagger]}$$

$$= \int \mathcal{D}\psi \mathcal{D}\psi^\dagger\, \mathcal{D}B\, e^{i\tilde{S}_f^{\mathrm{Th}}[\psi,\psi^\dagger,B]}$$

$$= \int \mathcal{D}B e^{-iN_F S_{\mathrm{wzw}}[u_{(+)}]-iN_F S_{\mathrm{wzw}}[u_{(-)}^{-1}]+i\int \mathrm{d}^2\sigma\left(a+G^{-1}\right)\mathrm{Tr}(B_+ B_-)}, \tag{5.4.23}$$

where a local counter term $\mathrm{Tr}(B_+ B_-)$ with a parameter a was added to the action.

The first bosonization duality arises by making a change of variables from $B_\pm$ to $u_{(\pm)}$, which generates a nontrivial Jacobian in the path integral [137]. Taking this Jacobian into account, we obtain two interacting WZW models as the dual theory, where $-N_F$ is shifted to $-N_F - \mathsf{h}^\vee$ due to the

---

Jacobian, i.e., we obtain

$$Z^{\mathrm{Th}} = \int \mathcal{D}u_{(+)}\mathcal{D}u_{(-)}$$

$$\times \exp\Bigg(-i(N_F + \mathsf{h}^{\vee})S_{\mathrm{WZW}}[u_{(+)}] - i(N_F + \mathsf{h}^{\vee})S_{\mathrm{WZW}}[u_{(-)}^{-1}]$$

$$+ i\int \mathrm{d}^2\sigma\,(a + G^{-1})\,\mathrm{Tr}(u_{(-)}^{-1}\partial_+ u_{(-)} u_{(+)}^{-1}\partial_- u_{(+)})\Bigg). \tag{5.4.24}$$

Alternatively, to derive another bosonic dual, we can reexpress the path integral (5.4.23) by multiplying it by a constant of the form

$$\int \mathcal{D}g_{(+)}\mathcal{D}g_{(-)}e^{-iN_F S_{\mathrm{WZW}}[g_{(+)}] - iN_F S_{\mathrm{WZW}}[g_{(-)}]}, \tag{5.4.25}$$

and by using the Polyakov–Wiegmann identity[16]

$$S_{\mathrm{WZW}}[ug] = S_{\mathrm{WZW}}[u] + S_{\mathrm{WZW}}[g] - \frac{1}{\pi}\int \mathrm{d}^2\sigma \mathrm{Tr}\left(u^{-1}\partial_- u\partial_+ gg^{-1}\right). \tag{5.4.26}$$

We then obtain

$$Z^{\mathrm{Th}} = \int \mathcal{D}B\mathcal{D}g_{(+)}\mathcal{D}g_{(-)} \exp\Bigg(iN_F S_{\mathrm{WZW}}[g_{(+)}] + iN_F S_{\mathrm{WZW}}[g_{(-)}]$$

$$- \frac{iN_F}{\pi}\int_{\Sigma} \mathrm{d}^2\sigma\,\mathrm{Tr}(B_-\partial_+ g_{(+)}g_{(+)}^{-1}) + \frac{iN_F}{\pi}\int_{\Sigma} \mathrm{d}^2\sigma\,\mathrm{Tr}(B_+ g_{(-)}^{-1}\partial_- g_{(-)})$$

$$+ i\int_{\Sigma} \mathrm{d}^2\sigma\,(a + G^{-1})\,\mathrm{Tr}(B_+ B_-)\Bigg). \tag{5.4.27}$$

Subsequently integrating the path integral (5.4.27) with respect to the auxiliary fields $B_{\pm}$ gives us

$$B_+ = \frac{N_F}{\pi}\left(a + G^{-1}\right)^{-1}\partial_+ g_{(+)}g_{(+)}^{-1},$$

$$B_- = -\frac{N_F}{\pi}\left(a + G^{-1}\right)^{-1}g_{(-)}^{-1}\partial_- g_{(-)}, \tag{5.4.28}$$

---

[16]This computation will be further elucidated when explaining the 4d Chern–Simons derivation below.

which leads us to the following bosonic dual of the massless Thirring model

$$Z^{\mathrm{Th}} = \int \mathcal{D}g_{(+)}\mathcal{D}g_{(-)} \exp\bigg(iN_F S_{\mathrm{WZW}}[g_{(+)}] + iN_F S_{\mathrm{WZW}}[g_{(-)}]$$

$$+ i\left(\frac{N_F}{\pi}\right)^2 (a + G^{-1})^{-1} \int_\Sigma \mathrm{d}^2\sigma \, \mathrm{Tr}\left(\partial_+ g_{(+)} g_{(+)}^{-1} g_{(-)}^{-1} \partial_- g_{(-)}\right)\bigg).$$

$$(5.4.29)$$

The relationship between the two bosonic models derived here can be understood to be a strong-weak duality, and this boson-boson duality has been discussed in various works, such as [138–140].

5.4.2.2   *Derivation of bosonization dualities from 4d Chern–Simons theory*

From the perspective of 4d Chern–Simons theory, we already know how to realize the multi-flavor massless Thirring model, that is, by using surface defects of the form

$$S_+^f[A_-] = \sum_{I=1}^{N_F} \int_{\Sigma \times \{z_+\}} \mathrm{d}^2\sigma \, \psi^\dagger{}_{L,I} iD_- \psi_{L,I},$$

$$S_-^f[A_+] = \sum_{I=1}^{N_F} \int_{\Sigma \times \{z_-\}} \mathrm{d}^2\sigma \, \psi^\dagger{}_{R,I} iD_+ \psi_{R,I}.$$

$$(5.4.30)$$

We shall be even more general in what follows, and consider multiple chiral and anti-chiral surface defects, labeled respectively by $\alpha = 1, \ldots, n_+$ and $\beta = 1, \ldots, n_-$.

As discussed in Section 5.3, anomaly inflow requires the holomorphic one-form ω to be corrected to

$$\omega_{\mathrm{eff}} = \left(1 - \frac{\hbar}{4\pi i}\left(\sum_{\alpha=1}^{n_+} \frac{k_{+,\alpha}}{z - z_{+,\alpha}} - \sum_{\beta=1}^{n_-} \frac{k_{-,\beta}}{z - z_{-,\beta}}\right)\right) \mathrm{d}z, \qquad (5.4.31)$$

where $k_{+,\alpha}$ and $k_{-,\beta}$ indicate the respective affine Kac–Moody algebra levels.

At this point, it is useful to recall the equivalence between free-fermion defects and an effective action of WZW-model form that was used in the 2d derivation. This uses an argument similar to that of [135, 136]. Recall from Section 5.3 that the anomalies associated with chiral and anti-chiral defects supporting affine Kac–Moody algebras (prior to the inclusion of

Pauli–Villars counter-terms) are

$$\delta W_+ = \frac{k_+}{2\pi} \int_\Sigma \mathrm{d}^2 w \partial_w \epsilon_a(w, \bar{w}) A^a_{\bar{w}}(w, \bar{w}) \,, \tag{5.4.32}$$

and

$$\delta W_- = \frac{k_-}{2\pi} \int_\Sigma \mathrm{d}^2 w \partial_{\bar{w}} \epsilon_a(w, \bar{w}) A^a_w(w, \bar{w}) \,. \tag{5.4.33}$$

In Minkowski signature these are

$$\delta W_+ = -\frac{k_+}{2\pi} \int_\Sigma \mathrm{d}\sigma^+ \mathrm{d}\sigma^- \epsilon_a \partial_+ A^a_- \,, \tag{5.4.34}$$

and

$$\delta W_- = -\frac{k_-}{2\pi} \int_\Sigma \mathrm{d}\sigma^+ \mathrm{d}\sigma^- \epsilon_a \partial_- A^a_+ \,. \tag{5.4.35}$$

Let us focus on the chiral defect. The variation of the effective action can also be described as

$$W_+[A_-] \to W_+[A_- - D_- \epsilon]$$

$$= W_+[A] - \int \mathrm{d}\sigma^+ \mathrm{d}\sigma^- \, \mathrm{tr} \left(D_- \epsilon \frac{\delta}{\delta A_-} W_+[A_-] \right) \tag{5.4.36}$$

$$= W_+[A] + \int \mathrm{d}\sigma^+ \mathrm{d}\sigma^- \, \mathrm{tr} \left(\epsilon D_- \frac{\delta}{\delta A_-} W_+[A_-] \right) .$$

We then have

$$\partial_- \frac{\delta W_+}{\delta A_-} + \left[A_-, \frac{\delta W_+}{\delta A_-} \right] = -\frac{k_+}{2\pi} \partial_+ A_- \,. \tag{5.4.37}$$

If we express A_- as

$$A_- = g^{-1} \partial_- g, \tag{5.4.38}$$

we can obtain the following expression for $\frac{\delta W_+}{\delta A_-}$ from (5.4.37)

$$\frac{\delta W_+}{\delta A_-} = -\frac{k_+}{2\pi} g^{-1} \partial_+ g \,. \tag{5.4.39}$$

If one assumes that the chiral defect has no modes that decouple from the 4d CS gauge field, an arbitrary variation of the effective action then takes the form

$$\delta W_+ = \int \mathrm{d}\sigma^+\mathrm{d}\sigma^- \, \mathrm{Tr}\left(\frac{\delta W_+}{\delta A_-}\delta A_-\right)$$

$$= -\frac{k_+}{2\pi}\int \mathrm{d}\sigma^+\mathrm{d}\sigma^- \mathrm{Tr}(g^{-1}\partial_+ g \delta(g^{-1}\partial_- g))$$

$$= \frac{k_+}{2\pi}\int \mathrm{d}\sigma^+\mathrm{d}\sigma^- \mathrm{Tr}(([g^{-1}\partial_- g, g^{-1}\partial_+ g] + \partial_-(g^{-1}\partial_+ g))g^{-1}\delta g)$$

$$= \frac{k_+}{2\pi}\int \mathrm{d}\sigma^+\mathrm{d}\sigma^- \mathrm{Tr}(\partial_+(g^{-1}\partial_- g)g^{-1}\delta g)\,.$$

$$(5.4.40)$$

Then, we find that

$$W_+ = -\frac{k_+}{4\pi}\int_\Sigma \mathrm{d}\sigma^+\mathrm{d}\sigma^- \mathrm{Tr}(g^{-1}\partial_+ g g^{-1}\partial_- g)$$

$$-\frac{k_+}{12\pi}\int_{\Sigma\times\mathbb{R}_+} \mathrm{Tr}(g^{-1}\mathrm{d}g \wedge g^{-1}\mathrm{d}g \wedge g^{-1}\mathrm{d}g)\,,$$

$$(5.4.41)$$

which is the aforementioned effective action. To ensure that the Wess–Zumino consistency condition is satisfied, one should also include the Pauli–Villars counter-term. The resulting effective action, regarded as a defect in 4d CS, was referred to as a chiral edge mode defect in [75], for reasons that will become apparent below. An analogous computation relates anti-chiral free-fermion defects and an anti-chiral edge mode action.

Explicitly, we have the following expressions for the chiral and anti-chiral edge mode defects located, e.g., at points z_+ and z_- on C. The actions of the defects are given by

$$S_+^{\mathrm{edge}}[g, A] = S_{\mathrm{WZW}}[g] + \frac{1}{2\pi}\int_{\Sigma\times\{z_+\}} \mathrm{d}^2\sigma \, \mathrm{Tr}\,(A_+ A_-) \qquad (5.4.42)$$

with the constraint

$$A_-|_{z_+} = g^{-1}\partial_- g \qquad (5.4.43)$$

and

$$S_-^{\mathrm{edge}}[g, A] = \widetilde{S}_{\mathrm{WZW}}[g] + \frac{1}{2\pi}\int_{\Sigma\times\{z_-\}} \mathrm{d}^2\sigma \, \mathrm{Tr}\,(A_+ A_-) \qquad (5.4.44)$$

with the constraint

$$A_+|_{z_-} = g^{-1}\partial_+ g \qquad (5.4.45)$$

where $S_{\mathrm{WZW}}[g]$ is the WZW action

$$S_{\mathrm{WZW}}[g] = -\frac{1}{2\pi} \int_{\Sigma} \mathrm{d}^2\sigma \, \mathrm{Tr} \left(g^{-1}\partial_+ g \, g^{-1}\partial_- g \right) + S_{\mathrm{WZ}}[g]$$

$$S_{\mathrm{WZ}}[g] = \frac{1}{12\pi} \int_{\Sigma \times \mathbb{R}_+} \mathrm{Tr} \left(g^{-1}\mathrm{d}g \wedge g^{-1}\mathrm{d}g \wedge g^{-1}\mathrm{d}g \right) ,$$

$$(5.4.46)$$

and $\widetilde{S}_{\mathrm{WZW}}[g] = S_{\mathrm{WZW}}\left[g^{-1}\right]$ is its anti-chiral counterpart.

Note that, although we have emphasized the duality between free-fermion defects and edge mode defects, it is clear from the anomaly-based derivation above that the duality to edge mode defects also holds more generally for other chiral and anti-chiral defects.

The derivation of the first bosonization from 4d CS then proceeds as follows. We shall show that the 4d CS action coupled to edge mode defects (that we obtained from dualizing the free fermion defects) can be interpreted as 4d CS with a meromorphic 1-form (5.4.31), with Dirichlet boundary conditions at the locations of poles of the 1-form, subject to a formal gauge transformation (described in Chapter 2). These shall be referred to as chiral and anti-chiral Dirichlet boundary conditions, and are located at $z = z_{+,\alpha}$ and $z = z_{-,\beta}$, and are given explicitly by

$$A_-|_{z_{+,\alpha}} = 0 , \qquad A_+|_{z_{-,\beta}} = 0 . \qquad (5.4.47)$$

Let us now apply the formal gauge transformation

$$A = -\mathrm{d}\widehat{g}\widehat{g}^{-1} + \widehat{g}\mathcal{L}\widehat{g}^{-1}. \qquad (5.4.48)$$

By using this gauge transformation, one can set $\mathcal{L}_{\bar{z}} = 0$, as in Chapter 2. This is possible since we can write the $\bar{z}$-component of the 4d CS gauge field as

$$A_{\bar{z}} = -\partial_{\bar{z}}\widehat{g}\widehat{g}^{-1}, \qquad (5.4.49)$$

where $\widehat{g}$ is determined by a map from Σ to the moduli space of G-bundles on $\mathbb{CP}^1$ where the boundary conditions (5.4.47) are imposed. As explained in Chapter 2, this is because in defining the 4d CS theory, we consider a topologically trivial G-bundle on $\Sigma \times \mathbb{CP}^1$, and a topologically trivial complex bundle on $\mathbb{CP}^1$ is generically holomorphically trivial as well, allowing us to express $A_{\bar{z}}$ as in (5.4.49).

The 4d CS action is transformed into

$$S_{\mathrm{CS}}[A] = \frac{1}{2\pi\hbar} \int_{\Sigma \times C} \omega_{\mathrm{eff}} \wedge \left(\mathrm{CS}(\mathcal{L}) + \mathrm{d}\left(\mathrm{Tr}(\widehat{g}^{-1}\mathrm{d}\widehat{g} \wedge \mathcal{L}) \right) + I_{\mathrm{WZ}}[\widehat{g}] \right) ,$$

$$(5.4.50)$$

where $I_{\mathrm{WZ}}[\widehat{g}]$ is the Wess–Zumino three-form defined as

$$I_{\mathrm{WZ}}[\widehat{g}] = \frac{1}{3}\mathrm{Tr}\left(\widehat{g}^{-1}\mathrm{d}\widehat{g} \wedge \widehat{g}^{-1}\mathrm{d}\widehat{g} \wedge \widehat{g}^{-1}\mathrm{d}\widehat{g}\right), \qquad (5.4.51)$$

while the boundary conditions (5.4.47) are transformed to

$$\begin{aligned}
\mathcal{L}_-|_{z_{+,\alpha}} &= g_{+,\alpha}^{-1}\partial_- g_{+,\alpha}|_{z_{+,\alpha}}\,, \\
\mathcal{L}_+|_{z_{-,\beta}} &= g_{-,\beta}^{-1}\partial_+ g_{-,\beta}|_{z_{-,\beta}}\,,
\end{aligned} \qquad (5.4.52)$$

where we have denoted the value of $\widehat{g}$ at $z_{+,\alpha}$ and $z_{-,\beta}$ to be $g_{+,\alpha}$ and $g_{-,\beta}$, respectively.

The second and third terms in the parenthesis in (5.4.50) can be shown, using the Cauchy–Pompeiu integral formula, to take the form

$$\begin{aligned}
&-\sum_{\alpha=1}^{n_+} \frac{k_{+,\alpha}}{4\pi}\left(\int_{\Sigma\times\{z_{+,\alpha}\}} \mathrm{d}\sigma^+\mathrm{d}\sigma^- \mathrm{Tr}(g_{+,\alpha}^{-1}\partial_+ g_{+,\alpha}\mathcal{L}_- - g_{+,\alpha}^{-1}\partial_- g_{+,\alpha}\mathcal{L}_+)\right. \\
&\qquad\qquad\qquad\left. + \int_{\Sigma\times\mathbb{R}_+\times\{z_{+,\alpha}\}} I_{\mathrm{WZ}}[g_{+,\alpha}]\right) \\
&+\sum_{\beta=1}^{n_-} \frac{k_{-,\beta}}{4\pi}\left(\int_{\Sigma\times\{z_{-,\beta}\}} \mathrm{d}\sigma^+\mathrm{d}\sigma^- \mathrm{Tr}(g_{-,\beta}^{-1}\partial_+ g_{-,\beta}\mathcal{L}_- - g_{-,\beta}^{-1}\partial_- g_{-,\beta}\mathcal{L}_+)\right. \\
&\qquad\qquad\qquad\left. + \int_{\Sigma\times\mathbb{R}_+\times\{z_{-,\beta}\}} I_{\mathrm{WZ}}[g_{-,\beta}]\right).
\end{aligned}$$
$$(5.4.53)$$

To relate this to edge mode defects, we then use the boundary constraints (5.4.52), to rewrite (5.4.53) into

$$\begin{aligned}
&-\sum_{\alpha=1}^{n_+} \frac{k_{+,\alpha}}{4\pi}\left(\int_{\Sigma\times\{z_{+,\alpha}\}} \mathrm{d}\sigma^+\mathrm{d}\sigma^- \mathrm{Tr}(g_{+,\alpha}^{-1}\partial_+ g_{+,\alpha}g_{+,\alpha}^{-1}\partial_- g_{+,\alpha} - \mathcal{L}_+\mathcal{L}_-)\right. \\
&\qquad\qquad\qquad\left. + \int_{\Sigma\times\mathbb{R}_+\times\{z_{+,\alpha}\}} I_{\mathrm{WZ}}[g_{+,\alpha}]\right) \\
&-\sum_{\beta=1}^{n_-} \frac{k_{-,\beta}}{4\pi}\left(\int_{\Sigma\times\{z_{-,\beta}\}} \mathrm{d}\sigma^+\mathrm{d}\sigma^- \mathrm{Tr}(g_{-,\beta}^{-1}\partial_+ g_{-,\beta}g_{-,\beta}^{-1}\partial_- g_{-,\beta} - \mathcal{L}_+\mathcal{L}_-)\right. \\
&\qquad\qquad\qquad\left. - \int_{\Sigma\times\mathbb{R}_+\times\{z_{-,\beta}\}} I_{\mathrm{WZ}}[g_{-,\beta}]\right).
\end{aligned}$$
$$(5.4.54)$$

Thus, the path integral has the form

$$\int \mathcal{DL}\, e^{iS_{\mathrm{CS}}[\mathcal{L}]} \prod_{\alpha=1}^{n_+} \exp(-ik_{+,\alpha} S_+^{\mathrm{edge}}[g_{+,\alpha}, \mathcal{L}]) \prod_{\beta=1}^{n_-} \exp(-ik_{-,\beta} S_-^{\mathrm{edge}}[g_{-,\beta}, \mathcal{L}]),$$

$$(5.4.55)$$

where $S_+^{\mathrm{edge}}[g_{+,\alpha}, \mathcal{L}]$ and $S_-^{\mathrm{edge}}[g_{-,\beta}, \mathcal{L}]$ are the actions for the chiral and anti-chiral edge mode defects. This is the path integral we obtained by dualizing free-fermion defects to edge mode defects, and picking the gauge where the $\bar{z}$-component of the 4d CS gauge field is zero.

We can thus derive the first bosonic dual of the multi-flavor massless Thirring model by deriving the integrable field theory corresponding to the appropriate specialization of the meromorphic one-form (5.4.31) to $n_+ = n_- = 1$, using techniques from Chapter 2. If we do not specialize (5.4.31), and consider multiple defects, we would in fact obtain a generalized bosonization duality.

Let us now proceed with this derivation. For simplicity, let us consider an equal number of chiral and anti-chiral defects, i.e., $n_+ = n_- = N$. We assume the anomaly cancellation condition (5.3.14) so that there is no pole at infinity, such that (5.4.31) can be written as

$$\omega_{\mathrm{eff}} = \prod_{\alpha=1}^{N} \frac{(z - \zeta_{-,\alpha})(z - \zeta_{+,\alpha})}{(z - z_{-,\alpha})(z - z_{+,\alpha})}\mathrm{d}z \qquad (5.4.56)$$

$$= \tilde{\varphi}_+(z)\tilde{\varphi}_-(z),$$

where

$$\tilde{\varphi}_+(z) = \prod_{\alpha=1}^{N} \frac{(z - \zeta_{-,\alpha})}{(z - z_{-,\alpha})}, \qquad \tilde{\varphi}_-(z) = \prod_{\alpha=1}^{N} \frac{(z - \zeta_{+,\alpha})}{(z - z_{+,\alpha})}. \qquad (5.4.57)$$

Now, we shall make the observation that the resulting integrable model should be a special limit of the coupled WZW models reviewed in Chapter 2. To this end, note that the twist function of these models correspond to the meromorphic one-form

$$\omega = \frac{\prod_{i=1}^{2N}(z - q_i^+)\prod_{j=1}^{2N}(z - q_j^-)}{\prod_{k=1}^{2N}(z - p_k)^2}\,\mathrm{d}z, \qquad (5.4.58)$$

where non-chiral Dirichlet boundary conditions

$$A_\pm|_{z=p_k} = 0 \qquad (5.4.59)$$

are imposed at the double poles at p_k, while at the zeroes q_i^+ and q_i^-, A_+ and A_- are respectively allowed to have poles, i.e.,

$$A_+ \sim \frac{1}{z - q_i^+}\,, \qquad A_- \sim \frac{1}{z - q_i^-}\,. \tag{5.4.60}$$

Let us factorize the denominator of (5.4.58) by expressing the meromorphic one-form (5.4.58) as

$$\omega = \varphi_+(z)\varphi_-(z)\mathrm{d}z\,, \tag{5.4.61}$$

where

$$\varphi_+(z) = \frac{\prod_{j=1}^{2N}\left(z - q_j^+\right)}{\prod_{k=1}^{2N}(z - p_k)}\,, \tag{5.4.62}$$

and

$$\varphi_-(z) = \frac{\prod_{j=1}^{2N}\left(z - q_j^-\right)}{\prod_{k=1}^{2N}(z - p_k)}\,. \tag{5.4.63}$$

We also make use of the expressions

$$\begin{aligned}
\varphi_{+,i}(z) &= (z - p_i)\,\varphi_+(z)\,,\\
\varphi_{-,i}(z) &= (z - p_i)\,\varphi_-(z)\,.
\end{aligned} \tag{5.4.64}$$

The action for these models as originally derived by [24] is given in (2.2.23). In what follows, we shall replace g_i^{-1} by g_i and replace $k/4\pi$ by $i/\hbar$ in (2.2.23).

Now, colliding a zero, q_i^- (q_i^+), of the twist function, where A_- (A_+) has a pole, with a double pole, p_k, of the twist function results in a simple pole where A_+ (A_-) satisfies a chiral Dirichlet boundary condition. We shall collide the zeroes labelled q_j^+ for $j = 1,\ldots,N$ with the double poles p_k for $k = N+1,\ldots 2N$, and collide the zeroes labelled q_j^- for $j = 1,\ldots,N$ with the double poles p_k for $k = 1,\ldots N$. The former results in N simple poles which can be identified with $z_{+,\alpha}$ and the latter results in N simple poles which can be identified with $z_{-,\alpha}$. Renaming the remaining N zeroes labelled q_j^- as $\zeta_{+,\alpha}$ and the remaining N zeroes labelled q_j^+ as $\zeta_{-,\alpha}$, we can retrieve a twist function of the form (5.4.56) with chiral and anti-chiral Dirichlet boundary conditions imposed as in (5.4.47). In particular, in this limit, we find that

$$\varphi_+(z) \to \tilde{\varphi}_+(z) \tag{5.4.65}$$

and

$$\varphi_-(z) \to \tilde{\varphi}_-(z) . \tag{5.4.66}$$

In addition, in this limit,

$$\varphi_{+,\alpha}(z_{+,\alpha}) \to 0 ,$$
$$\varphi_{-,\alpha}(z_{-,\alpha}) \to 0 , \tag{5.4.67}$$

and

$$\varphi'_{+,\alpha}(z_{+,\alpha}) \to \tilde{\varphi}_+(z_{+,\alpha}) ,$$
$$\varphi'_{-,\alpha}(z_{-,\alpha}) \to \tilde{\varphi}_-(z_{-,\alpha}) . \tag{5.4.68}$$

Once we understand the limit, we can write down the action of the bosonic dual of interest as a limit of the action of the coupled WZW models given in (2.2.23), leading to the action:

$$S[g_{+,\alpha}, g_{-,\beta}]$$

$$= -\frac{i}{\hbar} \sum_{\alpha=1}^{N} \left(-\int_{\Sigma} d\sigma^+ d\sigma^- \tilde{\varphi}_{-,\alpha}(z_{+,\alpha}) \tilde{\varphi}_+(z_{+,\alpha}) \mathrm{Tr}\left(g_{+,\alpha}^{-1} \partial_+ g_{+,\alpha} g_{+,\alpha}^{-1} \partial_- g_{+,\alpha} \right) \right.$$

$$+ \int_{\Sigma} d\sigma^+ d\sigma^- \tilde{\varphi}_{+,\alpha}(z_{-,\alpha}) \tilde{\varphi}_-(z_{-,\alpha}) \mathrm{Tr}\left(g_{-,\alpha}^{-1} \partial_+ g_{-,\alpha} g_{-,\alpha}^{-1} \partial_- g_{-,\alpha} \right)$$

$$+ \int_{\Sigma} d\sigma^+ d\sigma^- \sum_{\beta=1}^{N} \frac{\tilde{\varphi}_{+,\beta}(z_{-,\beta})}{z_{+,\alpha} - z_{-,\beta}} \tilde{\varphi}_{-,\alpha}(z_{+,\alpha}) \mathrm{Tr}\left(g_{-,\beta}^{-1} \partial_+ g_{-,\beta} g_{+,\alpha}^{-1} \partial_- g_{+,\alpha} \right)$$

$$- \int_{\Sigma} d\sigma^+ d\sigma^- \sum_{\beta=1}^{N} \frac{\tilde{\varphi}_{-,\beta}(z_{+,\beta})}{z_{-,\alpha} - z_{+,\beta}} \tilde{\varphi}_{+,\alpha}(z_{-,\alpha}) \mathrm{Tr}\left(g_{+,\beta}^{-1} \partial_- g_{+,\beta} g_{-,\alpha}^{-1} \partial_+ g_{-,\alpha} \right)$$

$$- \int_{\Sigma \times \mathbb{R}_+} \tilde{\varphi}_{-,\alpha}(z_{+,\alpha}) \tilde{\varphi}_+(z_{+,\alpha}) I_{\mathrm{WZ}}[\hat{g}_{+,\alpha}]$$

$$\left. - \int_{\Sigma \times \mathbb{R}_+} \tilde{\varphi}_{+,\alpha}(z_{-,\alpha}) \tilde{\varphi}_-(z_{-,\alpha}) I_{\mathrm{WZ}}[\hat{g}_{-,\alpha}] \right)$$

$$= - \sum_{\alpha=1}^{N} \left[k_{+,\alpha} S_{\mathrm{WZW}}[g_{+,\alpha}] + k_{-,\alpha} S_{\mathrm{WZW}}[g_{-,\alpha}^{-1}] \right.$$

$$\left. + \sum_{\beta=1}^{N} \rho_{\alpha\beta} \int_{\Sigma} d^2\sigma \, \mathrm{Tr}\left(j_+^{(\alpha)} j_-^{(\beta)} \right) \right], \tag{5.4.69}$$

where

$$\rho_{\alpha\beta} = \frac{2i}{\hbar} \frac{\tilde{\varphi}_{+,\alpha}(z_{-,\alpha})\tilde{\varphi}_{-,\beta}(z_{+,\beta}) + \tilde{\varphi}_{-,\beta}(z_{+,\beta})\tilde{\varphi}_{+,\alpha}(z_{-,\alpha})}{z_{-,\alpha} - z_{+,\beta}}, \qquad (5.4.70)$$

and

$$\tilde{\varphi}_{\pm,\alpha}(z) = (z - z_{\mp,\alpha})\tilde{\varphi}_{\pm}(z), \qquad j_{\pm}^{(\alpha)}(\tau,\sigma) = g_{\mp,\alpha}^{-1}\partial_{\pm}g_{\mp,\alpha}. \qquad (5.4.71)$$

In addition, the Lax connection is given by

$$\mathcal{L}_{\pm}(\tau,\sigma,z) = \sum_{\alpha=1}^{N} \frac{\tilde{\varphi}_{\pm,\alpha}(z_{\mp,\alpha})}{\tilde{\varphi}_{\pm,\alpha}(z)} j_{\pm}^{(\alpha)}(\tau,\sigma), \qquad (5.4.72)$$

and satisfies the boundary condition (5.4.52) since

$$\frac{1}{\varphi_{\pm,\alpha}(z_{\mp,\beta})} = \delta_{\alpha,\beta}\frac{1}{\varphi_{\pm,\alpha}(z_{\mp,\alpha})}. \qquad (5.4.73)$$

In particular, the Lax connection (5.4.72) arises in the limit described in the previous paragraph from the Lax connection derived in Ref. [24].

Note that at the path integral level, there are further corrections to the levels $k_{+,\alpha}$ and $k_{-,\alpha}$ in (5.4.69). This follows since the path integral measure for the 4d CS gauge fields in (5.4.55) can be split into measures for the gauge fields at the poles of ω and their bulk complement:

$$\int \mathcal{DL} = \int \mathcal{DL}_{\text{bulk}} \prod_{\alpha=1}^{N} \mathcal{DL}_{-}|_{z_{+,\alpha}} \prod_{\beta=1}^{N} \mathcal{DL}_{+}|_{z_{+,\beta}}. \qquad (5.4.74)$$

The imposition of the bulk equation of motion removes the 4d CS kinetic term from the path integral. The path integral over bulk components of the gauge field can then be evaluated to give a constant. Transforming the remaining path integral measures to path integral measures for the edge modes using (5.4.52) incurs a Jacobian, whose effect is to shift the levels of the WZW models by $-h^{\vee}$ [136].

The resulting path integral is the dual of a generalized multi-flavor massless Thirring model (the generalization coming from having multiple surface defects). Indeed, in the initial setup, taking $\hbar$ to be small such that ω_{eff} can be approximated by $\omega = dz$, and integrating out the 4d CS gauge fields in the $A_{\bar{z}} = 0$ gauge gives us an integrable field theory with the action

$$\frac{1}{\hbar_{2d}} \int_\Sigma \mathrm{d}^2\sigma \left(\sum_{\alpha=1}^{N} \sum_{I=1}^{k_{+,\alpha}} \psi^{*I}_{L,\alpha} i\partial_- \psi^I_{L,\alpha} + \sum_{\beta=1}^{N} \sum_{J=1}^{k_{-,\beta}} \psi^{*J}_{R,\beta} i\partial_+ \psi^J_{R,\beta} \right.$$

$$+ \frac{\hbar_{2d}}{4} \sum_{\alpha=1}^{N} \sum_{\beta=1}^{N} \frac{i\hbar}{z_{+,\alpha} - z_{-,\beta}} \left(\sum_{I=1}^{k_{+,\alpha}} \psi^{*I}_{L,\alpha} i\rho(T^a) \psi^I_{L,\alpha} \right)$$

$$\left. \times \left(\sum_{J=1}^{k_{-,\beta}} \psi^{*J}_{R,\beta} i\rho(T_a) \psi^J_{R,\beta} \right) \right). \qquad (5.4.75)$$

Hence, the bosonization duality of integrable quantum field theories that has been derived is an equivalence between the path integrals for the generalized multi-flavor massless Thirring model (5.4.75) and the coupled WZW models (5.4.69). We can check that in the case of $N = 1$, the bosonic dual is of the form given in (5.4.24), and we shall show this explicitly after deriving the second bosonization duality.

We shall now go on to derive the second bosonization duality from 4d Chern–Simons theory. We shall restrict ourselves here to the case of a single chiral and a single anti-chiral defect. To be precise, we shall utilize the fact that the edge mode defects can be further dualized to WZW defects, which can be defined as path integrals over gauged WZW actions.

Let us first explain how a WZW defect is equivalent to an edge mode defect, which is a classical result of Refs. [135, 136]. One type of WZW defect can be defined as the path integral

$$\widehat{Z}^b_+[A] = \int \mathcal{D}\mathcal{G}_{(+)} \exp\left(i S^b_+[A_-, \mathcal{G}_{(+)}] \right),$$

$$S^b_+[A, \mathcal{G}_{(+)}] = N_F S_{\mathrm{WZW}}[\mathcal{G}_{(+)}] - \frac{N_F}{\pi} \int_{\Sigma \times \{z_+\}} \mathrm{d}^2\sigma \, \mathrm{Tr}(A_- \partial_+ \mathcal{G}_{(+)} \mathcal{G}_{(+)}^{-1})$$

$$- \frac{N_F}{2\pi} \int_{\Sigma \times \{z_+\}} \mathrm{d}^2\sigma \, \mathrm{Tr}(A_+ A_-). \qquad (5.4.76)$$

If we parametrize the component A_- of the gauge field by

$$A_-|_{z_+} = g^{-1} \partial_- g \qquad (5.4.77)$$

and use the Polyakov–Wiegmann identity

$$S_{\mathrm{WZW}}[g_{(+)}\mathcal{G}_{(+)}] = S_{\mathrm{WZW}}[g_{(+)}] + S_{\mathrm{WZW}}[\mathcal{G}_{(+)}]$$

$$- \frac{1}{\pi} \int_{\Sigma \times \{z_+\}} \mathrm{d}^2\sigma \mathrm{Tr}\left(A_- \partial_+ \mathcal{G}_{(+)} \mathcal{G}_{(+)}^{-1} \right), \qquad (5.4.78)$$

we can then rewrite the path integral in (5.4.76) as

$$\widehat{Z}^b_+[A] = \int \mathcal{D}\mathcal{G}_{(+)} \exp\left(iN_F\, S_{\mathrm{WZW}}[g_{(+)}\mathcal{G}_{(+)}] - iN_F\, S_{\mathrm{WZW}}[g_{(+)}] \right.$$
$$\left. - \frac{N_F}{2\pi} \int_{\Sigma \times \{z_+\}} \mathrm{d}^2\sigma \, \mathrm{Tr}(A_+ A_-) \right)$$
$$= \exp\left(-iN_F\, S^{\mathrm{edge}}_+[g_{(+)}, A] \right) \int \mathcal{D}\mathcal{G}_{(+)} \exp\left(iN_F\, S_{\mathrm{WZW}}[\mathcal{G}_{(+)}] \right),$$

$$(5.4.79)$$

where in the second equality we utilized the invariance of the integration measure with respect to a left action, i.e. $\mathcal{D}(g_{(+)}\mathcal{G}_{(+)}) = \mathcal{D}\mathcal{G}_{(+)}$. In other words, the edge mode defect and the WZW defect are equivalent up to a constant that can be absorbed into the path integral measure.

We can also consider the anti-chiral edge mode defect action, and relate it to a WZW defect of the form

$$S^b_-\left[A, \mathcal{G}_{(-)}; N_F\right] = N_F S_{\mathrm{WZW}}\left[\mathcal{G}^{-1}_{(-)}\right]$$
$$+ \int_{\Sigma \times \{z_-\}} \mathrm{d}^2\sigma \, \mathrm{Tr}\left(A_+ \frac{N_F}{\pi} \partial_- \mathcal{G}_{(-)} \mathcal{G}^{-1}_{(-)} \right) \quad (5.4.80)$$
$$- \frac{N_F}{2\pi} \int_{\Sigma \times \{z_-\}} \mathrm{d}^2\sigma \, \mathrm{Tr}\left(A_+ A_- \right),$$

in an analogous manner.

We are now in a position to derive the second bosonization by considering 4d CS theory coupled to a chiral and anti-chiral WZW defect on $C = \mathbb{CP}^1$. The resulting 4d–2d system is

$$Z^{\mathrm{Th}}_b(z_\pm) = \int \mathcal{D}A\mathcal{D}\mathcal{G}_{(\pm)} \exp\left(iS^{\mathrm{4d-2d}}[A, \mathcal{G}_{(\pm)}] \right),$$

$$(5.4.81)$$

$$S^{\mathrm{4d-2d}}[A, \mathcal{G}_{(\pm)}] = S_{\mathrm{CS}}[A] + S^b_+[A, \mathcal{G}_{(+)}] + S^b_-[A, \mathcal{G}^{-1}_{(-)}].$$

To ensure gauge invariance, we require the usual mechanism of anomaly inflow whereby the meromorphic 1-form takes the form,

$$\omega_{\mathrm{eff}} = \mathrm{d}z - \frac{\hbar}{4\pi i}\left(\frac{k_+}{z - z_+} - \frac{k_-}{z - z_-} \right) \mathrm{d}z = \frac{(z - \zeta_1)(z - \zeta_2)}{(z - z_+)(z - z_-)} \mathrm{d}z, \quad (5.4.82)$$

where we set $k := k_+ = k_- = N_F$.

We shall now derive the bulk and the boundary equations of motion of (5.4.81). Taking a variation of the gauge field, we obtain the bulk equations of motion

$$F_{+-} = 0, \qquad \omega_{\text{eff}}\, F_{\bar{z}\pm} = 0, \tag{5.4.83}$$

and the boundary equations of motion localized at the poles $\{z_\pm, \infty\}$ of ω_{eff}

$$-\frac{i}{\hbar}\epsilon^{\alpha\beta}\xi_\infty \partial_{\xi_\infty} \text{Tr}(A_\alpha \delta A_\beta)|_{\xi_\infty} - \frac{k}{\pi}\text{Tr}(\delta A_-\,(A_+ + \partial_+ \mathcal{G}_{(+)}\mathcal{G}_{(+)}^{-1}))|_{z=z_+}$$
$$+ \frac{k}{\pi}\text{Tr}(\delta A_+\,(A_- - \mathcal{G}_{(-)}^{-1}\partial_-\mathcal{G}_{(-)}))|_{z=z_-} = 0, \tag{5.4.84}$$

where $F_{\mu\nu} = \partial_\mu A_\nu - \partial_\nu A_\mu + [A_\mu, A_\nu]$ and $\xi_\infty = 1/z$. We shall impose the following values of the gauge field A at the poles:

$$A|_{z=\infty} = 0, \qquad A_+|_{z=z_+} = -\partial_+ \mathcal{G}_{(+)}\mathcal{G}_{(+)}^{-1}, \qquad A_-|_{z=z_-} = \mathcal{G}_{(-)}^{-1}\partial_-\mathcal{G}_{(-)}. \tag{5.4.85}$$

The Lax connection can now be obtained by performing the formal gauge transformation,

$$A = \widehat{g}\mathcal{L}\widehat{g}^{-1} - \mathrm{d}\widehat{g}\widehat{g}^{-1}, \qquad \mathcal{G}_{(+)} = \widehat{g}|_{z=z_+}\cdot g_{(+)}, \qquad \mathcal{G}_{(-)} = g_{(-)}\cdot \widehat{g}^{-1}|_{z=z_-}, \tag{5.4.86}$$

where $\widehat{g}$ is a $G_{\mathbb{C}}$-valued function on $\Sigma \times C$. Here, we will make the usual choice of $\mathcal{L}_{\bar{z}} = 0$. The bulk equations of motion (5.4.83) then reduce to

$$\partial_+ \mathcal{L}_- - \partial_- \mathcal{L}_+ + [\mathcal{L}_+, \mathcal{L}_-] = 0, \qquad \omega_{\text{eff}}\, \partial_{\bar{z}}\mathcal{L}_\pm = 0, \tag{5.4.87}$$

and the boundary conditions (5.4.85) are translated to

$$\mathcal{L}|_{z=\infty} = 0, \qquad \mathcal{L}_+|_{z=z_+} = -\partial_+ g_{(+)}g_{(+)}^{-1}, \qquad \mathcal{L}_-|_{z=z_-} = g_{(-)}^{-1}\partial_- g_{(-)}. \tag{5.4.88}$$

The second equation of (5.4.87) can be interpreted to mean that $\mathcal{L}_\pm$ are meromorphic functions with poles at the zeroes of ω_{eff}, and the order of its poles is less than or equal to the order of the corresponding zeroes of ω_{eff}. We can thus find a solution to the second equation of (5.4.87) satisfying the conditions (5.4.88) given by

$$\mathcal{L} = -\frac{z_+ - \zeta_+}{z - \zeta_+}\partial_+ g_{(+)}g_{(+)}^{-1}\mathrm{d}\sigma^+ + \frac{z_- - \zeta_-}{z - \zeta_-}g_{(-)}^{-1}\partial_- g_{(-)}\mathrm{d}\sigma^-, \tag{5.4.89}$$

where $\zeta_\pm$ are simple zeroes of ω_{eff}

$$\zeta_\pm = \frac{1}{2}\left(z_+ + z_- \pm \sqrt{(z_+ - z_-)\left(z_+ - z_- + \frac{i\hbar k}{\pi}\right)}\right).\qquad(5.4.90)$$

We can then proceed to derive the corresponding 2d action by using the classical solution (5.4.89). The 4d-2d action (5.4.81) subject to the formal gauge transformation (5.4.86) is

$$S^{\text{4d–2d}} = S_{\text{CS}}[\mathcal{L}] + S^b_+[\mathcal{L}, g_{(+)}] + S^b_-[\mathcal{L}, g_{(-)}].\qquad(5.4.91)$$

Here, the first term vanishes due to the bulk equation of motion (5.4.83), i.e., we have

$$S_{\text{CS}}[\mathcal{L}] = \frac{1}{2\pi\hbar}\int_{\Sigma\times\mathbb{CP}^1}\omega_{\text{eff}}\wedge\text{Tr}\left(\mathcal{L}\wedge d\mathcal{L}\right) = 0.\qquad(5.4.92)$$

Note that, unlike the standard disorder defect case, there are no contributions from the 4d CS action to the 2d integrable field theory. The subtitution of (5.4.89) into (5.4.91) leads to the bosonic dual of the form

$$\begin{aligned}
S_{\text{2d}} &= S^b_+[\mathcal{L}, g_{(+)}] + S^b_-[\mathcal{L}, g_{(-)}^{-1}]\\[2mm]
&= S^{(k)}_{\text{WZW}}[g_{(+)}] + S^{(k)}_{\text{WZW}}[g_{(-)}] - \frac{k}{\pi}\int_{\Sigma\times\{z_+\}}d^2\sigma\,\text{Tr}(\mathcal{L}_-\partial_+ g_{(+)}g_{(+)}^{-1})\\[2mm]
&\quad + \frac{k}{\pi}\int_{\Sigma\times\{z_-\}}d^2\sigma\,\text{Tr}(\mathcal{L}_+ g_{(-)}^{-1}\partial_- g_R) - \frac{k}{2\pi}\int_{\Sigma\times\{z_+\}}d^2\sigma\,\text{Tr}(\mathcal{L}_+\mathcal{L}_-)\\[2mm]
&\quad - \frac{k}{2\pi}\int_{\Sigma\times\{z_-\}}d^2\sigma\,\text{Tr}(\mathcal{L}_+\mathcal{L}_-)\\[2mm]
&= S^{(k)}_{\text{WZW}}[g_{(+)}] + S^{(k)}_{\text{WZW}}[g_{(-)}]\\[2mm]
&\quad - \frac{8i}{\hbar}\rho_{+-}\int_\Sigma d^2\sigma\,\text{Tr}(\partial_+ g_{(+)}g_{(+)}^{-1}g_{(-)}^{-1}\partial_- g_{(-)}),
\end{aligned}$$

$$\qquad(5.4.93)$$

where ρ_{+-} is given by

$$\rho_{+-} = -\frac{(z_+ - \zeta_+)(z_- - \zeta_-)}{2(z_+ - z_-)} = \frac{(z_+ - z_- - (\zeta_+ - \zeta_-))^2}{8(z_+ - z_-)}.\qquad(5.4.94)$$

Having derived two bosonizations of the multi-flavor massless Thirring model from 4d Chern–Simons theory, we are now in a position to compare the results with those of the 2d derivation. Since the massless Thirring model (5.4.75) is obtained in the small $\hbar$ limit, we ought to study the bosonic models obtained via 4d Chern–Simons theory in this limit.

We thus study the coupling constants of (5.4.69) (for $N = 1$) and (5.4.93) in the limit where $\hbar$ is small. In (5.4.69), for the case of $N = 1$, the coupling takes the form

$$
\begin{aligned}
-\rho_{11} &= -\frac{4i}{\hbar} \frac{(z_- - \zeta_-)(z_+ - \zeta_+)}{z_- - z_+} \\
&= -\frac{8i}{\hbar} \left(-\frac{(z_- + \sqrt{(z_- - z_+)^2} - z_+)^2}{8(z_- - z_+)} \right. \\
&\quad + \frac{(z_- + \sqrt{(z_- - z_+)^2} - z_+)\, i\hbar k}{8\sqrt{(z_- - z_+)^2}} \frac{1}{\pi} + \frac{\left(\frac{(i\hbar k)}{\pi}\right)^2}{32\sqrt{(z_- - z_+)^2}} \left. + O(\hbar^3) \right) \\
&= \frac{4}{i\hbar}(z_+ - z_-) + O(\hbar^0),
\end{aligned}
\tag{5.4.95}
$$

if we choose the branch of the square root to be $\sqrt{(z_- - z_+)^2} = z_- - z_+$. The leading order term is thus G^{-1}, in agreement with (5.4.24) for $a = 0$. In (5.4.93), the coupling takes the form

$$
\begin{aligned}
-\frac{8i}{\hbar}\rho_{+-} &= -\frac{8i}{\hbar}\left(-\frac{(z_+ - \zeta_+)(z_- - \zeta_-)}{2(z_+ - z_-)} \right) \\
&= -\frac{8i}{\hbar}\left(-\frac{(z_- + \sqrt{(z_- - z_+)^2} - z_+)^2}{8(z_- - z_+)} \right. \\
&\quad + \frac{(z_- + \sqrt{(z_- - z_+)^2} - z_+)\, i\hbar k}{8\sqrt{(z_- - z_+)^2}} \frac{1}{\pi} + \frac{\left(\frac{(i\hbar k)}{\pi}\right)^2}{32\sqrt{(z_- - z_+)^2}} \left. + O(\hbar^3) \right) \\
&= \frac{k^2}{\pi^2} \frac{1}{4} \frac{i\hbar}{z_+ - z_-} + O(\hbar^2),
\end{aligned}
\tag{5.4.96}
$$

if we choose $\sqrt{(z_- - z_+)^2} = z_+ - z_-$. The leading order term is thus $(k^2/\pi^2)G$, in agreement with (5.4.29) for $a = 0$. Hence, there is agreement between the bosonization derivations performed using 4d CS and those in the literature (for $a = 0$) in the small $\hbar$ (semi-classical) limit, but only after making a particular choice of branch for $\sqrt{(z_- - z_+)^2}$. Such a choice could be justified by demanding the convergence of the path integral of the bosonized theories, which ought to restrict the form of the coupling constants.

Thus, the procedure of dualization of defects in 4d Chern–Simons theory well describes both the bosonizations of the multi-flavor massless Thirring model in the semi-classical limit, and moreover, furnishes a boson-boson duality between the two models.

Appendices

Appendix A

Numerical Factor for Two-Loop Gauge Anomaly for the Yangian

Following Ref. [?], we shall evaluate the numerical factor associated with the diagrams in Figure 1.12. To this end we consider the Wilson line positioned along a straight line at $y = z = 0$. We denote the positions of the vertices on the Wilson line as p_1, p_2, p_3, and the positions of the internal vertices as $v_1, v_2, v_3 \in \mathbb{R} \times \mathbb{C}$. The coordinates for vertex v_i are parametrized by x_i, y_i, z_i for $i = 1, 2, 3$. The labeling of the vertices for the first diagram in Figure 1.12 is depicted in Figure A.1.

The complete amplitude, including all the relevant diagrams and the color factor, can be expressed (with wedge products implicit) as

$$\frac{1}{2}\left(\frac{i}{2\pi}\right)^3 \int_{p_1,p_2,p_3 \in \mathbb{R}} \int_{v_1,v_2,v_3} P(0,v_1)\mathrm{d}z_1 A_1^a P(v_1,v_2)\mathrm{d}z_2 P(v_2,p_2)$$

$$\times P(v_2,v_3)\mathrm{d}z_3 A_3^b P(v_3,p_3) f^{afc} f^{fgd} f^{gbe}\{\rho(t^c),\rho(t^d),\rho(t^e)\}, \quad \text{(A.0.1)}$$

where A_2, A_3 represent the external gauge fields, and $P(a,b)$ denotes the propagator 2-form connecting two vertices.

To regulate the integral, a point-splitting regulator is adopted in which the domain of integration is confined to the region where $p_3 - p_1 \geq \epsilon$ under the assumption that the three points are arranged in ascending order such that $p_1 < p_2 < p_3$. To ensure that this is an effective regulator, it must be demonstrated that the integral converges absolutely within this domain. Since we are only interested in UV divergences, we can further narrow down the integration domain to a region where all vertices are enclosed within some ball around the origin.

To establish convergence, we note that the absolute value of the propagator is bounded as

$$|P| \leq (x^2 + y^2 + |z\bar{z}|)^{-\frac{3}{2}}, \quad \text{(A.0.2)}$$

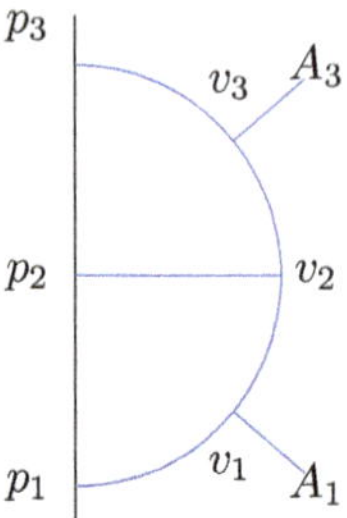

Figure A.1.   The Feynman diagram with vertices labeled. The A_i on the external lines indicate gauge fields.

which follows since the denominator of P is $(x^2 + y^2 + |z|^2)^{-2}$, while the numerator is a linear function of the variables $x, y, \bar{z}$, and a linear function on $\mathbb{R}^4$ is bounded in absolute value by a multiple of $(x^2 + y^2 + |z|^2)^{\frac{1}{2}}$.

To verify convergence, it suffices to demonstrate that the integral converges when each propagator $P(a, b)$ is replaced by $d(a, b)^{-3}$, where d represents the Euclidean distance

$$d(v_i, v_j) = \sqrt{x_{ij}^2 + y_{ij}^2 + |z_{ij}|^2}, \qquad (\text{A.0.3})$$

where $x_{ij} = x_i - x_j$, $y_{ij} = y_i - y_j$ and $z_{ij} = z_i - z_j$.

By bounding the external gauge fields with a constant, we can omit them from the integral. Hence, we are left with the following integral:

$$\int_{p_2, p_3 > \epsilon, v_1, v_2, v_3} \frac{1}{d(p_3, v_3)^3 d(v_2, v_3)^3 d(v_2, p_2)^3 d(v_2, v_1)^3 d(v_1, p_1 = 0)^3}, \qquad (\text{A.0.4})$$

where, due to overall translation invariance, we set $p_1 = 0$. The most divergent region in the integration domain occurs when v_1, v_2, v_3, p_2 are all near 0 (or all near p_3). Focusing on this region is the same as considering p_3 to be very distant from the other points, which permits us to drop the $d(p_3, v_3)^{-3}$ term since it is nonsingular. Consequently, we can consider the integral:

$$\int_{p_2, v_1, v_2, v_3} \frac{1}{d(0, v_1)^3 d(v_1, v_2)^3 d(v_2, p_1)^3 d(v_2, v_3)^3} \mathrm{d}^4 v_1 \mathrm{d}^4 v_2 \mathrm{d}^4 v_3 \mathrm{d} p_2. \qquad (\text{A.0.5})$$

Here, there are a total of 13 integration variables, and the integrand has a weight of -12 when all the variables are scaled. Thus, the integral converges absolutely in a domain where the integration variables v_i and p_2 are bounded from above.

Now, taking into account all three diagrams that contribute to the anomaly, the regularized amplitude turns out to be

$$\frac{1}{2}\left(\frac{i}{2\pi}\right)^3 \int_{\substack{p_i,v_j \\ p_{\min}<p_{\max}-\epsilon}} P(0,v_1)\mathrm{d}z_1 A_1^a P(v_1,v_2)\mathrm{d}z_2 P(v_2,p_2)P(v_2,v_3)\mathrm{d}z_3 A_3^b$$

$$\times P(v_3,p_3)f^{afc}f^{fgd}f^{gbe}\{\rho(t^c),\rho(t^d),\rho(t^e)\}, \tag{A.0.6}$$

where $p_{\min}, p_{\max}$ are the coordinates on the Wilson line with the minimum and maximum value, respectively. The factor of $\frac{1}{2}$ arises because there are 6 possible ways to arrange three points on a line, but we only need to consider 3 distinct diagrams.

To investigate the failure of gauge invariance in the integral, perform a linearized BRST transformation $A \mapsto A + \mathrm{d}c$ on the external gauge field A, where c denotes the fermionic ghost field used for gauge-fixing. By employing integration by parts, the resulting expression can be written as a sum of terms where we apply the exterior derivative d to one of the propagators or integrate over one of the boundary components of the integration domain. One can demonstrate that all such terms vanish except for the term where one integrates over the boundary component where $p_{\max} = p_{\min} + \epsilon$ (details can be found in [?]).

The integral describing the anomaly can then be written as:

$$\left(\frac{i}{2\pi}\right)^3 \int_{p_i,v_j\ p_{\min}=p_{\max}-\epsilon} P(p_1,v_1)\mathrm{d}z_1 c_1^a P(v_1,v_2)\mathrm{d}z_2$$

$$\times P(v_2,p_2)P(v_2,v_3)\mathrm{d}z_3 A_3^b P(v_3,p_3)f^{afc}f^{fgd}f^{gbe}\{\rho(t^c),\rho(t^d),\rho(t^e)\}, \tag{A.0.7}$$

where the external gauge field on the vertex v_1 has been replaced by the ghost field, c. The factor of $\frac{1}{2}$ has been canceled out by the factor of 2 originating from the two possible external lines connected to the ghost field.

The integral can be expressed as a sum of six terms corresponding to the six possible orderings of the points on the Wilson line. However, a reflection in a plane orthogonal to the Wilson line reveals that there are only three independent integrals. We can choose these three integrals to be the ones where the points p_1, p_2, p_3 are cyclically ordered. Hence, we need to compute:

$$2\left(\frac{i}{2\pi}\right)^3 \int_{p_1<p_2<p_3=p_1+\epsilon} \int_{v_1,v_2}$$

$$\times\, P(p_1,v_1)\mathrm{d}z_1 c_1^a P(v_1,v_2)\mathrm{d}z_2 P(v_2,p_2)P(v_2,v_3)\mathrm{d}z_3 A_3^b P(v_3,p_3)$$

$$+\,2\left(\frac{i}{2\pi}\right)^3 \int_{p_2<p_3<p_1=p_2+\epsilon} \int_{v_1,v_2}$$

$$\times\, P(p_1,v_1)\mathrm{d}z_1 c_1^a P(v_1,v_2)\mathrm{d}z_2 P(v_2,p_2)P(v_2,v_3)\mathrm{d}z_3 A_3^b P(v_3,p_3)$$

$$+\,2\left(\frac{i}{2\pi}\right)^3 \int_{p_3<p_1<p_2=p_3+\epsilon} \int_{v_1,v_2}$$

$$\times\, P(p_1,v_1)\mathrm{d}z_1 c_1^a P(v_1,v_2)\mathrm{d}z_2 P(v_2,p_2)P(v_2,v_3)\mathrm{d}z_3 A_3^b P(v_3,p_3).$$

$$(\mathrm{A.0.8})$$

The factor of 2 in this context arises because each integral represents the contribution from one of the two possible orderings of the points p_i.

To detect the anomaly, one utilizes the knowledge that the anomaly must involve a z-derivative for each external line while excluding other derivatives. In this regard, one can assume that the external ghost field c is z, and the external gauge field A is z multiplied by a delta function at $x=0$, i.e., $A=z\delta_{x=0}$.

By fixing the external gauge field, we effectively determine the position of the vertex v_3. By translation invariance, we can instead choose to fix the position of the vertex p_1 to be $p_1=0$.

In what follows, we shall verify that the anomaly is given by (1.5.20). Let us proceed with the evaluation of the integral for the diagram shown in Figure A.1, where $p_1<p_2<p_3$. We will integrate over the region where $p_3=p_1+\epsilon$, where $p_1=0$. Moreover, we will denote p_2 by p.

The integral can be expressed as

$$2\left(\frac{i}{2\pi}\right)^3 \int_{p=0}^{\epsilon} \int_{v_1,v_2,v_3} P(0,v_1)\wedge z_1\mathrm{d}z_1 \wedge P(v_1,v_2)$$

$$\wedge\, \mathrm{d}z_2 \wedge P(v_2,p)\wedge P(v_2,v_3)\wedge z_3\mathrm{d}z_3 \wedge P(v_3,\epsilon). \qquad (\mathrm{A.0.9})$$

Let us begin with the evaluation of the integral relevant for the v_1 variable in the expression (A.0.9):

$$\int_{v_1} P(v_0,v_1)\wedge \mathrm{d}z_1(z_1) \wedge P(v_1,v_2). \qquad (\mathrm{A.0.10})$$

Here, we have introduced a temporary vertex labeled v_0 with coordinates $v_0 = (x_0, y_0, z_0, \bar{z}_0)$. Later on, we will set these coordinates to zero.

Using the explicit expression for the propagator in (1.3.7), we have

$$P(v_i, v_j) = \frac{1}{2\pi} \frac{x_{ij}\mathrm{d}y_{ij} \wedge \mathrm{d}\bar{z}_{ij} - y_{ij}\mathrm{d}x_{ij} \wedge \mathrm{d}\bar{z}_{ij} + 2\bar{z}_{ij}\mathrm{d}x_{ij} \wedge \mathrm{d}y_{ij}}{d(v_i, v_j)^4}, \quad (A.0.11)$$

where the Euclidean distance $d(v_i, v_j)$ between two points v_i, v_j was defined in (A.0.3).

The wedge product of the propagators $P(v_0, v_1)$ and $P(v_1, v_2)$ can be calculated as:

$$P(v_0, v_1) \wedge P(v_1, v_2)$$

$$= \frac{1}{(2\pi)^2}\mathrm{d}x_1\mathrm{d}y_1\mathrm{d}\bar{z}_1 \Bigg(x_{01}y_{12}\mathrm{d}\bar{z}_2 - 2x_{01}\bar{z}_{12}\mathrm{d}y_2 + 2y_{01}\bar{z}_{12}\mathrm{d}x_2$$

$$- y_{01}x_{12}\mathrm{d}\bar{z}_2 + 2\bar{z}_{01}x_{12}\mathrm{d}y_2 - 2\bar{z}_{01}y_{12}\mathrm{d}x_2 \Bigg) \frac{1}{d(v_0, v_1)^4 d(v_1, v_2)^4},$$

$$(A.0.12)$$

where we have dropped the terms involving $\mathrm{d}x_0$, $\mathrm{d}y_0$, $\mathrm{d}\bar{z}_0$ since we will not be integrating over the vertex v_0. We thereby observe that

$$P(v_0, v_1) \wedge z_1\mathrm{d}z_1 P(v_1, v_2)$$

$$= \frac{1}{(2\pi)^2}\mathrm{d}x_1\mathrm{d}y_1\mathrm{d}z_1\mathrm{d}\bar{z}_1 \left[\partial_{\bar{z}_0} \frac{1}{d(v_0, v_1)^2 d(v_1, v_2)^4} \right]$$

$$\times (x_{01}y_{12}\mathrm{d}\bar{z}_2 - 2x_{01}\bar{z}_{12}\mathrm{d}y_2 + 2y_{01}\bar{z}_{12}\mathrm{d}x_2 - y_{01}x_{12}\mathrm{d}\bar{z}_2$$

$$+ 2\bar{z}_{01}x_{12}\mathrm{d}y_2 - 2\bar{z}_{01}y_{12}\mathrm{d}x_2). \quad (A.0.13)$$

having set the v_0 coordinates to zero. This is equivalent to

$$P(v_0, v_1) \wedge z_1\mathrm{d}z_1 P(v_1, v_2)$$

$$= \frac{1}{(2\pi)^2}\mathrm{d}x_1\mathrm{d}y_1\mathrm{d}z_1\mathrm{d}\bar{z}_1$$

$$\times \partial_{\bar{z}_0} \left[\frac{\begin{array}{c}(x_{01}y_{12}\mathrm{d}\bar{z}_2 - 2x_{01}\bar{z}_{12}\mathrm{d}y_2 + 2y_{01}\bar{z}_{12}\mathrm{d}x_2 - y_{01}x_{12}\mathrm{d}\bar{z}_2 \\ + 2\bar{z}_{01}x_{12}\mathrm{d}y_2 - 2\bar{z}_{01}y_{12}\mathrm{d}x_2)\end{array}}{d(v_0, v_1)^2 d(v_1, v_2)^4} \right]$$

$$- \frac{1}{(2\pi^2)^2}\mathrm{d}x_1\mathrm{d}y_1\mathrm{d}z_1\mathrm{d}\bar{z}_1 \frac{(2x_{12}\mathrm{d}y_2 - 2y_{12}\mathrm{d}x_2)}{d(v_0, v_1)^2 d(v_1, v_2)^4}. \quad (A.0.14)$$

We can now use integration by parts to show that the integral over $x_1, y_1, z_1, \bar{z}_1$ of the first term on the RHS of (A.0.14) vanishes. One such manipulation is given explicitly by

$$
\int_{v_1} (x_{01}y_{12} - y_{01}x_{12}) \frac{1}{d(v_0, v_1)^2} \frac{1}{d(v_1, v_2)^4}
$$

$$
= \frac{1}{2} \int_{v_1} (-x_{01}) \frac{1}{d(v_0, v_1)^2} \left(\partial_{y_1} \frac{1}{d(v_1, v_2)^2} \right)
$$

$$
+ \frac{1}{2} \int_{v_1} y_{01} \frac{1}{d(v_0, v_1)^2} \left(\partial_{x_1} \frac{1}{d(v_1, v_2)^2} \right)
$$

$$
= \frac{1}{2} \int_{v_1} x_{01} \left(\partial_{y_1} \frac{1}{d(v_0, v_1)^2} \right) \frac{1}{d(v_1, v_2)^2}
$$

$$
- \frac{1}{2} \int_{v_1} y_{01} \left(\partial_{x_1} \frac{1}{d(v_0, v_1)^2} \right) \frac{1}{d(v_1, v_2)^2}
$$

$$
= \int_{v_1} (x_{01}y_{01} - x_{01}y_{01}) \frac{1}{d(v_0, v_1)^4} \frac{1}{d(v_1, v_2)^2} = 0.
$$

Proceeding in this manner, we obtain

$$
-\frac{1}{(2\pi^2)^2} \int_{x_1, y_1, z_1} \mathrm{d}x_1 \mathrm{d}y_1 \mathrm{d}z_1 \mathrm{d}\bar{z}_1 \frac{(2x_{12}\mathrm{d}y_2 - 2y_{12}\mathrm{d}x_2)}{d(v_0, v_1)^2 d(v_1, v_2)^4}. \tag{A.0.15}
$$

This integral can be evaluated using the identity

$$
\frac{1}{A^\alpha B^\beta} = \frac{\Gamma(\alpha + \beta)}{\Gamma(\alpha)\Gamma(\beta)} \int_0^1 \mathrm{d}t \frac{t^{\alpha-1}(1 - t)^{\beta-1}}{(tA + (1 - t)B)^{\alpha+\beta}}. \tag{A.0.16}
$$

with $\alpha = 2, \beta = 1$. This leads us to

$$
-\frac{2}{(2\pi)^2} \int_0^1 \mathrm{d}t\, t \int \mathrm{d}x_1 \mathrm{d}y_1 \mathrm{d}z_1 \mathrm{d}\bar{z}_1 \frac{(2x_{12}\mathrm{d}y_2 - 2y_{12}\mathrm{d}x_2)}{\mathsf{X}^3}, \tag{A.0.17}
$$

where

$$
\mathsf{X} = (1 - t)(x_1^2 + y_1^2 + |z_1|^2) + t(x_{12}^2 + y_{12}^2 + |z_{12}|^2)
$$

$$
= (x_1 - tx_2)^2 + (y_1 - ty_2)^2 + |z_1 - tz_2|^2 + t(1 - t)(x_2^2 + y_2^2 + |z_2|^2). \tag{A.0.18}
$$

We now make the following shifts in integration variables:

$$
\begin{aligned}
x_1 &\to x_1 + tx_2 \\
y_1 &\to y_1 + ty_2 \\
z_1 &\to z_1 + tz_2.
\end{aligned} \tag{A.0.19}
$$

This results in the expression

$$\frac{4}{(2\pi)^2} \int_0^1 \mathrm{d}t\, t(1-t) \int \mathrm{d}x_1 \mathrm{d}y_1 \mathrm{d}z_1 \mathrm{d}\bar{z}_1$$

$$\times \frac{(x_2 \mathrm{d}y_2 - y_2 \mathrm{d}x_2)}{\left(x_1^2 + y_1^2 + |z_1|^2 + t(1-t)\left(x_2^2 + y_2^2 + |z_2|^2\right)\right)^3}, \quad (A.0.20)$$

where we have noted that terms that are odd in x_1 or y_1 do not contribute to the integral. We can then evaluate the integrals over x_1 and y_1 to obtain

$$\frac{4}{(2\pi)^2} \frac{\pi}{2} \left(x_2 \mathrm{d}y_2 - y_2 \mathrm{d}x_2\right)$$

$$\times \int_0^1 \mathrm{d}t\, t(1-t) \int \mathrm{d}z_1 \mathrm{d}\bar{z}_1 \frac{1}{\left(|z_1|^2 + t(1-t)\left(x_2^2 + y_2^2 + |z_2|^2\right)\right)^2} \quad (A.0.21)$$

We can then proceed to evaluate the integral over the z_1-plane using the polar coordinates $z_1 = re^{i\theta}$, as well as the integral over t, to obtain

$$\frac{4}{(2\pi)^2} \frac{\pi}{2} (-2i)\pi \left(x_2 \mathrm{d}y_2 - y_2 \mathrm{d}x_2\right) \frac{1}{\left(x_2^2 + y_2^2 + |z_2|^2\right)} \cdot \quad (A.0.22)$$

Equivalently, we have

$$\int_{v_1} P(v_0, v_1) \wedge z_1 \mathrm{d}z_1 \wedge P(v_1, v_2) = \frac{1}{i} \frac{(x_2 \mathrm{d}y_2 - y_2 \mathrm{d}x_2)}{d(v_0, v_2)^2}. \quad (A.0.23)$$

An analogous derivation leads to

$$\int_{v_3} P(v_2, v_3) \wedge z_3 \mathrm{d}z_3 \wedge P(v_3, p_3) = -\frac{1}{i} \frac{((x_2 - \epsilon)\mathrm{d}y_2 - y_2 \mathrm{d}x_2)}{d(v_0, v_2)^2}. \quad (A.0.24)$$

Using (A.0.23) and (A.0.24) in (A.0.9), we obtain

$$2 \left(\frac{i}{2\pi}\right)^3 \frac{1}{2\pi} \int_{p=0}^\epsilon \int_{x,y,z,\bar{z}} \frac{(x\mathrm{d}y - y\mathrm{d}x)(\mathrm{d}z)(y\mathrm{d}\bar{z}\mathrm{d}p)((x - \epsilon)\mathrm{d}y - y\mathrm{d}x)}{d(0, v)^2 d(p, v)^4 d(\epsilon, v)^2},$$

$$(A.0.25)$$

where $v_2 = (x_2, y_2, z_2, \bar{z}_2)$ has been replaced by $v = (x, y, z, \bar{z})$. This is equivalent to

$$2 \frac{i}{(2\pi)^4} \int_{p=0}^\epsilon \int_{x,y,z,\bar{z}}$$

$$\times \frac{\epsilon y^2 \mathrm{d}x \mathrm{d}y \mathrm{d}z \mathrm{d}\bar{z} \mathrm{d}p}{(x^2 + y^2 + |z|^2)((x - p)^2 + y^2 + |z|^2)^2((x - \epsilon)^2 + y^2 + |z|^2)}. \cdot$$

$$(A.0.26)$$

We also need to take into account the remaining two diagrams in Fig. 1.12, which contribute the expressions

$$2\frac{i}{(2\pi)^4}\int_{p=0}^{\epsilon}\int_{x,y,z,\bar{z}}$$

$$\times\frac{\epsilon y^2 \mathrm{d}x\mathrm{d}y\mathrm{d}z\mathrm{d}\bar{z}\mathrm{d}p}{(x^2+y^2+|z|^2)^2((x-p)^2+y^2+|z|^2)((x-\epsilon)^2+y^2+|z|^2)},$$

$$2\frac{i}{(2\pi)^4}\int_{p=0}^{\epsilon}\int_{x,y,z,\bar{z}}$$

$$\times\frac{\epsilon y^2 \mathrm{d}x\mathrm{d}y\mathrm{d}z\mathrm{d}\bar{z}\mathrm{d}p}{(x^2+y^2+|z|^2)((x-p)^2+y^2+|z|^2)((x-\epsilon)^2+y^2+|z|^2)^2}.$$

$$(A.0.27)$$

The sum of the amplitudes of the three diagrams can be written as the following expression:

$$-\frac{i}{(2\pi)^4}\int_{p=0}^{\epsilon}\int_{x,y,z,\bar{z}}(\epsilon y)\frac{\partial}{\partial y}$$

$$\times\frac{\mathrm{d}x\mathrm{d}y\mathrm{d}z\mathrm{d}\bar{z}\mathrm{d}p}{(x^2+y^2+|z|^2)((x-p)^2+y^2+|z|^2)((x-\epsilon)^2+y^2+|z|^2)},$$

$$(A.0.28)$$

which can be rewritten, after integration by parts, as

$$\frac{i}{(2\pi)^4}\int_{p=0}^{\epsilon}\int_{x,y,z,\bar{z}}$$

$$\times\epsilon\frac{\mathrm{d}x\mathrm{d}y\mathrm{d}z\mathrm{d}\bar{z}\mathrm{d}p}{(x^2+y^2+|z|^2)((x-p)^2+y^2+|z|^2)((x-\epsilon)^2+y^2+|z|^2)}.$$

$$(A.0.29)$$

The integration variables can be rescaled to remove the ϵ dependence, to give

$$\frac{i}{(2\pi)^4}\int_{p=0}^{1}\int_{x,y,z,\bar{z}}$$

$$\times\frac{\mathrm{d}x\mathrm{d}y\mathrm{d}z\mathrm{d}\bar{z}\mathrm{d}p}{(x^2+y^2+|z|^2)((x-p)^2+y^2+|z|^2)((x-1)^2+y^2+|z|^2)}.$$

$$(A.0.30)$$

To proceed, we shall employ angular coordinates on the $(y,z,\bar{z})$-plane, with the volume form

$$\mathrm{d}y\mathrm{d}z\mathrm{d}\bar{z}=-8\pi ir^2\mathrm{d}r\mathrm{d}\Omega_{S^2},\qquad(A.0.31)$$

where $d\Omega_{S^2}$ is the volume form on a two-sphere of unit volume. Integrating over this two-sphere gives us

$$\frac{8\pi}{(2\pi)^4} \int_{p=0}^1 dp \int_x dx \int drr^2 \frac{1}{(x^2+r^2)((x-p)^2+r^2)((x-1)^2+r^2)}.$$

$$(A.0.32)$$

Then, performing the r integral leads to

$$\frac{8\pi}{(2\pi)^4}\frac{\pi}{2} \int_{p=0}^1 dp \int_x dx \frac{1}{(|x|+|x-p|)(|x|+|x-1|)(|x-p|+|x-1|)}.$$

$$(A.0.33)$$

The integral (A.0.33) can be evaluated by considering the four cases $x < 0, 0 < x < p, p < x < 1, 1 < x$. In fact, there are only two independent integrals, corresponding to $p < x < 1$ and $1 < x$, which can be deduced by the change of variables $x \to 1 - x$ and $p \to 1 - p$. These two independent integrals, denoted A_{int} and B_{int}, can be evaluated to be

$$A_{int} = \int_{0<p<x<1} \frac{1}{(2x-p)(1-p)} = \frac{\pi^2}{8} , \qquad (A.0.34)$$

$$B_{int} = \int_{0<p<1<x} \frac{1}{(2x-p)(2x-1)(2x-1-p)} = \frac{\pi^2}{24} . \qquad (A.0.35)$$

Hence, we obtain

$$2 \times \frac{8\pi}{(2\pi)^4}\frac{\pi}{2}(A_{int} + B_{int}) = \frac{1}{12} . \qquad (A.0.36)$$

This is precisely the numerical factor of (1.5.20), up to a factor of $\hbar^2$. This factor of $\hbar^2$ arises since, in the Feynman diagrams in Figure 1.12, each propagator contributes a factor of $\hbar$, while each interaction vertex contributes a factor of $\frac{1}{\hbar}$.

Appendix B

Gauge Anomaly of Chiral Surface Defects

In this appendix, we shall describe the gauge anomaly of a two-dimensional chiral surface defect which supports an affine Kac–Moody algebra, that can be embedded in 4d Chern–Simons theory on $\Sigma \times C$, either along the Σ or C direction. The following derivation holds for surface defects supported on either $\Sigma = \mathbb{C}$ or $C = \mathbb{C}$, and we shall use $v, \bar{v}$ to denote complex coordinates on $\mathbb{C}$.

Consider a 2d chiral defect on the complex plane $\mathbb{C}$ with a classically gauge-invariant action of the form

$$S_{\text{defect}}[X, A] = S[X] + \int_{\mathbb{C}} \mathrm{d}^2 v A_{\bar{v}}^a \, \mathcal{J}_a(X) \,. \tag{B.0.1}$$

Here, we denote the field content of the surface defect collectively as X, $\mathcal{J}$ is the current for the G gauge symmetry, whose OPE gives rise to an affine Kac–Moody algebra, and $A_{\bar{v}}$ denotes the antiholomorphic component of a gauge field.

We can detect the gauge anomaly of such a chiral surface defect by performing an infinitesimal gauge transformation of its partition function $Z[A_{\bar{v}}]$ as a functional of $A_{\bar{v}}$, in order to obtain the gauge transformation of the effective action $W[A_{\bar{v}}] = -i \log Z[A_{\bar{v}}]$. Doing so, we obtain

$$\delta Z[A_{\bar{v}}] = \delta \left(\int \mathcal{D} X \, e^{iS[X] + i \int_{\mathbb{C}} \mathrm{d}^2 v A_{\bar{v}}^a \mathcal{J}_a} \right) \tag{B.0.2}$$

$$= -i \int \mathcal{D} X \left(\int_{\mathbb{C}} \mathrm{d}^2 v' D_{\bar{v}'} \epsilon^a(v', \bar{v}') \mathcal{J}_a(v') \right) e^{iS[X] + i \int_{\mathbb{C}} \mathrm{d}^2 v A_{a\bar{v}} \mathcal{J}^a} \tag{B.0.3}$$

$$= -i \int \mathcal{D}X \left(\int_{\mathbb{C}} \mathrm{d}^2 v' \, \partial_{\bar{v}'} \epsilon^a(v', \bar{v}') \mathcal{J}_a(v') \right) e^{iS[X] + i \int_{\mathbb{C}} \mathrm{d}^2 v A_{a\bar{v}} \mathcal{J}^a}$$

$$- i \int \mathcal{D}X \left(\int_{\mathbb{C}} \mathrm{d}^2 v' \, f^a{}_{bc} A^b_{\bar{v}} \epsilon^c(v', \bar{v}') \mathcal{J}_a(v') \right) e^{iS[X] + i \int_{\mathbb{C}} \mathrm{d}^2 v A_{a\bar{v}} \mathcal{J}^a} .$$

$$(\text{B.0.4})$$

Let us study the first term in (B.0.4). Taylor-expanding the exponential of the gauge field coupling in this term and contracting pairs of currents leads to

$$- \int \mathcal{D}X \left(i \int_{\mathbb{C}} \mathrm{d}^2 v' \, i \int_{\mathbb{C}} \mathrm{d}^2 v \, \partial_{\bar{v}'} \epsilon^a(v', \bar{v}') \right.$$

$$\left. \left(- \frac{f_{ab}{}^c \mathcal{J}_c(v)}{2\pi(v' - v)} + \frac{k \delta_{ab}}{(2\pi)^2 (v' - v)^2} + (\text{regular}) \right) A^b_{\bar{v}}(v, \bar{v}) \right) e^{iS_{\text{defect}}[X, A]} ,$$

$$(\text{B.0.5})$$

where we have used the OPE

$$\mathcal{J}_a(v') \mathcal{J}_b(v) \sim - \frac{1}{2\pi} \frac{f_{ab}{}^c \mathcal{J}_c(v)}{v' - v} + \frac{1}{(2\pi)^2} \frac{k \delta_{ab}}{(v' - v)^2} . \qquad (\text{B.0.6})$$

Integration by parts in (B.0.5) picks up both the first-order and second-order poles in the OPE (B.0.6), with the first-order pole contributing the expression

$$-i \int \mathcal{D}X \int_{\mathbb{C}} d^2 v \, \epsilon_a(v, \bar{v}) f^{ab}{}_c \mathcal{J}^c(v) A_{b\bar{v}}(v, \bar{v}) e^{iS_{\text{defect}}[X, A]} , \qquad (\text{B.0.7})$$

which cancels the second term in (B.0.4).

Using the identity

$$\frac{1}{2\pi i} \partial_{\bar{v}} \frac{1}{v^2} = \partial_v \delta^2(v) , \qquad (\text{B.0.8})$$

we find that the remaining expression is

$$k \int \mathcal{D}X \left(\int_{\mathbb{C}} d^2 v' \int_{\mathbb{C}} \frac{d^2 v}{2\pi i} \, \epsilon_a(v', \bar{v}') A^a_{\bar{v}}(v, \bar{v}) \right) \partial_{v'} \delta^2(v' - v) e^{iS_{\text{defect}}[X, A]} .$$

$$(\text{B.0.9})$$

Integrating by parts again moves the derivative onto $\epsilon_a(v')$, upon which we can integrate over the delta function, to obtain

$$-k \int \mathcal{D}X \left(\int_{\mathbb{C}} \frac{d^2 v}{2\pi i} \, \partial_v \epsilon_a(v, \bar{v}) A^a_{\bar{v}}(v, \bar{v}) \right) e^{iS_{\text{defect}}[X, A]} . \qquad (\text{B.0.10})$$

Hence, we deduce that the gauge variation of the effective action is given by

$$\delta W = \frac{k}{2\pi} \int_{\mathbb{C}} d^2 v \, \partial_v \epsilon_a(v, \bar{v}) A^a_{\bar{v}}(v, \bar{v}) . \qquad (\text{B.0.11})$$

In Chapter 5, we consider both chiral and anti-chiral defects on a complex plane parametrized by coordinates $w, \bar{w}$. The gauge anomaly we computed above for the chiral defect is denoted there as

$$\delta W_+ = \frac{k_+}{2\pi} \int_{\mathbb{C}} \mathrm{d}^2 w \, \partial_w \epsilon_a(w, \bar{w}) A^a_{\bar{w}}(w, \bar{w}) \,, \qquad (\text{B.0.12})$$

where k_+ is the level of the corresponding affine Kac–Moody algebra, and W_+ is the effective action of the defect. Analogously, the gauge variation of the effective action, W_-, for an anti-chiral defect which couples to A_w, with Kac–Moody symmetry at level k_-, is given by

$$\delta W_- = \frac{k_-}{2\pi} \int_{\mathbb{C}} \mathrm{d}^2 w \, \partial_{\bar{w}} \epsilon_a(w, \bar{w}) A^a_w(w, \bar{w}) \,. \qquad (\text{B.0.13})$$

As explained in the main text, when considering chiral/anti-chiral surface defects along Σ, the anomalies acquire corrections arising from counterterms in order to satisfy the Wess–Zumino consistency condition, while anomalies associated with chiral surface defects along C satisfy this consistency condition without such corrections.

We have thus observed how the correlation function of currents of a two-dimensional field theory (utilized as a defect in 4d CS) leads to a gauge anomaly. The relationship of this 2d chiral gauge anomaly to the correlation function of currents is analogous to the familiar relationship of the 4d chiral anomaly to the 1-loop triangle Feynman diagram involving axial and vector currents.

Appendix C

Wilson Lines from Quantum Mechanics

In this appendix, we shall review the description of Wilson lines/loops in gauge theories as the path integrals of one-dimensional quantum mechanical systems. This description arises since, for suitable periodic/anti-periodic boundary conditions, such path integrals can be expressed as the following trace of the time evolution operator over the Hilbert space, $\mathcal{H}$, of the theory:

$$\mathrm{Tr}_{\mathcal{H}} \, \mathcal{T} \exp \left(\int \mathrm{d}t H \right), \tag{C.0.1}$$

where H is the Hamiltonian of the theory, t indicates the Euclidean time coordinate, and $\mathcal{T}$ denotes time ordering. Indeed, interpreting time ordering as path ordering, if $\mathcal{H}$ is the representation space of a representation of some Lie group, and if H could be identified with a gauge field in that representation, then (C.0.1) would indeed take the form of a Wilson loop. In what follows, we shall review several examples of this correspondence.

C.1 Bosonic Quantum Mechanics and Symmetric Representations

Consider the partition function of a bosonic quantum mechanical system with gauge symmetry, that is,

$$Z = \int \mathcal{D}\beta \mathcal{D}\gamma \, \exp \left(i \int_{\mathbb{R}} \mathrm{d}t \, \left(\beta^i \partial_t \gamma_i + A_t^a \beta^i \rho(t_a)_i{}^j \gamma_j \right) \right), \tag{C.1.1}$$

where γ_i and β^i describe the position and momentum of a particle in N-dimensional space, respectively.[1] Here, $i = 1, \ldots, N$, while t_a are the generators of the gauge group $GL(N)$, in the fundamental representation,

---

[1] We allow for the possibility that γ and β are complex, whereby the particle would be propagating in $\mathbb{C}^N$.

305

ρ, while A_t is a background gauge field. The fields β^i and γ_i transform in the fundamental representation of $GL(N)$ and its dual, and the action of the particle can be shown to be gauge invariant.

Upon imposing periodic boundary conditions, the partition function (C.1.1) can be rewritten as the trace, over the Hilbert space, of the time-evolution operator of the system. This is equivalent to a Wilson loop

$$W_\rho = \mathrm{Tr}_{\mathcal{H}} \left(\mathcal{P} \exp \left(i \int_{S^1} \mathrm{d}t \, A^a \widehat{\rho}(t_a) \right) \right), \tag{C.1.2}$$

where $\widehat{\rho}(t_a)$ is the Schwinger representation of $t_a \in \mathfrak{gl}(N)$ given by

$$\widehat{\rho}(t_a) = \widehat{\beta}^i \rho(t_a)_i{}^j \widehat{\gamma}_j. \tag{C.1.3}$$

The Hilbert space $\mathcal{H}$ is generated by operators $\{\widehat{\gamma}_i, \widehat{\beta}^i\}$ $(i = 1, \ldots, N)$ satisfying the canonical commutation relations

$$[\widehat{\gamma}_i, \widehat{\beta}^j] = i\delta_i^j, \qquad [\widehat{\gamma}_i, \widehat{\gamma}_j] = 0, \qquad [\widehat{\beta}^i, \widehat{\beta}^j] = 0, \tag{C.1.4}$$

known as the Heisenberg-Weyl algebra, whereby (C.1.3) can be shown to satisfy the Lie algebra $\mathfrak{gl}(N)$.

The equivalence between (C.1.1) and (C.1.2) can be derived by utilizing coherent states for the Heisenberg-Weyl algebra. This involves rewriting (C.1.2) by partitioning S^1 into M infinitesimal segments, inserting the completeness relation for the coherent states at each partition point, and taking $M \to \infty$. Details on this procedure can be found in the literature, for example, in the work of Kondo [152] (also see the appendix of [75]).

We shall now focus on understanding the representations of $GL(N)$ that can be obtained from C.1.1. The Hilbert space, $\mathcal{H}$, is formed by acting with $\widehat{\beta}^i$ on the vacuum that satisfies $\widehat{\gamma}_i|0\rangle = 0$, that is, it consists of states of the form

$$|i_1 \ldots i_n\rangle = \widehat{\beta}^{i_1} \ldots \widehat{\beta}^{i_n}|0\rangle. \tag{C.1.5}$$

Since all the $\widehat{\beta}_i$ commute with each other, (C.1.5) corresponds to a symmetric representation of $GL(N)$. Thus, the Wilson loop (C.1.2) is associated with the direct sum of symmetric representations of $GL(N)$.

In order to obtain a Wilson loop in a particular irreducible representation, say, the l-th symmetric representation, we ought to project the Hilbert space, $\mathcal{H}$, to a restricted Hilbert space containing only l bosonic excitations. This can be achieved by demanding that the elements $|\Psi\rangle$ of the restricted

Hilbert space, $\mathcal{H}^{(l)}$, satisfy

$$\widehat{n}_b|\Psi\rangle = l|\Psi\rangle, \qquad \widehat{n}_b := \sum_{i=1}^{N} \widehat{\gamma}_i \widehat{\beta}^i. \qquad (C.1.6)$$

This projection can be implemented at the level of the path integral by generalizing (C.1.1) to

$$W_b^{(l)}[A] = \int \mathcal{D}\beta \mathcal{D}\gamma \mathcal{D}\tilde{A}^b \exp\left(i\,S_b^{(l)}\right) \exp\left(-il_{\text{eff}} \int dt\, \tilde{A}_t^b\right), \qquad (C.1.7)$$

$$S_b^{(l)} = \int dt\, \beta_i i(\delta_j^i \partial_t + A_t^a \rho(t_a)^i{}_j - i\delta_j^i \tilde{A}_t^b)\gamma^j, \qquad (C.1.8)$$

where $\tilde{A}^b$ is an auxiliary gauge field, and where l_{eff} is the integer l shifted by $N/2$ due to Weyl ordering, i.e.,

$$l_{\text{eff}} = l + \frac{N}{2}. \qquad (C.1.9)$$

Integrating over the auxiliary gauge field in the path integral implements the desired constraint via a delta functional.

The discussion above can be generalized to the case of arbitrary irreducible representations as discussed, for example, in [90, 91]. Let us consider the Wilson loop operator for an irreducible representation, $\tilde{\rho}$, of $GL(N)$, which is characterized by the following Young tableau with l_I boxes in the I-th row $(1 \leq I \leq L)$:

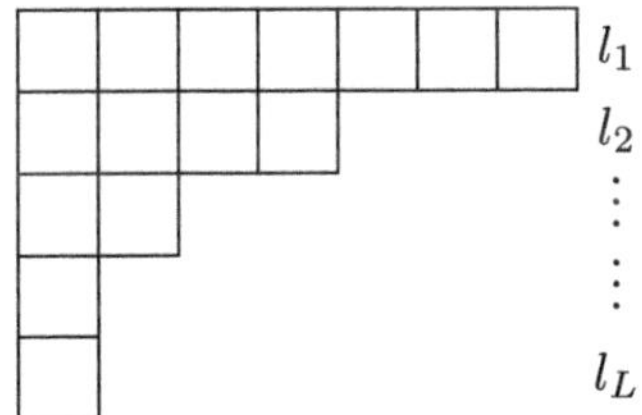

To construct a Hilbert space that realizes the irreducible representation $\tilde{\rho}$, one introduces operators $\widehat{\beta}_I^i$ and $\widehat{\gamma}_I^i$ with flavor indices $I, J = 1, \ldots, L$, where L is the number of rows of the Young tableau. These operators satisfy the commutation relations

$$[\widehat{\gamma}_i^I, \widehat{\beta}_J^j] = \delta_i^j \delta_J^I, \qquad [\widehat{\gamma}_i^I, \widehat{\gamma}_j^J] = 0, \qquad [\widehat{\beta}_I^i, \widehat{\beta}_J^j] = 0, \qquad (C.1.10)$$

whereby we can construct the Schwinger representation $\widehat{\rho}(t_a)$ of $t_a \in GL(N)$ defined by

$$\widehat{\rho}(t_a) = \sum_{I=1}^{K} \widehat{\beta}_I^i \rho(t_a)_i{}^j \widehat{\gamma}_j^I. \qquad (C.1.11)$$

Let $\mathcal{H}_b$ be the Hilbert space generated by acting with $\widehat{\beta}_I^i$ on the vacuum $|0\rangle$ which is defined by $\widehat{\gamma}_i^I|0\rangle = 0$, which corresponds to a reducible representation, as before. To project to the irreducible representation $\tilde{\rho}$, we require the following constraints

$$\widehat{L}^I{}_I|\Psi\rangle = l_I|\Psi\rangle\,, \qquad \widehat{L}^J{}_K|\Psi\rangle = 0\,, \qquad J > K\,, \tag{C.1.12}$$

where $|\Psi\rangle$ is an element of the restricted Hilbert space, and where $\widehat{L}^I{}_J$ are generators of $\mathfrak{gl}(L)$ defined by

$$\widehat{L}^I{}_J = \sum_{i=1}^{N} \widehat{\beta}^i_J \widehat{\gamma}^I_i\,, \tag{C.1.13}$$

that satisfy the commutation relations

$$[\widehat{L}^I{}_J, \widehat{L}^K{}_L] = \delta^I_L \widehat{L}^K{}_J - \delta^K_J \widehat{L}^I{}_L\,. \tag{C.1.14}$$

The first constraint of (C.1.12) specifies the length of each row in the Young tableau for $\tilde{\rho}$ as in the example of the l-th symmetric representation, and the second constraint realizes antisymmetrization between different rows.

For example, consider the second antisymmetric representation of $GL(N)$. If we impose the constraint

$$\widehat{\beta}^i_1 \widehat{\gamma}^2_i|\Psi\rangle = 0\,, \tag{C.1.15}$$

where $|\Psi\rangle$ is of the form $\tilde{\Psi}_{i,j}\widehat{\beta}^i_1\widehat{\beta}^j_2|0\rangle$ and satisfies $\widehat{L}^1{}_1|\Psi\rangle = |\Psi\rangle = \widehat{L}^2{}_2|\Psi\rangle$, we find the state corresponding to the second antisymmetric representation, described explicitly as

$$|\Psi\rangle_{[l_1=1,l_2=1]} = \frac{1}{2}\Psi_{i,j}\left(\widehat{\beta}^i_1\widehat{\beta}^j_2 - \widehat{\beta}^j_1\widehat{\beta}^i_2\right)|0\rangle\,. \tag{C.1.16}$$

This follows since, for fixed i and j,

$$\begin{aligned}
\widehat{\beta}^k_1\widehat{\gamma}^2_k(\widehat{\beta}^i_1\widehat{\beta}^j_2 - \widehat{\beta}^j_1\widehat{\beta}^i_2)|0\rangle &= \widehat{\beta}^k_1\widehat{\beta}^i_1\widehat{\gamma}^2_k\widehat{\beta}^j_2 - \widehat{\beta}^k_1\widehat{\beta}^j_1\widehat{\gamma}^2_k\widehat{\beta}^i_2|0\rangle \\
&= (\widehat{\beta}^{j1}\widehat{\beta}^{i1} - \widehat{\beta}^{i1}\widehat{\beta}^{j1})|0\rangle \tag{C.1.17} \\
&= 0\,.
\end{aligned}$$

In this way, we obtain the state corresponding to the Young diagram specified by $l_1 = 1, l_2 = 1$.

Let us now describe the path integral formulation of a Wilson loop associated with an irreducible representation specified by $\{l_I\}$, which generalizes (C.1.7). Note that the operators $\widehat{L}^I{}_I$ and $\widehat{L}^J{}_K$ ($J > K$) which enter the constraints (C.1.12) form the commutation relations of the Borel subalgebra $\mathfrak{b}_L$ of $\mathfrak{gl}(L)$,

$$[\widehat{L}^I{}_I, \widehat{L}^J{}_J] = 0\,,$$
$$[\widehat{L}^I{}_J, \widehat{L}^K{}_K] = -\delta_{JK}\widehat{L}^I{}_K\,, \qquad I > K\,, I > J\,, \qquad (\text{C.1.18})$$
$$[\widehat{L}^I{}_J, \widehat{L}^K{}_K] = \delta_{IK}\widehat{L}^K{}_J\,, \qquad K > J\,, I > J\,.$$

Thus, the path integral formulation of the Wilson loop in a fixed irreducible representation specified by $\{l_I\}$ is

$$W_{\tilde{\rho}}[A] = \int \mathcal{D}\beta \mathcal{D}\gamma \mathcal{D}\tilde{A}^b \exp\left(iS^b\right) \exp\left(-i\int_{S^1} dt \sum_{I=1}^{L} l_{\text{eff},I}\, (\tilde{A}^b_t)_{II}\right),$$

$$S^b = \int_{S^1} dt \sum_{I,J=1}^{L} \sum_{i,j=1}^{N} \beta_I^i \left(\delta^{IJ}\delta^i_j\partial_t + \delta^{IJ}A^a_t\rho(t_a)^i{}_j + \delta^{ij}(\tilde{A}^b_t)_{IJ}\right)\gamma_j^J\,,$$

$$(\text{C.1.19})$$

where the auxiliary gauge field $(\tilde{A}^b_t)_{IJ}$ is valued in the Borel subalgebra $\mathfrak{b}_L$ of $\mathfrak{gl}(L)$ and has non-zero components $(\tilde{A}^b_t)_{IJ}$ with $I \geq J$. In addition, $l_{\text{eff},I}$ measures the size of each row, where

$$l_{\text{eff},I} = l_I + \frac{N}{2}\,, \qquad (\text{C.1.20})$$

where the shift of $N/2$ is due to Weyl ordering. Integrating over the auxiliary gauge field furnishes the necessary constraints to obtain a Wilson loop in the irreducible representation $\tilde{\rho}$, via a delta functional.

Although we have focused on Wilson loops in the preceding discussion by considering quantum mechanics with periodic boundary conditions, one could instead show an equivalence between a path integral on $\mathbb{R}$ with fixed boundary states and a Wilson line that acts as an endomorphism on the Hilbert space of the system, and maps one boundary state to the other.

C.2 Fermionic Quantum Mechanics and Antisymmetric Representations

Wilson lines in a gauge theory can also be realized using quantum mechanics with fermionic degrees of freedom. By replacing the fields in (C.1.1) by fermionic fields denoted ψ^i and $\psi^\dagger_i$, we have

$$Z = \int \mathcal{D}\psi^\dagger \mathcal{D}\psi \exp\left(i\int_{\mathbb{R}} dt \left(\psi^\dagger_i \partial_t \psi^i + A^a_t \psi^\dagger_i \rho(t_a)^i{}_j \psi^j\right)\right). \qquad (\text{C.2.1})$$

Imposing antiperiodic boundary conditions, this partition function can be rewritten as the trace, over the Hilbert space, of the time-evolution operator, which is equivalent to a Wilson loop

$$W_\rho = \mathrm{Tr}_{\mathcal{H}} \left(\mathcal{P} \exp \left(i \int_{S^1} \mathrm{d}t \, A^a \widehat{\rho}(t_a) \right) \right). \tag{C.2.2}$$

Here, $\widehat{\rho}(t_a)$ is the Schwinger representation of $t_a \in \mathfrak{gl}(N)$ in terms of the fermionic operators $\widehat{\psi}^i$, $\widehat{\psi}^\dagger_i$ ($i = 1, 2, \dots, N$):

$$\widehat{\rho}(t_a) = \widehat{\psi}^\dagger_i \rho(t_a)^i{}_j \widehat{\psi}^j, \tag{C.2.3}$$

where $\rho(t_a)^i{}_j$ is the fundamental representation of $t_a \in \mathfrak{gl}(N)$, and the operators $\widehat{\psi}^i$, $\widehat{\psi}^\dagger_i$ satisfy the anticommutation relations

$$\{\widehat{\psi}^j, \widehat{\psi}^\dagger_k\} = \delta^j_k, \qquad \{\widehat{\psi}^j, \widehat{\psi}^k\} = 0, \qquad \{\widehat{\psi}^\dagger_j, \widehat{\psi}^\dagger_k\} = 0. \tag{C.2.4}$$

The trace $\mathrm{Tr}_{\mathcal{H}}$ is taken over the Hilbert space, $\mathcal{H}$, that is generated by acting with $\widehat{\psi}^\dagger_i$ on the vacuum, $|0\rangle$, satisfying $\widehat{\psi}^i|0\rangle = 0$ for all i.

The Wilson loop (C.2.2) is associated with the direct sum of all antisymmetric representations. This follows since states in the Hilbert space, $\mathcal{H}$, are of the form

$$|i_1 \dots i_n\rangle = \widehat{\psi}^\dagger_{i_1} \dots \widehat{\psi}^\dagger_{i_n} |0\rangle. \tag{C.2.5}$$

To obtain a Wilson loop in the k-th antisymmetric representation of $GL(N)$, we can follow the procedure of [93] (see also [90, 94]). This involves projecting the Hilbert space onto a restricted Hilbert space that contains only k fermionic excitations, i.e., an element $|\Psi\rangle$ of the restricted Hilbert space satisfies

$$\widehat{n}_f|\Psi\rangle = k|\Psi\rangle, \qquad \widehat{n}_f := \sum_{i=1}^{N} \widehat{\psi}^\dagger_i \widehat{\psi}^i. \tag{C.2.6}$$

This constraint can be realized at the path integral level by generalizing (C.2.1) to

$$W_f^{(l)}[A] = \int \mathcal{D}\psi^\dagger \mathcal{D}\psi \mathcal{D}\tilde{A}^f \exp \left(i S_f^{(l)} \right) \exp \left(-i k_{\mathrm{eff}} \int \mathrm{d}t \, \tilde{A}^f_t \right), \tag{C.2.7}$$

$$S_f^{(l)} = \int \mathrm{d}t \, \psi^\dagger_i i (\delta^i_j \partial_t + A^a_t \rho(t_a)^i{}_j - i\delta^i_j \tilde{A}^f_t)\psi^j, \tag{C.2.8}$$

where $\tilde{A}^f$ is an auxiliary gauge field and where k_{eff} is the integer k shifted by $-N/2$ due to Weyl ordering, i.e.,

$$k_{\mathrm{eff}} = k - \frac{N}{2}. \tag{C.2.9}$$

Just as in the case of bosonic quantum mechanics, the preceding discussion can be generalized to the case of arbitrary irreducible representations, using the results of [90, 91]. Consider the Wilson loop operator for an irreducible representation, $\tilde{\rho}$, of $GL(N)$ which is characterized by the following Young tableau with k_I boxes in the I-th column ($1 \leq I \leq K$):

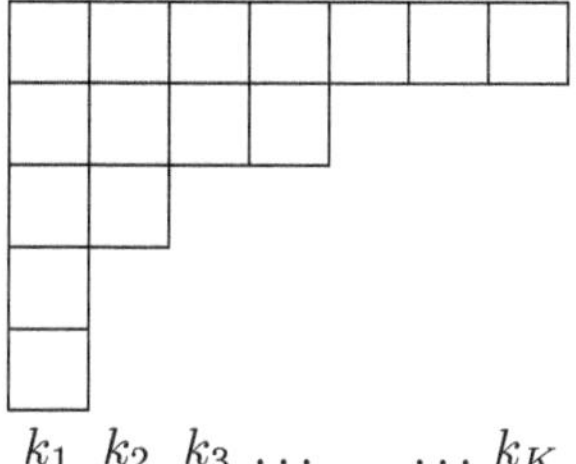

$$k_1 \ \ k_2 \ \ k_3 \ \ldots \qquad \ldots k_K$$

The Hilbert space that realizes the irreducible representation $\tilde{\rho}$ can be realized by introducing the fermionic operators $\widehat{\psi}^i_I$ and $\widehat{\psi}_i^{\dagger\,I}$ ($i = 1, \ldots, N$) with flavor indices $I, J = 1, \ldots, K$, which is equal to the number of columns of the Young tableau [90, 91]. These operators satisfy the anticommutation relations

$$\{\widehat{\psi}^i_I, \widehat{\psi}_j^{\dagger\,J}\} = \delta^i_j \delta^J_I \,, \qquad \{\widehat{\psi}^i_I, \widehat{\psi}^j_J\} = 0 \,, \qquad \{\widehat{\psi}_i^{\dagger\,I}, \widehat{\psi}_j^{\dagger\,J}\} = 0 \,, \qquad \text{(C.2.10)}$$

whereby we may construct the Schwinger representation $\widehat{\rho}(t_a)$ of $t_a \in \mathfrak{gl}(N)$ defined by

$$\widehat{\rho}(t_a) = \sum_{I=1}^{K} \widehat{\psi}_i^{I\dagger} \rho(t_a)^i{}_j \widehat{\psi}^j_I \,. \qquad \text{(C.2.11)}$$

The Hilbert space generated by acting with $\widehat{\psi}_i^{\dagger\,I}$ on the vacuum $|0\rangle$ which is defined by $\widehat{\psi}^i_I|0\rangle = 0$ corresponds to a reducible representation of the gauge group, $GL(N)$.

To project to the irreducible representation $\tilde{\rho}$, we impose the constraints

$$\widehat{L}^I{}_I|\Psi\rangle = k_I|\Psi\rangle \,, \qquad \widehat{L}^J{}_M|\Psi\rangle = 0 \,, \qquad J > M \,, \qquad \text{(C.2.12)}$$

where $\widehat{L}^I{}_J$ are generators of $\mathfrak{u}(K)$ defined by

$$\widehat{L}^I{}_J = \sum_{i=1}^{N} \widehat{\psi}_J^{i\dagger} \widehat{\psi}^I_i \,, \qquad \text{(C.2.13)}$$

that satisfy the commutation relations

$$[\widehat{L}^I{}_J, \widehat{L}^M{}_L] = \delta^I_L \widehat{L}^M{}_J - \delta^M_J \widehat{L}^I{}_L \,. \qquad \text{(C.2.14)}$$

The first constraint of (C.2.12) specifies the length of each column in the Young tableau for $\tilde{\rho}$, as in the case of the k-th antisymmetric representation, and the second constraint realizes symmetrization between different columns.

The generators $\widehat{L}^I{}_I$ and $\widehat{L}^J{}_M$ ($J > M$) that enter the constraints form the commutation relations of the Borel subalgebra $\mathfrak{b}$ of $\mathfrak{u}(K)$,

$$
\begin{aligned}
&[\widehat{L}^I{}_I, \widehat{L}^J{}_J] = 0 \,, \\
&[\widehat{L}^I{}_J, \widehat{L}^M{}_M] = -\delta_{JM}\widehat{L}^I{}_M \,, \qquad I > M \,, I > J \,, \\
&[\widehat{L}^I{}_J, \widehat{L}^M{}_M] = \delta_{IM}\widehat{L}^M{}_J \,, \qquad M > J \,, I > J \,.
\end{aligned}
\tag{C.2.15}
$$

Thus, introducing an auxiliary gauge field $\tilde{A}^f \in \mathfrak{b}$, whose non-zero components are $(\tilde{A}^f)_{IJ}$ with $I \geq J$, we can write down a path integral formulation of a Wilson loop associated with the representation $\tilde{\rho}$, which is

$$
W^f_{\rho}[A] = \int \mathcal{D}\psi^\dagger \mathcal{D}\psi \mathcal{D}\tilde{A}^f \, \exp\left(iS_f\right) \exp\left(-i \int dt \sum_{I=1}^{K} k_{\text{eff},I}\,(\tilde{A}^f_t)_{II}\right),
\tag{C.2.16}
$$

$$
S^f = \int_{S^1} dt \sum_{I,J=1}^{K} \sum_{i,j=1}^{N} \psi_i^{\dagger I}(i\delta^{IJ}\delta^i_j\partial_t + \delta^{IJ}A^a_t\rho(t_a)^i{}_j + \delta^{ij}(\tilde{A}^f_t)_{IJ})\psi^j_J \,,
\tag{C.2.17}
$$

where $k_{\text{eff},I}$ is the integer k_I shifted by $-N/2$ due to Weyl ordering,

$$
k_{\text{eff},I} = k_I - \frac{N}{2} \,.
\tag{C.2.18}
$$

C.3 Infinite-dimensional Representations

So far, we have considered quantum mechanical systems describing particles moving in flat spaces. Let us now consider Wilson lines obtained from quantum mechanical systems describing particles propagating in *curved* spaces. As we shall see, this will allow us to obtain Wilson lines in infinite-dimensional representations.

Consider the following action for a particle propagating on a curved target space, X, with field content consisting of $\gamma : \mathbb{R} \to X$ and $\beta \in \Omega^{1,0}(\mathbb{C}, \gamma^*T^*X)$ (with T^*X denoting the holomorphic cotangent bundle of X):

$$
\int_{\mathbb{R}} \left(\beta^i\partial_t\gamma_i + A_{a,t}\beta^i\rho^a_i(\gamma)\right) dt,
\tag{C.3.1}
$$

where $i = 1, \ldots, \dim_{\mathbb{C}} X$, and $\rho : \mathfrak{g} \to \mathrm{Vect}(X)$ is a Lie algebra homomorphism from $\mathfrak{g}$ to the Lie algebra of holomorphic vector fields on X, which gives rise to the infinitesimal G-action on X.

The Hamiltonian of this system is equal to $A_t^a \beta^i \rho_{a,i}(\gamma)$, and it is possible to quantize the action such that the composite field $\beta^i \rho_{a,i}(\gamma)$ acts on the Hilbert space in some representation of $\mathfrak{g}$.

Let us focus on the specific example of $X = \mathbb{CP}^1$ and $G = SL(2, \mathbb{C})$ for simplicity, so that $\dim_{\mathbb{C}} X = 1$. In this case, the holomorphic vector fields $\rho_a(\gamma)$ are described in terms of a local holomorphic coordinate, γ, as

$$\rho_1 = i\gamma \partial_\gamma, \quad \rho_2 = \frac{1}{2}\left(-\partial_\gamma - \gamma^2 \partial_\gamma\right), \quad \rho_3 = \frac{1}{2}\left(i\partial_\gamma - i\gamma^2 \partial_\gamma\right). \tag{C.3.2}$$

These vector fields satisfy $[\rho_i, \rho_j] = \sum_{k=1}^{3} \epsilon_{ijk} \rho_k$, where $i, j = 1, 2, 3$ and the antisymmetric tensor ϵ_{ijk} is normalized as $\epsilon_{123} = 1$. We can rewrite the Hamiltonian as

$$A_t^H \beta \rho_H + A_t^F \beta \rho_F + A_t^E \beta \rho_E, \tag{C.3.3}$$

where $A^H = -iA^1$, $A^F = \frac{1}{2}(iA^3 - A^2)$, $A^E = \frac{1}{2}(iA^3 + A^2)$, and

$$\rho^H = -\gamma, \qquad \rho^F = 1, \qquad \rho^E = -\gamma^2. \tag{C.3.4}$$

It shall also be convenient to define $\sigma_E = \beta \rho^E$, $\sigma_F = \beta \rho^F$ and $\sigma_H = \beta \rho^H$.

Quantization in a local coordinate chart of $\mathbb{CP}^1$ can be achieved by imposing the canonical commutation relations

$$[\widehat{\gamma}, \widehat{\beta}] = 1, \qquad [\widehat{\gamma}, \widehat{\gamma}] = 0, \qquad [\widehat{\beta}, \widehat{\beta}] = 0, \tag{C.3.5}$$

so that $\widehat{\beta}$ can be regarded as the differential operator $-\partial/\partial\gamma$. In particular, the quantities σ_E, σ_F and σ_H correspond to the representation of $SL(2, \mathbb{C})$ given by

$$\sigma_E = \gamma^2 \frac{\partial}{\partial\gamma} + 2\gamma a,$$

$$\sigma_F = -\frac{\partial}{\partial\gamma}, \tag{C.3.6}$$

$$\sigma_H = \gamma \frac{\partial}{\partial\gamma} + a,$$

where E, F, H are the Chevalley generators of $SL(2, \mathbb{C})$ satisfying the commutation relations $[E, F] = 2H$, $[H, E] = E$, $[H, F] = -F$ and a is in general complex, and accounts for the ambiguity in operator ordering.[2]

---

[2] This representation is also known as the Dyson–Maleev representation.

For $a \neq 0$, these are known as *twisted* differential operators. Note that a non-zero value of a is needed to obtain a non-trivial Casimir, i.e.,

$$\sigma_H \sigma_H + \frac{\sigma_E \sigma_F}{2} + \frac{\sigma_F \sigma_E}{2} = a(a+1)\,. \tag{C.3.7}$$

For generic a, the representation generated from a highest-weight state annihilated by σ_E is an infinite-dimensional Verma module, with σ_F acting as the lowering operator.

To define these operators globally, consider $\mathbb{CP}^1$ with charts parametrized by γ and γ', where $\gamma = 1/\gamma'$. For twisted differential operators defined on $\mathbb{CP}^1$ to be globally well-defined (see, e.g., page 273 of [96]), the transition function

$$\partial_{\gamma'} = -\gamma^2 \partial_\gamma \tag{C.3.8}$$

ought to be deformed to

$$\partial_{\gamma'} = -\gamma^2 \partial_\gamma - 2a\gamma\,, \tag{C.3.9}$$

whereby

$$
\begin{aligned}
\sigma_E &= \gamma^2 \frac{\partial}{\partial \gamma} + 2\gamma a &&\to -\frac{\partial}{\partial \gamma'}\,, \\
\sigma_F &= -\frac{\partial}{\partial \gamma} &&\to \gamma'^2 \frac{\partial}{\partial \gamma'} + 2\gamma' a\,, \\
\sigma_H &= \gamma \frac{\partial}{\partial \gamma} + a &&\to -\gamma' \frac{\partial}{\partial \gamma'} - a\,.
\end{aligned}
\tag{C.3.10}
$$

These transformation rules define a sheaf of twisted differential operators on $\mathbb{CP}^1$. We remark that the gluing equations in (C.3.10) correspond to an automorphism $(E, F, H) \to (F, E, -H)$ of $\mathfrak{sl}_2$.

Algebras of twisted differential operators defined on $\mathbb{CP}^1$, and more generally, flag varieties, were analyzed by Bernstein and Beilinson [97, 98] (also see [99]) precisely for the purpose of studying the representation theory of simple complex Lie algebras. Their result can be stated as follows. For a flag variety, X, of a reductive group, G, and a choice of weight λ, let χ_λ be the corresponding central character of the universal enveloping algebra $U(\mathfrak{g})$. Defining the associated central reduction $U_\lambda(\mathfrak{g}) = U(\mathfrak{g})/U(\mathfrak{g})\mathrm{Ker}\,\chi_\lambda$, there exists a twisted sheaf of differential operators D_λ on X such that

$$\Gamma(X; D_\lambda) \cong U_\lambda(\mathfrak{g})\,. \tag{C.3.11}$$

For regular (anti)dominant λ, coherent modules over D_λ can be identified through an equivalence of categories with representations of $U(\mathfrak{g})$ having central character χ_λ. In particular, the fields of the quantum mechanical system describe an affine chart of X (for $\mathbb{CP}^1$, one of the two standard affine charts). Quantizing on this chart yields a module that is generally infinite-dimensional (that is a Verma module when the chart is a Bruhat cell). In this way, we find that the Hilbert space of the quantum mechanical system with action (C.3.1) and flag manifold target space can be identified with infinite-dimensional representations of the gauge group.

Appendix D

Vertex Algebras and Chiral Differential Operators

D.1 Vertex Algebras and Zhu Algebras

Let us briefly review standard facts regarding vertex algebras and their corresponding Zhu algebras.

Define a *field* on a vector space V to be a formal series

$$a(z) = \sum_{n \in \mathbb{Z}} a_{(n)} z^{-n-1} \in (\mathrm{End}V)[[z, z^{-1}]]$$

such that for any $v \in V$ one has $a_{(n)}v = 0$ if $n \gg 0$. We shall use $\mathcal{F}(V)$ to denote the space of all fields on V.

A *vertex algebra* is a vector space V appended with the following data:

- a linear map (the state-operator correspondence) $Y : V \to \mathcal{F}(V)$, $V \ni a \mapsto a(z) = \sum_{n \in \mathbb{Z}} a_{(n)} z^{-n-1}$,
- an even vector $\mathbf{1} \in V$, known as the *vacuum vector*,
- a linear operator $\partial : V \to V$, known as the *translation operator*.

These data satisfy the following axioms:

1. (Translation Covariance) $(\partial a)(z) = \partial_z a(z)$.
2. (Vacuum) $\mathbf{1}(z) = \mathrm{id}$; $a(z) |\emptyset\rangle \in V[[z]]$ and $a_{(-1)} |\emptyset\rangle = a$.
3. (Locality) For all $a, b \in V$ there exists a positive integer N such that

$$(z - w)^N [Y(a, z), Y(b, w)] = 0. \tag{D.1.1}$$

To each vertex algebra V, one can assign an associative algebra known as the Zhu algebra $A(V)$. As a vector space, it is a quotient space of V. Let

317

$O(V)$ denote the linear span of elements $\mathrm{Res}_z(Y(a,z)((\hbar_\star z+1)^{\deg(a)}/z^2)b)$. One can define a multiplication in V as follows: For $a, b \in V$

$$a \star b := \mathrm{Res}_z\left(Y(a,z)\frac{(\hbar_\star z + 1)^{\deg(a)}}{z}b \right). \tag{D.1.2}$$

The $\star$-product induces the multiplication on the quotient $A(V) := V/O(V)$, and can be shown to be associative in $A(V)$.

In physics terminology, the Zhu algebra corresponds to the zero modes of the CFT Hilbert space. This is due to a theorem by Zhu [104], which can be stated as follows: For a graded module, M, over a vertex algebra V, the top component M_0 is a module over the Zhu algebra $A(V)$. The assignment $M \mapsto M_0$ gives rise to a one-to-one correspondence between isomorphism classes of graded irreducible V-modules and those of irreducible $A(V)$ modules.

D.2   2d Curved $\beta\gamma$ Systems and Twisted Chiral Differential Operators

We now describe the obstruction to defining a global sheaf of chiral differential operators on a Kähler manifold, X. We first compute the commutator of Laurent modes of currents associated with holomorphic vector fields that generate the G symmetry of X, whose OPE is given in equation (5.2.88), which includes the Lie algebra extension (5.2.89). We then review how a global sheaf of chiral differential operators can be defined on a general Kähler target space, following Ref. [106], emphasizing how this can be obstructed by the extension to $\mathfrak{g}$, and why this obstruction vanishes when $p_1(X) = 0$.

D.2.1   *Current-Current OPE in Curved $\beta\gamma$ Systems*

Let us first write the Laurent expansion of the currents as

$$J_V(w) = \sum_{n \in \mathbb{Z}} w^{-n-1} J_{Vn}, \qquad J_{Vn} = \frac{1}{2\pi i}\oint w^n J_V(w). \tag{D.2.1}$$

The contribution of the first term to the commutator $[J_{Vn}, J_{Wm}]$ can be computed using the formula

$$[A, B] = \oint_0 dw' \oint_w dw\, a(w)b(w') \tag{D.2.2}$$

to be

$$\frac{1}{(2\pi i)^2} \oint_0 dw' \oint_{w'} dw\, w^n w'^m \left(-\frac{\partial_j V^i \partial_i W^j(w')}{(w-w')^2} \right)$$

$$= \frac{1}{2\pi i} \oint_0 dw' \lim_{w\to w'} \frac{d}{dw} w^n (w')^m (-\partial_j V^i \partial_i W^j(w'))$$

$$= \frac{1}{2\pi i} \oint_0 dw' \quad n \quad (w')^{n-1}(w')^m \sum_{p\in\mathbb{Z}} (-\partial_j V^i \partial_i W^j)_p (w')^p$$

$$= n\delta_{n+m+p,0}(-\partial_j V^i \partial_i W^j)_p \,. \tag{D.2.3}$$

This is non-zero, but it does not contribute to the commutator $[J_{V0}, J_{W0}]$ that is our main object of interest.

Defining $\tilde{J}(w) = -(\partial_k \partial_j V^i)(\partial_i W^j \partial\gamma^k)$, the remaining terms in the commutator are computed to be

$$\frac{1}{(2\pi i)^2} \oint_0 dw' \oint_w dw\, w^n (w')^m \left(\frac{J_{[V,W]}(w')}{w-w'} + \frac{\tilde{J}(w')}{w-w'} \right)$$

$$= \frac{1}{2\pi i} \oint_0 dw' (w')^n (w')^m \left(\sum_{m+n\in\mathbb{Z}} (w')^{-m-n-1} J_{[V,W]m+n} \right.$$

$$\left. + \sum_{m+n\in\mathbb{Z}} (w')^{-m-n-1} \tilde{J}_{m+n} \right)$$

$$= J_{[V,W]m+n} + \tilde{J}_{m+n}, \tag{D.2.4}$$

where $[V,W]^j = (V^i \partial_i W^j - W^i \partial_i V^j)$. The full commutator arising from the current-current OPE is thus

$$[J_{Vn}, J_{Wm}] = n\delta_{n+m+p,0}(-\partial_j V^i \partial_i W^j)_p + J_{[V,W]m+n} + \tilde{J}_{m+n} \,. \tag{D.2.5}$$

The zero mode algebra is thus a Lie algebra extension defined via the exact sequence

$$0 \to \mathfrak{c} \to \mathfrak{g} \to \mathfrak{v} \to 0 \,. \tag{D.2.6}$$

D.2.2 *Defining global sheaves of twisted chiral differential operators*

To define a global sheaf of twisted chiral differential operators on a Kähler manifold, X, we first cover X by open sets U_a. For each a, b, choose an open set $U_{ab} \subset U_a$, and also an open set $U_{ba} \subset U_b$, and a holomorphic diffeomorphism f_{ab} between these open sets $f_{ab} : U_{ab} \cong U_{ba}$ such that

$f_{ba} = f_{ab}^{-1}$. In order to identify a point $P \in U_{ab}$ with a point $Q \in U_{ba}$ when $Q = f_{ab}(P)$, we would require that for any U_a, U_b, and U_c, we have

$$f_{ca} f_{bc} f_{ab} = 1 \tag{D.2.7}$$

on the triple intersection U_{abc}.

Analogously, to obtain a consistent global sheaf of chiral algebras defined over the entire space X, we ought to glue the chiral algebras using symmetries of the associated conformal field theory. For example, given two open sets U_a and U_b, we can use a CFT symmetry, denoted $\widehat{f}_{ab}$, to glue them, with the consistency condition on triple overlaps given by

$$\widehat{f}_{ca} \widehat{f}_{bc} \widehat{f}_{ab} = 1 \,. \tag{D.2.8}$$

Now, since we have a Lie algebra extension and thus a group extension as shown in Section 5.2.4.1, given such a consistent gluing of the sheaf of CDOs, one can always map the CFT symmetry group elements to geometrical symmetry group elements, i.e., (D.2.8) implies (D.2.7), thereby defining the complex manifold that is the target space of the CFT.

However, the $\widehat{f}_{ab}$ are not uniquely determined by the f_{ab}, as we can always pick an element $C_{ab} \in H^0\left(U_{ab}, \Omega^{2,cl}(X)\right)$ (where $\Omega^{2,cl}(X)$ is the sheaf of $(2,0)$-forms on X that are annihilated by ∂) representing an element of $\mathfrak{c}$, and perform the transformation $\widehat{f}_{ab} \to \widehat{f}'_{ab} = \exp\left(C_{ab}\right) \widehat{f}_{ab}$. If the CFT gluing condition on triple intersections is to still hold, we would then require that

$$C_{ab} + C_{bc} + C_{ca} = 0 \,,$$

and hence the $C._\cdot$'s define an element of the Čech cohomology group $H^1\left(X, \Omega^{2,cl}(X)\right)$.

Now, let us understand the anomaly that is the obstruction to gluing. If we wish to lift the gluing condition (D.2.7), in general we will not obtain (D.2.8), but rather

$$\widehat{f}_{ca} \widehat{f}_{bc} \widehat{f}_{ab} = \exp\left(C_{abc}\right) \tag{D.2.9}$$

for some $C_{abc} \in H^0\left(U_{abc}, \Omega^{2,cl}(X)\right)$ which represents an element of $\mathfrak{c}$. Note that $\widehat{f}_{ab}$ is not unique, and transforming $\widehat{f}_{ab} \to \exp\left(C_{ab}\right) \widehat{f}_{ab}$, we get

$$C_{abc} \to C'_{abc} = C_{abc} + C_{ab} + C_{bc} + C_{ca} \,. \tag{D.2.10}$$

In addition, in quadruple overlaps $U_a \cap U_b \cap U_c \cap U_d$, we have the relation

$$C_{abc} - C_{bcd} + C_{cda} - C_{dab} = 0 \,. \tag{D.2.11}$$

The relations (D.2.10) and (D.2.11) imply that the $C's$ define an element of the sheaf cohomology group $H^2\left(X, \Omega^{2,cl}(X)\right)$. There is no obstruction to obtaining a globally-defined sheaf of chiral algebras if it is possible to pick the C_{ab} to set all $C'_{abc} = 0$. This can be identified with the vanishing of $p_1(X)$.

Bibliography

[1] E. Witten, Quantum field theory and the Jones polynomial, *Communications in Mathematical Physics* **121**, 351–399 (1989).

[2] E. Witten, Gauge theories and integrable lattice models, *Nuclear Physics B* **322**, 629–697 (1989).

[3] E. Witten, Gauge theories, vertex models, and quantum groups, *Nuclear Physics B* **330**, 285–346 (1990).

[4] N. A. Nekrasov, *Four-Dimensional Holomorphic Theories.* Princeton University, 1996.

[5] K. Costello, Supersymmetric gauge theory and the Yangian, arXiv:1303.2632 [hep-th].

[6] K. Costello, Integrable lattice models from four-dimensional field theories, *Proceedings of Symposia in Pure Mathematics* **88**, 3–24 (2014), arXiv:1308.0370 [hep-th].

[7] K. Costello, E. Witten, and M. Yamazaki, Gauge theory and integrability, I, *ICCM Notices* **06**, 46–119 (2018), arXiv:1709.09993 [hep-th].

[8] K. Costello, E. Witten, and M. Yamazaki, Gauge theory and integrability, II, *ICCM Notices* **06**, 120–146 (2018), arXiv:1802.01579 [hep-th].

[9] A. A. Belavin and V. G. Drinfeld, Solutions of the classical Yang–Baxter equation for simple lie algebras, *Funktsional'nyi Analiz i ego Prilozheniya* **16**, 1–29 (1982).

[10] A. Polyakov, One physical problem with possible mathematical significance, *Journal of Geometry and Physics* **5**, 595–600 (1988).

[11] A. M. Polyakov, Fermi-Bose transmutations induced by gauge fields, *Modern Physics Letters A* **3**, 325–328 (1988).

[12] E. Guadagnini, M. Martellini, and M. Mintchev, Chern-Simons holonomies and the appearance of quantum groups, *Physics Letters B* **235**, 275–281 (1990).

[13] E. Guadagnini, M. Martellini, and M. Mintchev, Wilson lines in Chern-Simons theory and link invariants, *Nuclear Physics B* **330**, 575–607 (1990).

[14] M. Kontsevich, Vassiliev's knot invariants, *Advances in Soviet Mathematics* **16**, 137–150 (1993).

[15] V. V. Fock and A. A. Rosly, Poisson structure on moduli of flat connections on Riemann surfaces and r-matrix (1998), *arXiv preprint* math/9802054.

[16] A. Morozov and A. Smirnov, Chern–simons theory in the temporal gauge and knot invariants through the universal quantum r-matrix, *Nuclear Physics B* **835**, 284–313 (2010).

[17] V. G. Drinfel'd, Quantum groups, *Journal of Soviet Mathematics* **41**, 898–915 (1988).

[18] V. Chari and A. N. Pressley, *A Guide to Quantum Groups*. Cambridge University Press, 1995.

[19] A. Z. Khan, Holomorphic surface defects in four-dimensional Chern-Simons theory, arXiv:2209.07387 [hep-th].

[20] H. P. McKean Jr and I. M. Singer, Curvature and the eigenvalues of the laplacian, *Journal of Differential Geometry* **1**, 43–69 (1967).

[21] V. V. Bazhanov, T. Lukowski, C. Meneghelli, and M. Staudacher, A short-cut to the Q-operator, *Journal of Statistical Mechanics* **1011**, P11002 (2010), arXiv:1005.3261 [hep-th].

[22] K. Costello, D. Gaiotto, and J. Yagi, Q-operators are 't Hooft lines, arXiv:2103.01835 [hep-th].

[23] E. Witten, Dyons of charge e theta/2 pi, *Physics Letters B* **86**, 283–287 (1979).

[24] K. Costello and M. Yamazaki, Gauge theory and integrability, III, arXiv:1908.02289 [hep-th].

[25] A. L. Retore, Introduction to classical and quantum integrability, *Journal of Physics A* **55**, 173001 (2022), arXiv:2109.14280 [hep-th].

[26] L. D. Faddeev and N. Y. Reshetikhin, Integrability of the principal chiral field model in (1+1)-dimension, *Annals of Physics* **167**, 227 (1986).

[27] F. Delduc, M. Magro, and B. Vicedo, Alleviating the non-ultralocality of coset sigma models through a generalized Faddeev-Reshetikhin procedure, *JHEP* **08**, 019 (2012), arXiv:1204.0766 [hep-th].

[28] V. E. Zakharov and A. V. Mikhailov, Relativistically invariant two-dimensional models in field theory integrable by the inverse problem technique (in Russian), *Soviet Physics JETP* **47**, 1017–1027 (1978).

[29] V. O. Rivelles, On the S-matrix of the Faddeev-Reshetikhin model, *AIP Conference Proceedings* **1031**, 130–137 (2008), arXiv:0804.1523 [hep-th].

[30] O. Fukushima, J.-I. Sakamoto, and K. Yoshida, Faddeev-Reshetikhin model from a 4D Chern-Simons theory, *JHEP* **02**, 115 (2021), arXiv:2012.07370 [hep-th].

[31] M. Benini, A. Schenkel, and B. Vicedo, Homotopical analysis of 4d Chern-Simons theory and integrable field theories, *Communications in Mathematical Physics* **389**, 1417–1443 (2022), arXiv:2008.01829 [hep-th].

[32] F. Delduc, S. Lacroix, M. Magro, and B. Vicedo, A unifying 2D action for integrable σ-models from 4D Chern–Simons theory, *Letters in Mathematical Physics* **110**, 1645–1687 (2020), arXiv:1909.13824 [hep-th].

[33] B. Hoare, Integrable deformations of sigma models, *Journal of Physics A* **55**, 093001 (2022), arXiv:2109.14284 [hep-th].

[34] B. Vicedo, Deformed integrable σ-models, classical R-matrices and classical exchange algebra on Drinfel'd doubles, *Journal of Physics A* **48**, 355203 (2015), arXiv:1504.06303 [hep-th].

[35] B. Vicedo, Holomorphic Chern-Simons theory and affine Gaudin models, arXiv:1908.07511 [hep-th].

[36] J. M. Maillet, New integrable canonical structures in two-dimensional models, *Nuclear Physics B* **269**, 54–76 (1986).

[37] R. Bittleston and D. Skinner, Twistors, the ASD Yang-Mills equations and 4d Chern-Simons theory, *JHEP* **02**, 227 (2023), arXiv:2011.04638 [hep-th].

[38] T. Adamo, Lectures on twistor theory, *PoS* **Modave2017**, 003 (2018), arXiv:1712.02196 [hep-th].

[39] R. Boels, L. J. Mason, and D. Skinner, Supersymmetric gauge theories in twistor space, *JHEP* **02**, 014 (2007), arXiv:hep-th/0604040.

[40] L. T. Cole, R. A. Cullinan, B. Hoare, J. Liniado, and D. C. Thompson, Integrable deformations from twistor space, *SciPost Physics* **17**, 008 (2024), arXiv:2311.17551 [hep-th].

[41] L. T. Cole, R. A. Cullinan, B. Hoare, J. Liniado, and D. C. Thompson, Gauging the diamond: Integrable coset models from twistor space, arXiv:2407.09479 [hep-th].

[42] M. Yamazaki, New T-duality for Chern-Simons theory, *JHEP* **12**, 090 (2019), arXiv:1904.04976 [hep-th].

[43] K. Costello, M-theory in the Omega-background and 5-dimensional non-commutative gauge theory, arXiv:1610.04144 [hep-th].

[44] E. Witten, Chern-Simons gauge theory as a string theory, *Progress in Mathematics* **133**, 637–678 (1995), arXiv:hep-th/9207094.

[45] H. Ooguri and C. Vafa, Knot invariants and topological strings, *Nuclear Physics B* **577**, 419–438 (2000), arXiv:hep-th/9912123.

[46] K. Costello and J. Yagi, Unification of integrability in supersymmetric gauge theories, *Advances in Theoretical and Mathematical Physics* **24**, 1931–2041 (2020), arXiv:1810.01970 [hep-th].

[47] E. Witten, Topological quantum field theory, *Communications in Mathematical Physics* **117**, 353–386 (1988).

[48] N. Nekrasov and E. Witten, The Omega deformation, branes, integrability, and Liouville theory, *JHEP* **09**, 092 (2010), arXiv:1002.0888 [hep-th].

[49] S. Hellerman, D. Orlando, and S. Reffert, The Omega deformation from string and M-theory, *JHEP* **07**, 061 (2012), arXiv:1204.4192 [hep-th].

[50] M. Bershadsky, C. Vafa, and V. Sadov, D-branes and topological field theories, *Nuclear Physics B* **463**, 420–434 (1996), arXiv:hep-th/9511222.

[51] N. A. Nekrasov and S. L. Shatashvili, Quantization of integrable systems and four dimensional gauge theories, in *16th International Congress on Mathematical Physics*, pp. 265–289, August 2009. arXiv:0908.4052 [hep-th].

[52] M. Ashwinkumar, M.-C. Tan, and Q. Zhao, Branes and categorifying integrable lattice models, *Advances in Theoretical and Mathematical Physics* **24**, 1–24 (2020), arXiv:1806.02821 [hep-th].

[53] M. Ashwinkumar and M.-C. Tan, Unifying lattice models, links and quantum geometric Langlands via branes in string theory, *Advances in Theoretical and Mathematical Physics* **24**, 1681–1721 (2020), arXiv:1910.01134 [hep-th].

[54] A. Kapustin and E. Witten, Electric-magnetic duality and the geometric Langlands program, *Communications in Number Theory and Physics* **1**, 1–236 (2007), arXiv:hep-th/0604151.

[55] E. Witten, Fivebranes and knots, arXiv:1101.3216 [hep-th].

[56] E. Witten, A new look at the path integral of quantum mechanics, arXiv:1009.6032 [hep-th].

[57] E. Witten, Analytic continuation of Chern-Simons theory, *AMS/IP Studies in Advanced Mathematics* **50**, 347–446 (2011), arXiv:1001.2933 [hep-th].

[58] C. Elliott and V. Pestun, Multiplicative Hitchin systems and supersymmetric gauge theory, *Selecta Mathematica* **25**, 64 (2019), arXiv:1812.05516 [math.AG].

[59] M. Ashwinkumar, K.-S. Png, and M.-C. Tan, 4d Chern-Simons theory as a 3d Toda theory, and a 3d-2d correspondence, *JHEP* **09**, 057 (2021), arXiv:2008.06053 [hep-th].

[60] N. Ishtiaque, S. Faroogh Moosavian, and Y. Zhou, Topological holography: The example of the D2-D4 brane system, *SciPost Physics* **9**, 17 (2020), arXiv:1809.00372 [hep-th].

[61] M. Ashwinkumar, Integrable lattice models and holography, *Journal of High Energy Physics* **2021**, 1–22 (2021).

[62] A. Giveon and D. Kutasov, Brane dynamics and gauge theory, *Reviews of Modern Physics* **71**, 983–1084 (1999), arXiv:hep-th/9802067.

[63] L. Rozansky and E. Witten, HyperKahler geometry and invariants of three manifolds, *Selecta Mathematica* **3**, 401–458 (1997), arXiv:hep-th/9612216.

[64] O. Aharony, M. Berkooz, and S.-J. Rey, Rigid holography and six-dimensional $\mathcal{N} = (2,0)$ theories on AdS$_5 \times \mathbb{S}^1$, *JHEP* **03**, 121 (2015), arXiv:1501.02904 [hep-th].

[65] J. Yagi, Ω-deformation and quantization, *JHEP* **08**, 112 (2014), arXiv:1405.6714 [hep-th].

[66] S. Elitzur, G. Moore, A. Schwimmer, and N. Seiberg, Remarks on the canonical quantization of the chern-simons-witten theory, *Nuclear Physics B* **326**, 108–134 (1989).

[67] E. Witten, Non-abelian bosonization in two dimensions, *Communications in Mathematical Physics* **92**, 455–472 (1984).

[68] S. K. Donaldson, Anti self-dual yang-mills connections over complex algebraic surfaces and stable vector bundles, *Proceedings of the London Mathematical Society* **3**, 1–26 (1985).

[69] V. Nair and J. Schiff, A kähler-chern-simons theory and quantization of instanton moduli spaces, *Physics Letters B* **246**, 423–429 (1990).

[70] A. Losev, G. Moore, N. Nekrasov, and S. Shatashvili, Four-dimensional avatars of two-dimensional RCFT, *Nuclear Physics B-Proceedings Supplements* **46**, 130–145 (1996).

[71] T. Inami, H. Kanno, T. Ueno, and C.-S. Xiong, Two-toroidal lie algebra as current algebra of the four-dimensional kähler wzw model, *Physics Letters B* **399**, 97–104 (1997).

[72] E. Witten, Integrable lattice models from gauge theory (2016), arXiv preprint arXiv:1611.00592.

[73] N. Levine, Equivalence of 1-loop RG flows in 4d Chern-Simons and integrable 2d sigma-models, arXiv:2309.16753 [hep-th].

[74] D. M. Schmidtt, A generalized 4d Chern-Simons theory, *JHEP* **11**, 144 (2023), arXiv:2307.10428 [hep-th].

[75] M. Ashwinkumar, J.-I. Sakamoto, and M. Yamazaki, Dualities and discretizations of integrable quantum field theories from 4d Chern-Simons theory, arXiv:2309.14412 [hep-th].

[76] N. Levine, Universal 1-loop divergences for integrable sigma models, *JHEP* **03**, 003 (2023), arXiv:2209.05502 [hep-th].

[77] C. Beasley and E. Witten, Non-Abelian localization for Chern-Simons theory, *Journal of Differential Geometry* **70**, 183–323 (2005). arXiv:hep-th/0503126.

[78] L. Faddeev and N. Reshetikhin, Integrability of the principal chiral field model in $1 + 1$ dimension, *Annals of Physics* **167**, 227–256 (1986).

[79] H. B. Nielsen and M. Ninomiya, Absence of neutrinos on a lattice. 1. Proof by homotopy theory, *Nuclear Physics B* **185**, 20 (1981). [Erratum: *Nuclear Physics B* 195, 541 (1982)].

[80] H. B. Nielsen and M. Ninomiya, Absence of neutrinos on a lattice. 2. Intuitive topological proof, *Nuclear Physics B* **193**, 173–194 (1981).

[81] H. B. Nielsen and M. Ninomiya, No go theorem for regularizing chiral fermions, *Physics Letters B* **105**, 219–223 (1981).

[82] S. Hellerman, D. Orlando, and M. Watanabe, Quantum information theory of the gravitational anomaly, arXiv:2101.03320 [hep-th].

[83] C. Destri and H. J. de Vega, Light cone lattice approach to fermionic theories in 2-D: The massive thirring model, *Nuclear Physics B* **290**, 363–391 (1987).

[84] C. Destri and H. J. de Vega, Integrable quantum field theories and conformal field theories from lattice models in the light cone approach, *Physics Letters B* **201**, 261–268 (1988).

[85] C. Destri and H. J. de Vega, Light cone lattices and the exact solution of chiral fermion and σ models, *Journal of Physics A* **22**, 1329 (1989).

[86] L. D. Faddeev, How algebraic Bethe ansatz works for integrable model, in *Les Houches School of Physics: Astrophysical Sources of Gravitational Radiation*, pp. 149–219, May, 1996. arXiv:hep-th/9605187.

[87] V. P. Nair, Elements of geometric quantization and applications to fields and fluids, arXiv:1606.06407 [hep-th].

[88] V. Caudrelier, M. Stoppato, and B. Vicedo, On the Zakharov-Mikhailov action: 4d Chern-Simons origin and covariant Poisson algebra of the Lax connection, *Letters in Mathematical Physics* **111**, 82 (2021), arXiv:2012.04431 [hep-th].

[89] J. Gomis and F. Passerini, Wilson loops as D3-branes, *JHEP* **01**, 097 (2007), arXiv:hep-th/0612022.

[90] J. Gomis and F. Passerini, Holographic Wilson loops, *JHEP* **08**, 074 (2006), arXiv:hep-th/0604007.

[91] O. Corradini and J. P. Edwards, Mixed symmetry tensors in the worldline formalism, *JHEP* **05**, 056 (2016), arXiv:1603.07929 [hep-th].

[92] K. Costello, Supersymmetric gauge theory and the Yangian, arXiv:1303.2632 [hep-th].

[93] F. Bastianelli, R. Bonezzi, O. Corradini, and E. Latini, Particles with non abelian charges, *JHEP* **10**, 098 (2013), arXiv:1309.1608 [hep-th].

[94] D. Tong and K. Wong, Monopoles and Wilson lines, *JHEP* **06**, 048 (2014), arXiv:1401.6167 [hep-th].

[95] M.-A. Sato, Operator ordering and perturbation expansion in the path integration formalism, *Progress of Theoretical Physics* **58**, 1262 (1977).

[96] R. Hotta, K. Takeuchi, and T. Tanisaki, *D-Modules, Perverse Sheaves, and Representation Theory*, Vol. 236. Springer Science & Business Media, 2007.

[97] J. Bernstein and A. Beilinson, Localisation de g-modules, *Comptes rendus de l'Académie des sciences (Paris)* **292**, 15–18 (1981).

[98] A. Beilinson and J. Bernstein, A proof of jantzen conjectures, *Advances in Soviet Mathematics* **16**, 1–50 (1993).

[99] M. Kashiwara, *Representation theory and D-modules on flag varieties.* Kyoto University, Research Institute for Mathematical Sciences [RIMS], 1988.

[100] W. Taylor, D-brane field theory on compact spaces, *Physics Letters B* **394**, 283–287 (1997), arXiv:hep-th/9611042.

[101] Y. Zhu, Modular invariance of characters of vertex operator algebras, *Journal of the American Mathematical Society* **9**, 237–302 (1996).

[102] D. Brungs and W. Nahm, The Associative algebras of conformal field theory, *Letters in Mathematical Physics* **47**, 379–383 (1999), arXiv:hep-th/9811239.

[103] M. Dedushenko, From VOAs to short star products in SCFT, *Communications in Mathematical Physics* **384**, 245–277 (2021), arXiv:1911.05741 [hep-th].

[104] Y. Zhu, *Vertex Operator Algebras, Elliptic Functions and Modular Forms.* Yale University, 1990.

[105] F. Malikov, V. Schechtman, and A. Vaintrob, Chiral de Rham complex, *Communications in Mathematical Physics* **204**, 439–473 (1999), arXiv:math/9803041.

[106] E. Witten, Two-dimensional models with (0,2) supersymmetry: Perturbative aspects, *Advances in Theoretical and Mathematical Physics.* **11**, 1–63 (2007), arXiv:hep-th/0504078.

[107] T. Arakawa, D. Chebotarov, and F. Malikov, Algebras of twisted chiral differential operators and affine localization of g-modules, *Selecta Mathematica* **17**, 1–46 (2011).

[108] E. Abdalla, M. Forger, and M. Gomes, On the origin of anomalies in the quantum nonlocal charge for the generalized nonlinear σ models, *Nuclear Physics B* **210**, 181–192 (1982).

[109] I. Affleck, D. Bykov, and K. Wamer, Flag manifold sigma models: Spin chains and integrable theories, *Physics Reports* **953**, 1–93 (2022), arXiv:2101.11638 [hep-th].

[110] D. Bykov, The geometry of antiferromagnetic spin chains, *Communications in Mathematical Physics* **322**, 807–834 (2013), arXiv:1206.2777 [hep-th].

[111] E. Abdalla, M. C. B. Abdalla, and M. Gomes, Anomaly in the Nonlocal Quantum Charge of the $\mathrm{CP}^{(n-1)}$ Model, *Physical Review D* **23**, 1800 (1981).

[112] J. Oh and J. Yagi, Chiral algebras from Ω-deformation, *JHEP* **08**, 143 (2019), arXiv:1903.11123 [hep-th].

[113] S. Jeong, SCFT/VOA correspondence via Ω-deformation, *JHEP* **10**, 171 (2019), arXiv:1904.00927 [hep-th].

[114] N. Ishtiaque, S. F. Moosavian, S. Raghavendran, and J. Yagi, Super-spin chains from superstring theory, *SciPost Physics* **13**, 083 (2022), arXiv:2110.15112 [hep-th].

[115] R. Dijkgraaf, L. Hollands, P. Sulkowski, and C. Vafa, Supersymmetric gauge theories, intersecting branes and free fermions, *JHEP* **02**, 106 (2008), arXiv:0709.4446 [hep-th].

[116] C. P. Bachas, M. B. Green, and A. Schwimmer, (8,0) quantum mechanics and symmetry enhancement in type I' superstrings, *JHEP* **01**, 006 (1998), arXiv:hep-th/9712086.

[117] N. Itzhaki, D. Kutasov, and N. Seiberg, I-brane dynamics, *JHEP* **01**, 119 (2006), arXiv:hep-th/0508025.

[118] R. Emparan, Born-Infeld strings tunneling to D-branes, *Physics Letters B* **423**, 71–78 (1998), arXiv:hep-th/9711106.

[119] R. C. Myers, Dielectric branes, *JHEP* **12**, 022 (1999), arXiv:hep-th/9910053.

[120] D. Mateos and P. K. Townsend, Supertubes, *Physical Review Letters* **87**, 011602 (2001), arXiv:hep-th/0103030.

[121] E. Antonyan, J. A. Harvey, and D. Kutasov, Chiral symmetry breaking from intersecting D-branes, *Nuclear Physics B* **784**, 1–21 (2007), arXiv:hep-th/0608177.

[122] A. Kapustin, Holomorphic reduction of N=2 gauge theories, Wilson-'t Hooft operators, and S-duality, arXiv:hep-th/0612119.

[123] L.-Y. Hung, Comments on I1-branes, *JHEP* **05**, 076 (2007), arXiv:hep-th/0612207.

[124] M. B. Green, J. A. Harvey, and G. W. Moore, I-brane inflow and anomalous couplings on d-branes, *Classical and Quantum Gravity* **14**, 47–52 (1997), arXiv:hep-th/9605033.

[125] D. Diakonov and V. Petrov, NonAbelian Stokes theorems in Yang-Mills and gravity theories, *Journal of Experimental and Theoretical Physics* **92**, 905–920 (2001), arXiv:hep-th/0008035.

[126] O. DeWolfe, D. Z. Freedman, and H. Ooguri, Holography and defect conformal field theories, *Physical Review D* **66**, 025009 (2002), arXiv:hep-th/0111135.

[127] D. Gaiotto and E. Witten, Supersymmetric boundary conditions in N=4 super Yang-Mills theory, *Journal of Statistical Physics* **135**, 789–855 (2009), arXiv:0804.2902 [hep-th].

[128] N. Ishtiaque and Y. Zhou, Line operators in 4d Chern-Simons theory and Cherkis bows, arXiv:2211.00049 [hep-th].

[129] A. Hanany and E. Witten, Type IIB superstrings, BPS monopoles, and three-dimensional gauge dynamics, *Nuclear Physics B* **492**, 152–190 (1997), arXiv:hep-th/9611230.

[130] D. Gaiotto and E. Witten, S-duality of boundary conditions in N=4 super Yang-Mills theory, *Advances in Theoretical and Mathematical Physics* **13**, 721–896 (2009), arXiv:0807.3720 [hep-th].

[131] D. Gaiotto, J. H. Lee, and J. Wu, Integrable Kondo problems, *JHEP* **04**, 268 (2021), arXiv:2003.06694 [hep-th].

[132] D. Gaiotto, J. H. Lee, B. Vicedo, and J. Wu, Kondo line defects and affine Gaudin models, *JHEP* **01**, 175 (2022), arXiv:2010.07325 [hep-th].

[133] J. Wess and B. Zumino, Consequences of anomalous Ward identities, *Physics Letters B* **37**, 95–97 (1971).

[134] H. J. de Vega, H. Eichenherr, and J. M. Maillet, Canonical charge algebras for integrable fermionic theories, *Physics Letters B* **132**, 337–341 (1983).

[135] A. M. Polyakov and P. B. Wiegmann, Theory of nonabelian Goldstone bosons, *Physics Letters B* **131**, 121–126 (1983).

[136] A. M. Polyakov and P. B. Wiegmann, Goldstone fields in two-dimensions with multivalued actions, *Physics Letters B* **141**, 223–228 (1984).

[137] E. Alvarez, L. Alvarez-Gaume, and Y. Lozano, On nonabelian duality, *Nuclear Physics B* **424**, 155–183 (1994), arXiv:hep-th/9403155.

[138] G. Georgiou, K. Sfetsos, and K. Siampos, Double and cyclic λ-deformations and their canonical equivalents, *Physics Letters B* **771**, 576–582 (2017), arXiv:1704.07834 [hep-th].

[139] G. Georgiou, K. Sfetsos, and K. Siampos, All-loop anomalous dimensions in integrable λ-deformed σ-models, *Nuclear Physics B* **901**, 40–58 (2015), arXiv:1509.02946 [hep-th].

[140] R. C. B. Santos, C. A. Hernaski, and P. R. S. Gomes, Bosonization, duality, and the C-theorem in the non-abelian thirring model, arXiv:2305.11113 [hep-th].

[141] O. Gwilliam and B. R. Williams, A one-loop exact quantization of Chern-Simons theory, arXiv:1910.05230 [math-ph].

[142] R. F. Penna, Twistor Actions for Integrable Systems, *JHEP* **09** (2021) 140, arXiv:2011.05831 [hep-th].

[143] L. J. Mason and N. M. J. Woodhouse, *Integrability, self-duality, and twistor theory.* Oxford University Press, 1996.

[144] Y.-J. He, J. Tian, and B. Chen, Deformed integrable models from holomorphic Chern-Simons theory, *Sci. China Phys. Mech. Astron.* **65** (2022) 100413, arXiv:2105.06826 [hep-th].

[145] Y. Luo, M.-C. Tan, J. Yagi, and Q. Zhao, Ω-deformation of B-twisted gauge theories and the 3d-3d correspondence, *JHEP* **02** (2015) 047, arXiv:1410.1538 [hep-th].

[146] N. Ishtiaque and J. Yagi, Disk, interval, point: on constructions of quantum field theories with holomorphic action functionals, *JHEP* **06** (2020) 180, arXiv:2002.10488 [hep-th].

[147] K. Costello and D. Gaiotto, Twisted holography, *JHEP* **01** (2025) 087, arXiv:1812.09257 [hep-th].

[148] K. Costello and N. M. Paquette, Twisted Supergravity and Koszul Duality: A case study in AdS$_3$, *Commun. Math. Phys.* **384** (2021) 279–339, arXiv:2001.02177 [hep-th].

[149] M. Bershadsky, S. Cecotti, H. Ooguri, and C. Vafa, Kodaira-Spencer theory of gravity and exact results for quantum string amplitudes, *Commun. Math. Phys.* **165** (1994) 311–428, arXiv:hep-th/9309140.

[150] K. Costello and S. Li, Quantization of open-closed BCOV theory, I, arXiv:1505.06703 [hep-th].

[151] M. Bershadsky and V. Sadov, Theory of Kahler gravity, *Int. J. Mod. Phys. A* **11** (1996) 4689–4730, arXiv:hep-th/9410011.

[152] K.-I. Kondo, Wilson loop and magnetic monopole through a non-Abelian Stokes theorem, *Phys. Rev. D* **77** (2008) 085029, arXiv:0801.1274 [hep-th].